Breeding in Insects

BREEDING IN INSECTS

By

MANJU YADAV

Lecturer

Department of Zoology

M.M.H. College

Ghaziabad (U.P.)

(India)

DISCOVERY PUBLISHING HOUSE

NEW DELHI-110002

First Published-2003

ISBN 81-7141-737-X

Published by

DISCOVERY PUBLISHING HOUSE
4831/24, Ansari Road, Prahlad Street,
Darya Ganj, New Delhi-110002 (India)
Phone: 23279245 • Fax: 91-11-23253475
E-mail:dphtemp@indiatimes.com

Printed at: **Tarun Offset Printers, Delhi-53**

Preface

The fundamentals of Breeding in Insects are presented within the framework of scientific discovery. Researches in Entomology have been made almost increadible strides in the past few decades. Consequently, existing concepts of Insect biology have been expanded. These has been a revolution indeed in this direction.

The text integrates the descriptive, experimental and biochemical approaches into a conceptual framework. All important points are illustrated diagramatically. The title is not intended to be comprehensive nor could it be at length, but it concentrates as putting across the basic principles of the subject as briefly and lucidly as possible. It does this with the aid of careful selected examples–some recent and other classic of the field and with numerous illustrations. The aim is to enthuse the reader with this active and exciting area of research and to lay a solid foundation on which further study of its various facets may be based.

The author has freely consulted various articles, discussion notes, reviews and extracts from the various scientific journals in the preparation of the present book, in able to make it comprehensive and upto date.

Though every care has been taken by the printer, publisher and me, it is quite likely that some errors might have found their way into the book but I hope these are of very insignificant nature. However, sugestions to improve book and pointing out of errors and mistakes, if any, will gratefully be appreciated.

The author expresses grateful to her friends and colleagues whose constant inspiration have initiated her in bringing out this book.

Special thanks are given to Mr. Vasan and staff of M/s Discovery publishers for their whole hearted co-operation in the publication of this book.

Author

Contents

1

INTRODUCTION

Watch a cloud of mosquitoes dancing up and down over a stream, or a group of black-flies buzzing around under a river-side tree, and you are looking at insects ensuring the survival of their kind. At these gatherings, males and females find each other to mate. Yet most of the mosquitoes and black-flies you see among these swarms are males; the females merely visit to pick up or get picked up by a male.

When a female mosquito enters the swarm, the males find her by listening for the whining of her wings which beat at a different rate from their wings. They detect the vibrations from her wing beats passing through the air with their long bottle-brush style antennae. (In comparison females have much smoother antennae with fewer whorls of hairs). The first male to reach her, grasps her with his back legs and they either mate in mid-air or fall to the ground to mate.

In contrast to mosquitoes, male and female black-flies have smaller antennae and more developed eyes, so vision plays an important role in how they find each other. The male black-flies have larger eyes than the female, because they not only need to spot females flying into the swarm but have to orientate in relation to other males and keep their position under a tree (or where the swarm formed). The females are also attracted to the males by the markings on their back legs which dangle down as they fly along. Once a male blackfly grabs hold of the female, the pair falls to the ground to mate.

For the majority of animals mating is the most important part of their lives because they can pass on their genes to the next generation. Normally, offspring produced by mating get half their genes from their mother and half from their father. Yet some insects reproduce without mating and so pass on all their genes to their offspring. Producing young without mating is common among stick insects, something people who keep them as pets should be warned about because they may end up with scores of baby stick insects from just one female. All the babies are identical because they have more or less the same genes as their mother.

But why bother to mate if you can pass all your genes on to your offspring instead of only half? The major drawback is that if all your offspring are genetically identical and all their offspring and so on, they cannot adapt to changes in their environment. If a new disease should break out, all the identical individuals would be equally susceptible and the entire lot then wipe out. On the other hand, not bothering with mating saves a lot of time.

Some insects, such as aphids and gall wasps, get the best of both worlds by producing offspring with or without mating at different times in their life cycle. Queen honeybees to mate, but they choose whether or not to fertilize their own eggs from the sperm they acquired from males on their nuptial flight. The choice is important because it partly determines the role an individual plays in the hive. Fertilized eggs hatch into females that become either workers or, if fed a special diet, another queen. The fertilized eggs become males (drones) which are fertile even though they have only half the genetic material. Although the female workers are unable to produce their own offspring, they help to raise their sisters who share most of their genes.

Smelling Attractive

As the perfume industry knows, smell is an important part of attracting partner. For the vapourer moth female it is essential because she only has stubby wings and cannot fly so she must attract males to her. As she sits on a tree trunk at night, she pumps out smells from glands on her posterior end. Any male flying in the vicinity, picks up the scent molecules on his antennae and homes in on the female by flying in a zig-zag path upwind in the direction of her scent. Among the moths with the most sensitive sense of smell are some of the emperor moths, whose males can detect a female's scent from several kilometres (miles) away. Male emperor moths have highly branched antennae which provide a big surface on which to collect scent molecules.

Butterflies normally use scent at a closer range than moths. This is more like aftershave than perfume, because it is usually the males not the females who produce scents. In some butterfly species the male's scent persuades the female to land or dissuades her from flying off so he can then try to mate. In others the male's scent is an aphrodisiac stimulating her to mate. Male butterflies (and some moths) have a great variety of scent-producing structures on the wings, legs and abdomen. Some are simple scales or hairs associated with scent producing glands, others consist of hidden patches of scent glands such as those in folds on the wing or where the back wing overlaps the front wing. Because the scent gland is enclosed or covered the scent does not leak out at the wrong time or is wasted by evaporation. The most complex male scent organs are the hair pencils. These are a pair of brush-like structures that distribute scent particles or "love dust", scattering them over the female. This "love dust" can be produced in the hair pencils at the tip of the abdomen, or the hair-pencil brushes pick up scent from another part of the male butterfly's body. In the Queen butterfly from the southeast USA, the male first pursues the female and then overtakes her, scattering love dust from his hair pencils onto her antennae. She then settles on a plant pushing out his hair pencils. This encourages her to fold her wings. Finally, he lands beside her and they mate.

Male bumblebees also use scents to persuade females to land so they can mate. The males patrol a route along which there are spots marked with their scent. Several males share a route and the one flying over the spot at the time a female has landed mates with her. Orchid bee males do not make their own perfumes but scrape scent from the orchids they visit in the rainforests of Central and South America where they live. These are not used to attract females directly; instead it seems that the scent attracts other males to gather together. Females are probably attracted to the group of males where they choose the fittest one to mate with. Female bees also produce their own smells. For example, the newly emerged queen honeybee draws many males to her by producing a scent as she flies through a gathering of males.

Love Songs

Seldom can we smell the scents insects use in courtship (among the most noticeable are those used by some butterflies and moths), but we can listen to some of their courtship songs, often whether we want to or not. Among the noisiest is a species of mole cricket from southern Europe, whose males can be heard 600 metres (656 yards) away. Each male lives in a burrow with two openings which act like amplifiers in a hi-fi-music system. He makes a long, loud whirring noise by rubbing his wings together. On hearing this noise, the female heads towards the male and, as he continues to sing, he sways from side to side in the hope that the female may be impressed enough to climb onto him so they can mate. Male grasshoppers also make noises to attract a mate and to alert other males that their patch of grass is occupied. The chirruping is produced by the male rubbing small pegs on his back legs against a hard vein on his front wings, rather like making a noise by rubbing a coin along the teeth of a comb. Female grasshoppers are able to distinguish males of their own species by their distinctive songs, even when there are several species of grasshopper living in the same area which look almost identical. (Scientists can also tell grasshopper species apart by listening to their songs). The females hear the song through a pair of eardrums positioned one on each side of the abdomen.

Male cicadas can collectively produce extremely loud noises, up to 100 decibels, from about 20 metres (22 yards) away. They produce sound by clicking a lid-like structure on each side of the abdomen, which in some species is amphlified further by a pair of air sacs in the abdomen. Many kinds of male moths have a series of tiny clicking structures which produce high frequency sounds (ultrasound) often for defence, but those of the polka-dot wasp moth are used by males and females as a means to recognize each other and are just audible to us.

Another quiet courtship is that of the death-watch beetle which lives in the timbers of old houses and churches in southern England. They scared people in the past because the sound of males and females tapping could only be heard when the house was in silence, such as during a night-time vigil around someone's deathbed. Undaunted by the human goings-on, the adult beetles emerge from holes in timbers where they have spent their young as grubs and then pupae. The males immediately start knocking their heads on the wood to alert any females of their presence. The females then respond by doing likewise. The gentle tapping sounds are picked up by the beetles' feet as the vibrations pass through the wood they stand on.

Looking Good

Many insects have good sight and some are able to see in colour as well as detect ultraviolet patterns that are invisible to us and other vertebrates. Some male butterflies fly at any object thrown into the air that is approximately the same size as a female. Closer to, the colours of the females' wings are often attractive to males, though the intricate patterns are not necessarily important. Sometimes it is only the colours of the second pair of wings that produce a response. It is also important whether males view females with their wings open or folded at rest.

Among the most spectacular visual mating displays are those of fireflies which signal to each other at night by emitting flashes of light from their abdomen. In some species the females are wingless and called glow-worms. They rest among plants and signal their presence to males flying by. In other species, males also emit flahes of light to which the female responds. In North America there can be more than five species of fireflies living in the same area all trying to mate on the same night. To avoid confusion, each species has a unique sequence of flashes and the female must respond at the right point during the male's repertoire before he will approach her. When close to, either the pattern of bands on the female's abdomen or the scent of the female acts as a final cue for making. Sneaky males can try to disrupt the mating display of other males by flying close by and adding more flashes, so that the female fails to recognize the first male's display. The sneak then completes the full sequence and mates successfully. A worse misfortune can befall the unwary suitor; some predatory female fireflies attract males of other species only to devour them when they make their approach.

Male praying mantids can risk being eaten even when they attempt to mate with females of their own species. Male mantids are smaller than females so can be easily caught in the female's sharp spined front legs. Observations of the Chinese praying mantis (which was introduced to North America at the end of the last century on an attempt to control insect pests) shows that this rarely happens unless the female's is hungry. Still, the male if approaching his chosen female head-on, must do so with caution. He moves slowly towards her and makes all the right signals. First, he waggles his antennae, then he either stamps his feet or wiggles his abdomen in a figure-of-eight. Once within striking distance the male makes a flying leap onto her back. So for this species and the few others observed mating in the wild, the females do not make a habit of eating their males.

Some male insects facing fierce females offer their intended a nutritious gift. Hanging flies are predators, and the male offers the female juicy flies and other insects. With his gift ready, he releases a scent to draw the female to him. She assesses how good a match she is getting by the size of the gift and may even turn down a suitor if she deems it to be too small or not to her liking. Not only does the size of the gift indicate the quality of the male, but she benefits more if it is a big meal, especially if it is not easy for her to hunt for her own flies because she is weighed down with eggs. Dance-fly males present their gifts wrapped in silk (these are true flies unlike hanging flies), but some of these male flies give just symbolic gifts like a petal wrapped in silk or even an empty parcel. Some male crickets produce large packages of sperm with a supply of food attached which they deposit into the female's genital opening. The female then devours the store of food as it sticks out from her genital opening. The males of thynnine wasps in Australia also feed their wingless females and in some species the males carry her around so she can feed on the nectar from flowers.

Fighting for Females

Some male insects fight each other for females. Staghorn beetle males do battle on a branch or log in the woods of Europe where they live. They have outsized jaws which they use to wrestle with males attempting to take over their breeding and feeding site. The opponents try to seize each other in their jaws. The male which succeeds picks up his rival and drops him over the edge of the branch or log. Rarely do the battling beetles cause each other injury, although the toothed jaws are capable of piercing the armour-like exoskeleton. Atlas beetle males fight in a similar manner. These beetles are often found in coconut and oil-palm plantation in southeast Asia. Instead of outsized jaws, the males have long horns on the thorax and head which they use to lever opponents off a log. The males are attracted to a female by the scent she releases. The first male to reach her then fights off any rivals which arrive later. The horns are said to be used to carry off the female to a safer spot. The male picks her up by placing his head horn under her body and then clamping her against the long rigid horns on his thorax. Male rhinoceros beetles also have prongs on their heads and thorax. They compete by trying either to shove each other out of the way or to overturn each other. The strongest beetle wins the trial strength so the females then mate with the best male. Among many kinds of beetles armed with horns or big jaws there are smaller, cunning males lacking these weapons who sneak in and mate with the waiting female while the well-armed males do battle.

Scientist Chris Lyal observed another kind of beetle, wood-boring weevils, called *Mecopus*, fighting in the rainforest of Sulawesi. These beetles resemble flies in the way they fly and run in sudden jerks. *Mecopus* males home in on recently fallen trees and patrol the trunk looking for females drilling holes with their snouts in preparation for laying an egg. If they come into contact with other males, they tussle using their long snouts and spikes on the thorax. The males also have long front legs to gain maximum purchase on the tree trunk. Male bottle-brush weevils in Central America spar with rivals using their bushy snouts like miniature swords. Some brentid weevils guard the females they have mated with by actually standing over them on a tree trunk as they lay the eggs. If another male comes too close, the male jabs his snout under his rival and, by suddenly raising his head and pushing up on his front legs, flings his opponent off the trunk.

Male stalk-eyed flies in Australia defend their chosen females from the attentions of other males. They sometimes wrestle with other males but only if the competition is evenly matched. Male stalk-eyed flies assess the size, and therefore the strength of their opponent, by pushing against each other and aligning their stalked eyes. If the competitor has eyes set further apart, the male gives up. If they are evenly matched they shove hard against each other, until they are standing just on the tips of the now upright back two pairs of legs. The beautiful golden dung-flies are also competitive. The females come to fresh cow pats to mate and lay their eggs. The males hang around the edge of the cow pat and grab any female flying in. If there are not enough females to go round, males try to wrestle females away from other males. Even if the first male has succeeded in mating it does not matter as long as she still has eggs to lay. It is the last male to mate before she lays an egg who fertilizes it.

Sounds are sometimes used by male insects to show their annoyance to other males if they get too close to their female or trespass on their territory. Male cracker butterflies in Central and South America make snapping sounds with their wings as they duel in mid-air.

Mating Matters

Most male insects deposit either a package of sperm or free sperm directly inside the female through her genital opening. Bed bugs and their relatives are unusual because the male cuts through the female's exoskeleton and the sperm migrate through the female's blood, or through special tissues, and then to the ovaries where they fertilize the eggs. The males of various species of strepsipterans (twisted-wing insects) have a hard time because their females lie hidden away inside bees, wasps, plant bugs and other insects which they parasitize. The females are also cloaked by their last larval skin. However, the fused head and thorax of the female pokes out of the host's abdomen and here the larval skin has an opening to receive sperm which then migrate to the genital opening at the tip of her abdomen.

Insects have a variety of mating positions–sometimes one sits on top of the other or sometimes they mate side to side and sometimes end to end. The length of time insects remain paired varies enormously from a few seconds in honeybees up to 10 hours for some species of damselflies. Sperm transfer does not always come about by direct contact between male and females. Male springtails (tiny soil-dwelling insects) deposit little packages of sperm on the ground which the females pick up to fertilize their eggs. Some male springtails help the females find the packages by holding their antennae and leading them to the right spot.

Once males have mated it does not mean that their sperm fertilizes the eggs. In many insects, the last male to mate just before the female lays her eggs ends up being the father. For this reason males often keep guard and fight off rivals and this is also the reason for protracted mating sessions. Some male insects, including certain butterflies and mosquitoes, produce mating plugs which block the female's genital opening after mating so that, at least for a while, no more sperm can get in. Male heliconid butterflies add a strong scent which repels other suitors. Males of dragonflies and damselflies actually scoop out the sperm from any earlier couplings a female may have had, before mating with her. Some prevent the same thing happening to them by staying with the female until she drops eggs over the surface of the water.

Still, it is wise for a male to be the first to get to a female and essential if a female mates only once. Some male butterflies, such as those of heliconids, cling onto the pupal case (chrysalis) of a female awaiting the moment for it to emerge so they can be first to mate. Some take things one step further actually penetrating the pupal case and fertilizing the female before she emerges. The males of one mosquito species that lives in rock pools in New Zealand, fly over the water waiting for the moment that females start to emerge from their pupae. A male grabs a partially emerged female and hauls her over to the edge of the pool where they mate. Occasionally, he may grab a male by mistake but does not find out until his "intended" is almost clear of his pupal skin.

Digger bee males also like to be the first to reach a female. These bees spend their larval and pupal lives in burrows. The males emerge first and congregate around the entrance of the female's burrow hoping to be the first to get to her and mate.

Some female insects receive sperm from a number of males, and her offspring therefore have several different fathers. A queen honeybee mates with several males on her nuptial flight and obtains enough sperm to fertilize all the eggs she produces in her life time. For these successful males it is their last act because they die within a few hours of mating. When a drone mates with a female his sperm explode into the female's genital opening and in the process he ruptures his abdomen. The sperm is stored in pockets inside the female ready to fertilize the eggs as required. There is plenty of sperm to spare; for example, a queen honey-bee can receive over 80 million sperm during her nuptial flight but only has room to store about five million.

Getting it Right

Males and females may waste their effort if they end up mating with the wrong species. All the razzmatazz of courtship is designed to ensure no mistakes are made. A sequence of courtship behaviour is often required and if one or other partner fails to make the right response mating may not occur. Some beetles have the ultimate way to get it right. The male and female genital organs are like a lock and key and simply do not fit together unless they are of the same species. Yet there have been many instances where insects have got it wrong in their hurried attempts to mate.

An astonishing example of how things can go wrong is seen in certain buprestid beetles in Australia, whose males were found attempting to mate with beer bottles thrown into the bush by litter louts. The male beetles check whether they have the right female by noting her colour and the pits on her back. Unfortunately for the frustrated beetles, the colour and bumps on the beer bottles matched the colour and reflections from the pits of the female beetles. Smooth wine bottles and tin cans attracted no beetles at all.

A common mistake is males trying to mate with other males, especially if females are in short supply. Locusts bred in the laboratory are often seen trying to mate with other males. Bed bugs and their relatives are more prone to mate with males, because the males only have to pierce the exoskeleton of the female and do not need to find a specific genital opening. Once of the most bizarre phenomena in the insect world occurs in a species of bug where the males sometimes copulate with other males and inject sperm into them by piercing their exoskeleton. The sperm is not destroyed but migrates in the recipient's blood to his testes. When the recipient mates with a female he impregnates her not only with his own sperm but with that of his male attacker. So it seems in nature there are ever more ingenious ways to ensure that one's genes are passed on to the next generation.

Making more Insects

Awesome Swarm

The rains have come for another year to the fringes of the Sahara desert the land is refreshed. Newly planted crops are growing well and there is good grazing for the livestock. But the much welcomed rain brings with it insects which threaten to destroy all this good green–desert locusts are gathering together to breed. The males have already mated with the females but some stay perched on their mate's back as she walks around looking for a good spot to lay her eggs close to vegetation. The females probe the soil with the tip of their abdomen, assessing how much moisture is present, how hard the surface is, its temperature and whether too much salt has accumulated in the soil due to evaporation of water in the hot sun. More and more females gather together in the most suitable places for egg-laying where there is some moisture at least below the top 6 centimetres (2.4 inches) of the soil. The females are also attracted by each other's scent.

Each female pushes her egg-laying tube into the soil making a hole to take her eggs. Then she stretches her abdomen down into the hole and begins to lay the eggs. These emerge glued together by a frothy mixture to form an egg pod which is about half her length. The pod is wedged deep into the hole and topped with more froth which forms a plug sealing the hole. Under swarm conditions, each female can produces up to 80 eggs per pod but the numbers decline if she produces a second or third pod. Females lay their egg pods close to where others have laid theirs, with densities of up to 100 per metre^2 (!20 per yard^2). In the right place the eggs absorb almost their own weight in water from the soil. If the eggs are laid in soil which is too dry, they will not develop, nor will eggs laid on the surface of the ground. This happens not because the female is careless but because there are not suitable places available. Once she has fully developed eggs inside her, the female has to lay them within three days, whether the conditions are right or not.

After the eggs are laid, the females and the males have nothing more to do with them. Depending on the temperature the tiny first stage hoppers hatch out about two weeks to a month later (longer in very cold conditions). All the hoppers in an egg pod usually hatch at the same time, often just as the sun begins to rise. The burst out of the egg shell and work their way through the plug blocking the exit. As soon as they reach the surface they moult, shedding a thin outer skin. Only a few egg pods in an area hatch on the first day, the majority hatch on the next day, while the stragglers emerge on the third day. Because the hoppers emerge from egg pods laid close together they soon form groups, and then the groups join together to form bands which eat and march together. They will moult five times before turning into winged adults some 25 days or more later–so the dreaded swarm is formed.

All Alone

Like deserts locusts, many other insect parents do little else for their offspring after the female has laid her eggs in an appropriate spot. Flies lay their eggs on dung or carrion so that when the maggots hatch they have a good source of food. Before they lay their egg female butterflies often test the surface of a plant by stamping on it with their feet, to feel if it is the right one for the caterpillars to feed on when they hatch. Gall wasps choose the right part of the plant to inject their eggs into and it is these eggs, not the adult, which cause the plant tissue to produce the protective gall.

Acorn weevil females lay their eggs in young acorns, so that their grubs have food available when they hatch. The female gets some food herself when she drills a hole into the acorn with her long snout, feeding on the acorn tissue as she goes. Then she turns around and lays an egg down the hole with her egg-laying tube (ovipositor). According to beetle specialist Peter Hammond, she may risk her life when she drills this hole, because he has come across fallen acorns which have female weevils with their mouthpart stuck into the acorns and their legs waving helplessly in the air. If a female loses her grip on the acorn's surface while her snout is firmly inserted, she may be unable to regain her footing and withdraw her snout. By comparison, some stick and leaf insect mothers seem careless with their eggs because these simply shoot out of the abdomen and are scattered over the ground. But, this may be more careful than it seems, for predator would have to search quite a large area before it found all the eggs. Another maternal strategy is to hide the eggs where predators or parasites cannot get to them. Some water bugs insert their eggs into the stems of water plants, while some mantids produce a frothy substance which hardens around their eggs to protect them.

Parental Care

Some insects are more devoted parents and stay with and care for their offspring. Females of some species of cockroach, such as the common or oriental cockroach, carry the egg case containing their eggs around with them. Some species of fungus-feeding thrips stay with their eggs and care for the young when they hatch. The adults of one species that lives on trees in Panama accompany the young when they go out to feed, then herd them back into the safety of a crack in the bark each night. In many bug species, the mother remains with her eggs and nymphs, protecting them with her body–this behaviour helps toward off insects trying to parasitize the eggs. Often it is the eggs on the outside of the batch, which poke out from under the mother's body, that are attacked by parasites. Male giant water bugs, which live in ponds and sluggish flowing rivers, carry the fertilized eggs around on their back. The male strokes them regularly, not only to keep a good supply of freshly oxygenated water over them, but to keep them clean and also to prevent any fungus growing on them. The task is onerous since he cannot move about easily with the load. Until the eggs hatch some two to four weeks later he cannot feed and has to stay hanging onto weeds close to the water's surface.

Food and Shelter

Common earwigs make a little nest cavity in the soil, under stones or under bark, to protect their eggs. If the eggs are moved, the female scurries about picking up each egg in her jaws and depositing them all together in one place. She turns the eggs over in her mouthparts to clean them of any debris or fungi on their surface. When disturbed during her maternal duties she flexes her pincers and waves them at an intruder. The eggs of the common earwig are usually laid in winter and hatch the following spring, and they are guarded by their mother the entire time. When the earwig nymphs emerge, she feeds them by regurgitating food she collected on short night-time trips. The earwig young stay with their mother for about two weeks. She may then go on to produce two further broods using sperm stored from mating at the end of the previous summer.

Most hunting wasps do not stay with their eggs or young, but they do provide them with food and shelter. Among the most impressive are the spider-hunting wasps that catch spider even as large as tarantulas for their young to feed on. They sting the spider to paralyze it, but must avoid getting bitten by the spider's fangs. Most of these wasps catch their prey first and then make a burrow or use the spider's own burrow or another ready-made crevice. The prey is dragged down into the burrow where the female lays an egg on it. On hatching the wasp grubs feed on the spider.

Among the sphecid wasps there are also species which hunt spiders, while others specialize in hunting certain kinds of insects, such as caterpillars, weevils, crickets, or aphids, to feed their young. They make nests in a variety of places including within plant stems, in the ground or in rotten wood. Sand wasps make their nests in sandy soil. It is a laborious process for the female and takes her most of one day, after which she flies off to a night-time roost. After inspecting the nest the next day she goes off in search of caterpillars. She remembers the exact position of the nest hole by taking a reconnaissance flight when leaving the burrow for the first time, during which she memorizes the landmarks. Switching the position of a twig or stone near the burrow entrance can cause her much confusion. Once she catches a caterpillar she stings it several times, then she either flies back to the nest with the prey or, if it is too heavy, drags it along the ground. The prey is dragged into the burrow where she lays an egg on it. Depending on the species, the female either stocks the burrow with just one large caterpillar or she goes in search of more caterpillars. In any event, the female ensures that her grub has enough to feed on when it hatches. When fully provisioned the nest is closed off with soil which is compacted by the female vibrating her head against it or, in other species, the female vibrates a small pebble held in her jaws. In some sphecid species the female returns to the nest with new supplies of food for her grubs.

Some of the mason bees make their nest of mud, but they do not start from scratch. The females often make use of holes in old walls, or else dig new holes in old mortar. The female makes a cell at the far end of the hole using mud collected in her jaws. She stocks the cell with pollen and some nectar, and then lays an egg there before sealing it with mud and beginning the next in a series of up to about six cells. Once they have made and provisioned their nests, the mason bee females have nothing more to do with their offspring.

Carpenter bee females are territorial and guard the nests they have made by tunnelling into wood. Each cell in the nest is stocked with pollen and an egg laid on the food store. The female takes up a position nearby, from where she will attack any marauders, including parasites, predators and inquisitive scientists. When living in Malaysia, Laurence Mound found it most unnerving to be watched by a female carpenter bee who perched on the end of a washing line in his garden. Whenever he walked within a metre of her nest, she would take off and buzz angrily around his head.

House Hunting

While most bees and wasps lead a solitary life, others are social, living in colonies and helping each other to rear the young. Many species of bumblebees live in colonies which are begun afresh each year. The fertilized females spend the winter somewhere warm and dry. When the warm days of spring come, they forage for nectar and pollen from the first flowers. They seek out a warm spot to make a nest, usually in an abandoned mouse or shrew burrow. The female bumblebee makes two chambers using wax from her abdomen. She stores honey (processed nectar) in one and fills the other with pollen and a little nectar and then lays eggs on the mixture before sealing the chamber with wax. When the grubs hatch they feed from the food in their chamber, but she also has to leave the nest to forage for more food. Eventually the grubs pass through the pupa stage and emerge as workers. These then forage for food while the queen bumblebee stays at home and lays more eggs.

At the end of the season, males and new queen bumblebees are produced and these leave the nest to mate. The old queen, males and workers die off in the frosty cold of winter, while the newly fertilized queens hibernate ready to start the whole cycle again in the following year. Bumblebee nests can be plagued by cuckoo bumblebees, the females of which lay their eggs in a nest and trick the workers into rearing the cuckoo's young along with their queen's young. Some species of cuckoo bumblebee actually kill the queen bumblebee and any of her brood, leaving the workers to raise only the cuckoo bumblebee's young. There are also cuckoo wasps and bees which eat the stored food and sometimes the grubs in the nests of solitary bees and wasps.

Common wasps (yellowjackets) have a similar yearly cycle to bumblebees. The fertilized females or queen wasps find a warm spot to spend the winter, and it is these extra-large females which are seen inside during the autumn as they try to find somewhere to hibernate. They are more commonly seen in spring when, like the bumblebee queens, they are looking for a site to start a new colony. Unlike bumblebees, they feed their young on chewed up insects, such as caterpillars, and construct their nest out of chewed up wood. By summer, a large colony of wasps in the roof of your house can contain over 10,000 workers.

Honeybee colonies can survive the winter because they generate warmth in the hive by contracting their flight muscles without flapping their wings. This requires energy so the stock of honey in the hive is essential. Honey is stored in hexagonal cells made of wax which are built on either side of a comb by young workers. Once each cell is full it is sealed with wax. If honey is taken from the hive for human consumption, the bees may have to be fed syrup so they can last the winter. So that there are fewer mouths to feed, any drones (males) that are left after the new queen has been fertilized are pushed out of the hive and left to die before the onset of winter. The workers, who are all females, are allowed to stay.

In late winter and early spring the queen speeds up the rate of egg-laying so that there will be enough workers to forage for nectar and pollen from the first flush of flowers. During peak laying periods the queen can lay 2000 eggs per day. Each egg is laid in a brood cell. When the grub hatches it is fed mostly brood food, a nutritious substance worker bees produce from glands in their heads mixed with other substances, although older larvae are also fed pollen mixed with honey. After about five days of feeding, the workers seal the brood cell, and the grub spins a cocoon and transforms itself into a pupa. About two weeks later a new worker emerges from the cell. Her first chores will be within the hive where she works as a cleaner, a nursery maid caring for the young, and a builder constructing the wax comb and sealing gaps with a resinous substance. Young workers also receive nectar from older workers returning to the hive and help process it into honey by letting water evaporate from it while it is in their mouth. The honey is concentrated further in the wax cells by bees fanning it with their wings. Before the worker leaves the nest to forage, she spends a period as a guard at the nest entrance.

Royal Insects

New queens are produced either when the old queen begins to lay fewer eggs and makes less of the queen substance or when the numbers of workers build up. The queen substance is a chemical which is taken from the queen by the worker and spread through the hive when the workers feed each other. Normally, the queen produces enough queen substance to stop the workers rearing new queens. The queen substance also inhibits the development of the worker's ovaries so they cannot lay eggs themselves. To make new queens, the workers feed a number of grubs entirely on royal jelly (a more nutritious version of the brood food fed to the workers). The aspiring queens also get fed more food than worker grubs and inhabit large queen cells. Before the first of the new queens comes out of her cell, the old queen usually leaves with a retinue of workers in a swarm to start a new colony. The new queen kills the remainder of the still-developing queens so the hive only has one leader. The remaining workers become very excited when the new queen emerges and push her to the hive entrance, where she takes off on her nuptial flight.

All species of ants are more or less social, living in colonies and sharing the task of rearing the young. The newly fertilized ant queens usually have to set up colonies on their own. During this time they live off food reserves in their bodies including the wing muscles which are no longer needed, since by this time the female has shed her wings. The queen must have enough energy reserves to lay eggs and feed the first larvae when they hatch. These are much smaller than normal-sized larvae produced later on and do not require so much food. Once these develop into tiny workers, they immediately take over her chores and feed her. The workers are sterile females and lack wings.

Sometimes several newly mated queens start a colony, though once the workers are produced, they may kill the extra queens or the queens may quarrel and eventually set up separate nests. In other species of ant, the colonies are huge and contain many queens. The workers can help start up a new colony by leaving with a queen to make a nest at a new site. More intriguing still, a queen may enter the nest of another species and have her young cared for entirely by that species, and eventually her offspring take over the nest and oust the original queen. The queens of some species of ant that invade other species' nests do not bother to produce their own workers but just produce new queens and males. Once the new queens are fertilized they find other nests to invade. These are social parasites which depend on the social behaviour of other species in order to survive.

The workers do all the jobs in the nest apart from laying the eggs which, once colony is established, is the queen's sole responsibility. The workers feed the larvae regurgitated food but sometimes they lay infertile eggs which they feed to the older larvae and the queen. In some species of ants, the workers capture the workers and brood of other species (or sometimes even those of the same species, but from different colonies) and make them work as slaves.

Nests can be constructed underground, in a mound, or in vegetation. Weaver ants in the tropics are famous for their nests made by sewing leaves together. The worker weaver ants hold the leaves together while other workers stick the leaf edges together with a sticky substance produced by the larvae which they hold in their jaws. In many species of ants, the nest has separate chambers for the queen, eggs, larvae and pupae. As the young ants develop the workers transport them about the nest to another chamber. Not all the workers are the same size or do the same job. Some of the smallest leaf-cutter ant workers never leave the nest and are responsible for maintaining the fungal gardens. In some species the largest workers are soldiers that patrol the nest, but these can have other jobs as well, such as finding food. The difference in size between the smallest and largest workers can be tenfold. The control of the nest is highly complex with both workers and queens playing a part.

Termites are also social insects living in large colonies like ants. However, their young undergo direct development, so the eggs hatch into nymphs which look like small adults. Unlike ants and the social bees and wasps–where the workers are females–the termite workers can be of either sex. The males are always produced from fertilized eggs and have their full complement of genetic material. The termite colony is founded by a queen and a king (or sometimes several) and mating continues after the nests has been established. The queen's body becomes grossly distorted as she turns into an egg-laying machine, producing over 30,000 eggs per day.

The first nymphs to hatch from the eggs are workers, which tend the queen and king, forage for food, take care of the young, and begin to build the nest. When there are enough workers, soldiers are produced whose sole role is defending the colony. Workers are usually blind, which is not surprising since they spend most of their days hidden in the nest. Even on foraging trips worker termites construct mud tunnels to reach new sources of plants or wood to feed on, and they cover their food source with a layer of mud. However, it does seem strange that the soldiers are also blind, because they must defend the colony when the walls of the nest are breached. When the colony is fully established, winged males and females are produced, and they exit in great numbers through holes made by the workers and mix with those from other nests to mate.

The most impressive termite nests are those on the grasslands, such as those fungus-growing termites which live on the African savannas. These termites build some of the biggest nests known, over 7 metres high, to give them protection against the burning African sun and space to grow their fungal gardens. The nest is made of mud with a central chimney that draws air through the nest to keep it cool. In the rainforests, heavy rain is the problem and some termites living here build nests with a series of roofs shaped like umbrellas stacked on top of each other which the rain drains off. The Australian magnetic termites align their nests on the dry grasslands in a north-south direction. This means the flat sides of the wedge-shaped nest are exposed to morning and evening sun helping to keep the colony warm but at mid day–when the nest could overheat–the sun beats down only on the small surface of the blade-like top edge.

Mega Societies

The complex nests of the social insects are wonderful creations, providing protection for their young and allowing enormous populations to develop. A hive of honeybees can contain 80,000 individuals; termites' nests have been known to house seven million individuals; while a supercolony of ants in Japan consisted of over 300 million workers and one million queens housed in 45,000 interconnecting nests. The ultimate aim of these massive societies is to breed more pairs so that there are generations to follow. These reproductive pairs in turn set up more colonies which produce more breeding pairs.

The elaborate nests of social insects also provide new habitats for yet more insect species. There are moth species only found in bees' nests, bugs and fly maggots in termites' nests, and many species of beetles found in ant's nests. The uninvited guests may steal food or even eat their host's eggs and young. Some guests are welcome, such as certain species of caterpillar which give the ants food in return for their protection. Even invited guests can turn nasty, such as the caterpillars of the large blue butterfly from Europe, which the ants take down into the nest, but the caterpillars then feast on its brood. The larger the insect colony and the longer its total life-span, the greater the diversity of insects and other creatures sharing its space. So, insects themselves help to divide the world's living space into smaller and smaller units creating more places for further species of insects.

2

Reproductive Organs

Reproduction is the multiplication of organisms and the organs involved in it are the *reproductive organs.* The reproductive organs differ form all the other organs of the body in that their functions do not contribute primarily to the welfare of the individual of which they are a part; their chief concern lies with the succeeding generation. On the other hand, many of the activities of the organism must be correlated with the reproductive functions. This correlation is largely brought about in vertebrate animals by the secretion of hormones in the gonads, but with insects there is no evidence that the reproductive organs have a regulatory effect on the activities of the body organs or on the development of secondary sexual characters.

In insects, the reproductive system is a complex of organs derived from three anatomical sources. Its parts, therefore, may be classed in three morphological groups as follows: (1) primary internal mesodermal organs, (2) secondary ectodermal parts, produced from invaginations of the body wall, forming the usual exit apparatus; and (3) external appendicular structures. In a study of the reproductive system, however, it will be found more convenient to divide the subject into *internal genitalia* and *external genitalia,* each division including some of the parts classed above in the second group. The internal genitalia serve to lodge the germ cells, to provide for their nutrition, and to furnish them a protected space within the body where they may undergo a part or all of the development that brings them to a stage ready for conjugation; they include devices for insuring fertilization, and glands for the production of mucous or an adhesive or protective medium; and, finally, they discharge the germ cells from the body at the proper time.

The external genitalia accomplish the union between the sexes and enable the female to deposit the eggs according to the manner fixed in the instinct of each particular species. Insects, with rare exceptions, are bisexual, in so far as the male and female germ cells are matured in separate individuals; but in some species males are not known to exist, and parthenogenesis is of frequent occurrence, the unfertilized eggs in some cases producing males, in other females. A condition approaching hermaphroditism is said to occur in a species of Plecoptera, *perla marginata.* In the male organs of this insect, illustrated by Schoenemund (1912), the genital ducts are united in a transverse arch, as they are also in the female, and in young males the median part of he arch bears, between two lateral groups of sperm tubules, a number of small tubules containing egglike cells. These "male eggs," however, in the diploid stage, according to Junker (1923) have the same chromosome formula as the sperm cells, each being provided with two unlike heterochromosomes, and, though the undergo a partial development like that of a normal egg cell, they never reach maturity, and the tubules containing them degenerate when the insect approaches the time for transformation to the adult. The occurrence of functional hermaphoditism has been demonstrated by Hughes-Schrader to occur in the cottony-cushion scale *Icerya purchasi,* in which, she says (1930), the so-called females "are in reality hermaphrodites capable of self-fertilization of their own eggs by their own sperm."

The gonad of these hermaphroditic females, which are diploid in chromosome constitution, is primarily a pair of ovaries united anteriorly above the alimentary canal. The organ at an early stage contains no lumen, and in the first instar some of its central cells become reduced to the haploid condition and proliferate a solid central core of haploid cells. The outer cells of the gonad remain diploid and form the ovariolos with their contained oocytes and nurse cells. The central haploid cells develop into spermatozoa. The oocytes undergo normal maturation, resulting in haploid ova, and when the latter are fertilized they give rise to diploid hermaphrodites; unfertilized eggs develop parthenogenetically into haploid males. Males of this species are rare; they have been observed to copulate with the females, but it is not known that they accomplish fertilization. The majority of insects are amphigonous and oviparous. The spermatozoa are stored in a receptacle of the female at the time of mating.

The eggs in most cases are fertilized as they are extruded from the oviduct and thus undergo their entire development outside the body of the female. Viviparity, however, is of frequent occurrence among insects of various orders, the eggs, or even the larvae, being retained within the body of he female where they complete a part or all of their development. Usually, in such cases, the egg is deposited just before the time of hatching and almost immediately gives issue to an active young insects. Hatching may take place, however, within the egg passage of the female, where the larva then spends a varying length of time before its extrusion, as in the viviparous Diptera that give birth to living maggots. In the tsetse fly and Hippoboscidae the larva completes its development within the body of the parent female and pupates on emergence. In the viviparous Diptera the developing egg or larva is retained in a dilation (uterus) of the vaginal section of the median egg passage, where in pupiparous forms the larva is nourished from special glands.

In other cases, however, as in viviparous Aphididae and Coccidae, and in the viviparous beetle *Chrysomela varians,* the embryos are developed within the egg tubes of the ovaries. Probably in most cases development within the ovaries is parthenogenetic, but the eggs of *Chrysomela varians,* according to Rethfeldt (1924), are fertilized in the egg follicles by spermatozoa that travel upward into the ovarial tubes. The female of this beetle has no sperm receptacle, and copulation takes place during immature stages.

Female Reproductive Organs

The essential parts of the reproductive system in female insects consists of a pair of *ovaries,* two *lateral oviducts* converging posteriorly from the ovaries, and generally a *median oviduct,* or *oviductus communis,* receiving the lateral ducts anteriorly, and opening posteriorly to the exterior at the *gonopore.* In addition to these primary parts, there are usually a sac like *receptaculum seminis,* or *spermatheca,* for the reception and storage of the spermatozoa, a pair of *accessory glands* having various functions, and a copulatory pouch of the body wall, which is either an open *genital chamber,* or a tubular exit passage from the median oviduct, known as the *vagina.*

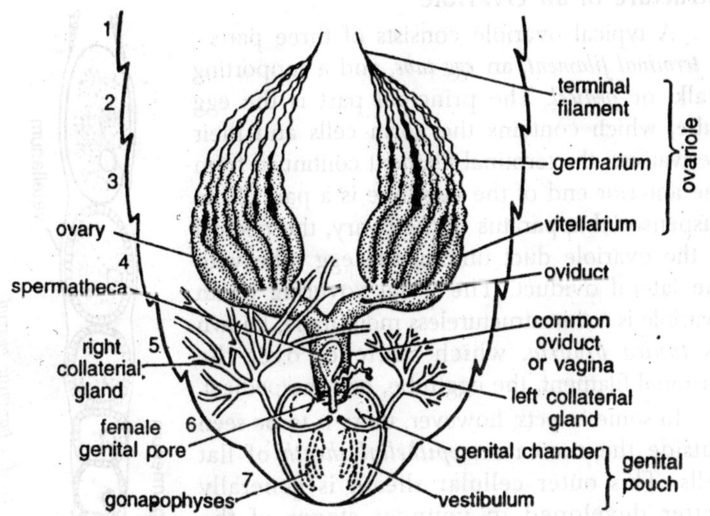

Fig. 2.1. P. americana. Female reproductive organs in dorsal view.

The Ovaries

Each ovary, in most insects, consists of a group of cylindrical or tapering units, the *ovarioles,* which ordinarily coverage upon the anterior end of the corresponding lateral oviduct, though in many of the more generalized insects the ovarioles of each ovary arise serially from one side of the oviduct. The anterior parts of the ovarioles consist of threadlike filaments, and usually all the ovariole filaments in each ovary are united distally with one another in a *suspensory ligament.* The ligament may end in the neighbouring fat tissue, but generally it is attached either to the body wall or to the dorsal diaphragm. In some cases, the ligaments from the two ovaries are combined in a single *median ligament,* which is inserted in the ventral wall of the dorsal blood vessel.

In young stages the entire ovary is usually incased in a *peritoneal sheath* of adventitious connective tissue; in the adult this sheath is generally absent, though it may be retained, as in some Diptera, in the form of a membrane enveloping the ovarioles. The number of ovarioles in an ovary varies greatly in different insects. Usually it is not large, four, six, or eight ovarioles being perhaps typical, but in some Hymenoptera and Diptera the number may be increased to 100 or even 200, and in the Isoptera it is said to reach 2,400 or more. On the other hand, the number of ovarioles may be reduced to two or to one.

Structure of an Ovariole

A typical ovariole consists of three parts : *a terminal filament,* an *egg tube,* and a supporting stalk, or *pedicel.* The principal part is the egg tube, which contains the germ cells and their derivatives; the terminal filament continued from the anterior end of the egg tube is a part of the suspensorial apparatus of the ovary, the pedicel is the ovariole duct uniting the egg tube with the lateral oviduct. The usual covering of an ovariole is a thin structureless membrane, known as *tunica propria,* which stretches over the terminal filament, the egg tube, and the pedicel.

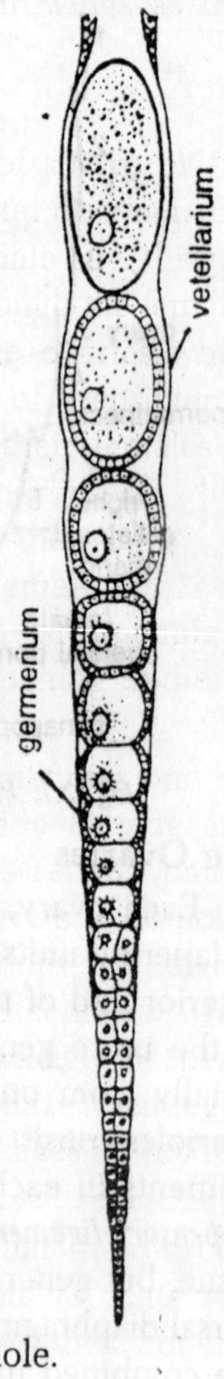

In some insects, however, there is to be seen outside the tunica an *epithelial sheath* of flat cells. This outer cellular sheath is generally better developed in younger stages of the ovary. It is sometimes regarded as a connective tissue layer, but according to K. Schneider (1917) its cells in the moth *Deilephila* originate within the body of the embryonic gonad. The epithelial sheat, therefore, is probably of mesodermal origin and represents the original mesodermal wall of the gonad. In early developmental stages it is continuous over the entire ovarial rudiment; but when the latter becomes divided into compartments that are to form the definitive ovarioles, the epithelial sheath is inflexted between the lobes and eventually forms a covering about each ovariole.

Fig. 2.2. Parts of a panoistic ovariole.

The terminal filament

The slender, threadlike filament that forms the anterior part of an ovariole is a solid strand of cells ensheathed in the tunica propria. The group of terminal filaments in each ovary is derived from the embryonic sheet of mesoderm by which the primitive ovarial rudiment was suspended from the splanchnic wall of the coelom and, as we have seen, the terminal filaments form in the adult the suspensory apparatus of the mature ovary. In some insects the terminal filaments are not united with one another, and in a few cases they are absent.

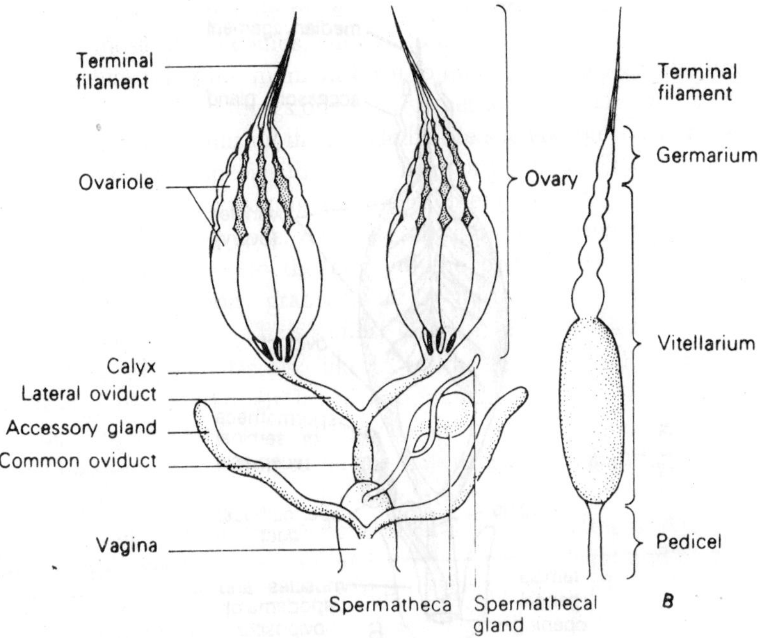

Fig. 2.3. Generalized female reproductive system. A–Principal female organs. B–Single ovariole.

The egg tube

The middle section of an ovariole, the egg tube, represents the intermediate part of the embryonic gonad, which contains the germ cells. In the adult organ it is an elongate sac filled with a mass of cells derived from the germinal elements, some of which become the ova. The wall of the egg tube is formed of the tunica propria, outside which there may be a cellular epithelial sheath, but the latter, as we have observed, is often not evident in adult insects. Most of the length of the egg tube is lined inside the tunica propria by a layer of *follicular epithelium.* In each mature egg tube there are to be distinguished two principal parts.

At the anterior end is a region containing the germ cells in an active state of division and incipient differentiation. This region is the *end chamber,* or *germarium.* Beyond the germarium is the region in which the egg cells grow and attain their mature size. This part of the egg tube is the *zone of growth,* or *vitellarium.* The germarium represents the primary egg tube of the young ovary in which the

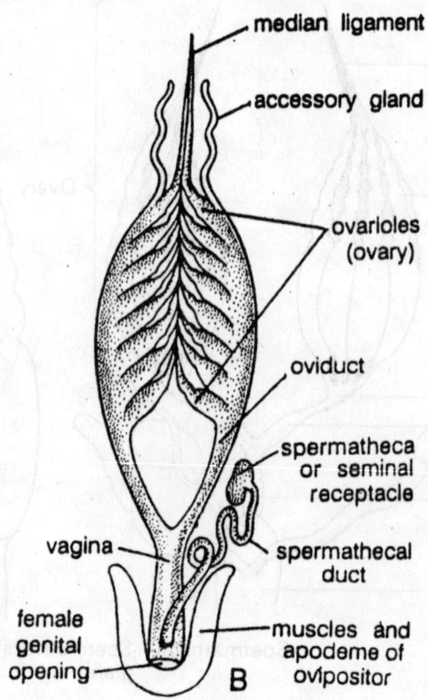

Fig. 2.4. Grasshopper. Female reproductive system.

germ cells are lodged. The germ cells, however, soon develop into *oogonia,* and from the latter are formed the *oocytes,* or young egg cells, which are generally accompanied by nutritive *nurse cells.* In the germarium there are usually also smaller cells, which become the *follicle cells* in the lower part of the egg tube.

In more general terms we may designate the follicle cells *cystocytes,* and the nutritive cells *trophocytes,* since both types of cells accompanying the true reproductive cells are found in various forms in both the ovary and the testis. The vitellarium is formed during the development of the insect as an extension of the egg tube beyond the germarium. It enlarges rapidly as the oocytes multiply and mature and varies in length according to the number and size of the eggs it contains. At the upper end of the vitellarium the follicle cells take a peripheral position in the tube and show the beginning of a definite epithelial arrangement, while the oocytes and nurse cells assume an axial position.

The growth of the oocytes distends the vitellarium into a series of *egg chambers,* or *follicles,* which become successively larger toward the posterior end of the tube. The follicular walls are formed by the follicle cells, which enclose each egg, or each egg and its accompanying nurse cells, in a cystlike sac. Beyond the last egg chamber a mass of follicle cells forms a "plug" that closes the rear end of the egg tube. Different types of ovarioles result from the presence or absence of nutritive cells or from differences in their position in the egg tube, but these modifications will be noted later. Since the oocytes are produced continuously from the oogonia, the first encysted becomes the lowermost and the first mature oocyte in the vitellarium. With each successive addition of an oocyte to the series the egg tube becomes lengthened between the last formed egg chamber and the germarium by a rapid multiplication of the follicle cells in this region.

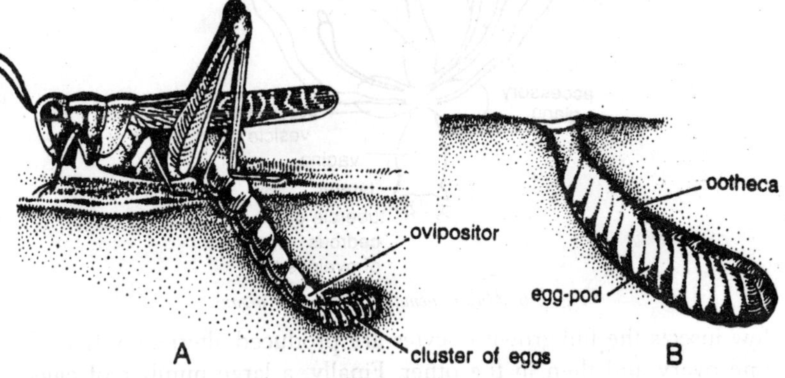

Fig. 2.5. Grasshopper. Oviposition. A–Female grasshopper laying eggs in ground; B–Egg-pod of grasshopper.

The ovarioles thus undergo a great increase in size during the growth period of the eggs, which is usually the early part of the imaginal stage of the insect. The last egg in each egg tube, when fully formed, is usually abruptly larger than any of those preceding it and is enclosed in the chorion, but it is still in the oocyte stage, since maturation does not ordinarily take place until the egg is laid. Hence, it is not strictly proper to speak of the full-grown ovarial oocytes as "mature" or "ripe" eggs. If an oocyte is fully formed in all the ovarioles at the same time, the insect may deposite at one laying as many eggs as there are ovarioles in the ovaries. Some insects, however, such as the queen been and the queen termite, eject the eggs singly in a continuous succession. With the

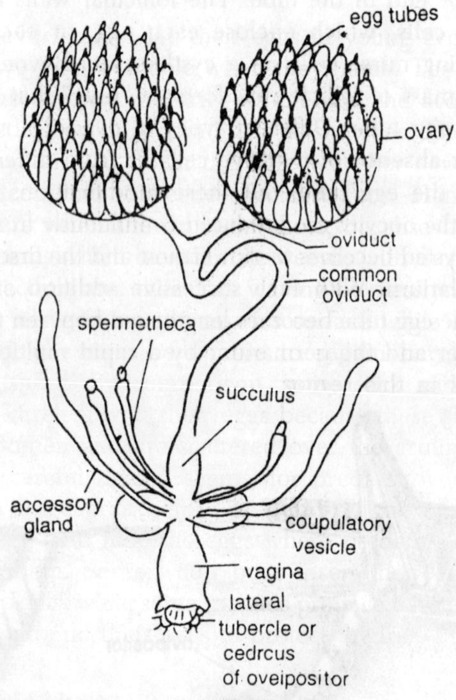

Fig. 2.6. Musca. Female reproductive organs.

few insects the full-grown oocytes are produced alternately first in one ovary and then in the other. Finally, a large number of eggs may accumulate in the exit passages before they are discharged. Thus in Ephemerida and Acrididae the elongate calyx and wide lateral ducts, prior to oviposition, become distended with eggs ready to be laid, and in some of the Lepidoptera the greatly lengthened pedicels of the ovarioles form storage tubes filled with long series of eggs, in which the latter are held until the moth is inseminated.

The ovariole pedicel

The pedicels, or stalks, of the ovarioles are short ducts connecting the egg tubes with the lateral oviduct. They are formed from the strand or mesodermal cells along the lower margin of the embryonic gonad, which is continuous posteriorly with the primitive oviduct. Each pedicel develops a lumen communicating that of the duct but closed at the upper end beneath the epithelial plug of the egg tube. At the time the first egg is ready to be laid the cells of both the plug and the retaining wall of the pedicel are dissolved to open a passage through the pedicel from the egg tube into the oviduct. The walls of the mature pedicels consists of a simple elastic epithelium, and in some insects the muscle sheath of the oviduct is continued upon them.

The Tropical Function of the Egg Tubes

The growth of the eggs in the ovary, which is mostly the accumulation of yolk, involves the utilization of much nutrient material. This material is supplied either from the daily food of the insect absorbed into the blood or from the food reserves stored in the body, principally in the fat tissue. With insects that do not feed in the imaginal stage, all the added egg material must be drawn from the latter source. Since the eggs in the egg tubes do not have direct access to the nutrient elements of the blood, an important function of the ovary is that of an intermediate trophic organ between the blood and the eggs. It discharges this function in various ways.

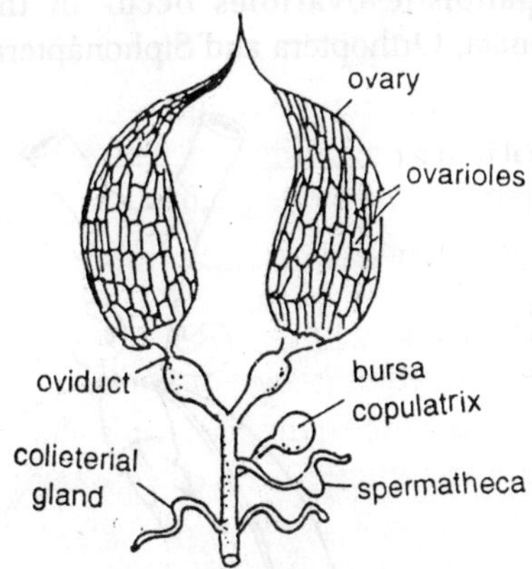

Fig. 2.7. Apis. Female reproductive organs.

It is possible that the eggs in some cases utilize material that passes by diffusion directly through the walls of the ovarioles; but observations on the histology of the epithelial layer of follicle cells indicate that more generally the walls of the egg chambers play an active physiological role in the nutrition of the eggs. On the one hand, the follicular cells apparently absorb food material from the blood and elaborate it in their cytoplasm, while, on the other hand, either they discharge the products into the egg tubes, where the material is directly or indirectly taken up by the egg cells, or the highly nutritious plasma itself of the follicle cells is absorbed by the egg cells. The first process is the more usual one. The structure of the egg tubes differ somewhat according to the manner in which the oocytes and nourished. In general two principal types of structure are distinguished by the presence or absence of the special nutritive cells within the follicular tubes. When there are no nutritive cells, an egg tube is classed as *panoistic*, when nutritive cells are present the tube in *meroistic*. The meroistic type is again subdivided into two groups known as *polytrophic* and *acrotrophic* according to the position of the trophocytes in the egg tube.

The panoistic type of egg tube

In an egg tube of their type there are no special nutritive cells differentiated from the egg cells; the food products elaborated by the follicular epithelium are absorbed directly by the oocytes. There is within the vitellarium, therefore, only a series of egg cells, each of which is generally contained in a distinct follicular egg chamber. Insects having panoistic ovarioles occur in the Apterygota, Ephemerida, Odonata, Orthoptera and Siphonaptera.

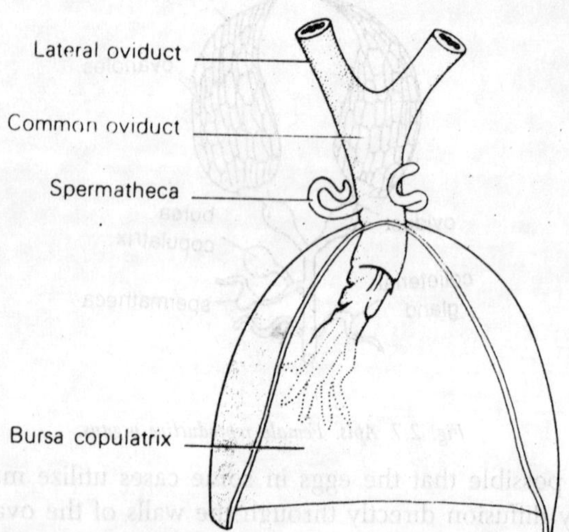

Fig. 2.8. Duct of a female triatomid bug, Rhodnius sp. (Hemiptera; Reduviidae), showing opened bursa copulatrix and entrance to the common oviduct.

The polytrophic type of egg tube

An egg tube of the polytrophic type contains an alternating succession of oocytes and trophocytes. In most cases the trophocytes, or nurse cells, are descendants along with the oocytes from the oogonia, but in some insects they are said to be derived from the follicle cells. Where the trophocytes are of germ-cells origin, the cells produced by the division of a single oogonium adhere in compact groups and maintain connections with one another in the form of protoplasmic strands. Only the most posterior cell in each group becomes a functional oocyte; the others specialize in the trophic function and become differentiated as the nutritive cells. Usually there is a definite number of nurse cells to each oocyte.

In muscoid Diptera, according to Verhein (1921), the proportion is 16 to 1, while in the honey bee, according to Paulcke (1901), the ratio is 48 to 1. In the first case, it would seem that the egg cells and the nurse cells are differentiated from the cells of the third generation produced from a single oogonium; in the second, the egg is one of the daughter cells of the first division, and the nurse cells the descendants of the other by four succeeding divisions. In *Deilephila euphorbiae,* it is said by K. Schneider (1917) that, of the five nurse cells accompanying each oocyte, four daughter cells of the oogonium and one a sister cell of the oocyte. On the other hand, the number of nurse cells accompanying each egg may be highly variable, as in Carabidae, described by Kern (1912). In *Carabus violaceus* Kern found a maximum of 127 nurse cells with a single egg cell, and the total of 128 cells he presume to have been produced by seven successive divisions from one oogonium. The polytrophic type of egg tube is characteristic of Anoplura, Neuroptera, Coleoptera, Lepidoptera, Hymenoptera and Diptera.

In some insects with polytropic ovarioles, such as Coleoptera and Lepidoptera, each oocyte and its accompanying nurse cells occupy the same ovariole chamber; in the others there is an alternation of egg chambers and nutritive chambers the nutritive chambers in such cases being the larger in the upper parts of the tubes, while in the lower parts the size relation is the reverse owing to the growth of the egg at the expense of its nurse cells. The nurse cells at first increase in size presumably by absorbing material elaborated by the cells of the oocytes are nourished by an active streaming of the plasmatic contents of the nurse cells into the cytoplasm of the eggs along the strands originally connecting the cells in the group produced form a primary oogonium. When the oocyte is mature, its nurse cells are exhausted and reduced to mere remnants in a state of degeneration. The production of special groups of nutritive cells from the follicular epithelium has been described in Apterygota by Willem (1900), in the May beetle *Melolontha vulgaris* by Mollison (1904), in a moth, *Deilephila euphorbiae,* by K. Schneider (1917), and in Tenthredinidae by Peacock and Gresson (1928).

In some of the Apterygota, according to Willem, the epithellial nurse cells form large protoplasmic masses of nutritive material in the egg tubes alternating with groups of oocytes. Earlier students of these insects, Willem claims, mistook the egg cells for the nurse cells and regarded the masses of epithelial nutritive cells as the oocytes. In *Melolontha,* as described by Mollison, the follicle cells of the egg tubes form a mass of nutritive cells in the upper part of each egg chamber, many of which become connected with the oocyte by protoplasmic strands, through which their contents are passed into the cytoplasm of the oocyte. A similar condition is reported by Peacock and Gresson in Tenthredinidae, where certain cells of the egg follicles appear to become nutritive cells since the chromatin of their nuclei is discharged into the oocytes.

The acrotrophic type of egg tube

In a few insects, particularly in the Hemiptera and some Coleoptera, the cells produced with the oocytes from the oogonia,

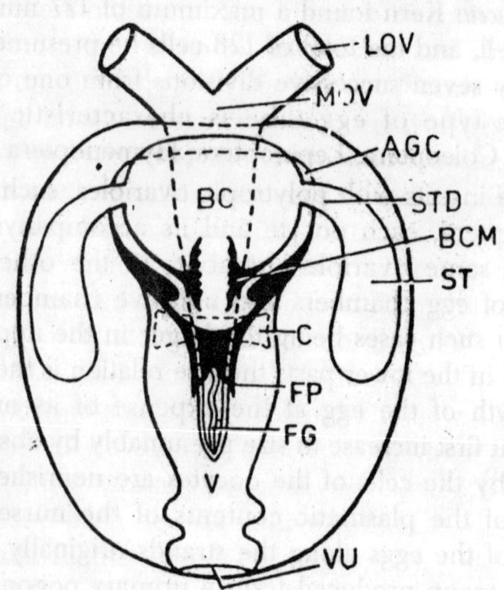

Fig. 2.9. The female genital complex in the drogonfly, Tramea. AGC–anterior genital chamber (anterior vaginal pouch); BC–bursa copulatrix; BCM–bursa communis; C–collar; FG–female gonopore; FP–fertilization pore; LOV–lateral oviduct; MOV–median oviduct; ST–spermathecae; STD–spermathecal duct; V–vagina; VU–vulva.

but which are destined to become nurse cells, remain in the upper part of the egg tube, while the oocytes become removed from them as the series of egg cells increases in the vitellarium. The original protoplasmic connections between the two sets of cells are maintained, however, as long plasmatic strands through which the oocytes in the egg tube continue to receive the yolk-forming material from the nurse cells. The germarium in the acrotrophic type of egg tube, therefore, is also an apical feeding chamber for the oocytes. A good illustration of the acrotrophic type of egg tube in Hemiptera is given by Malouf (1933).

Origin and Relation of the Cellular Elements of the Egg Tubes

There has been much difference of opinion as to the derivation of the various cell groups composing an ovariole and found within the egg tube. The terminal filament, the outer epithelial sheath of the egg tube (which may be absent in the adult organ), and the

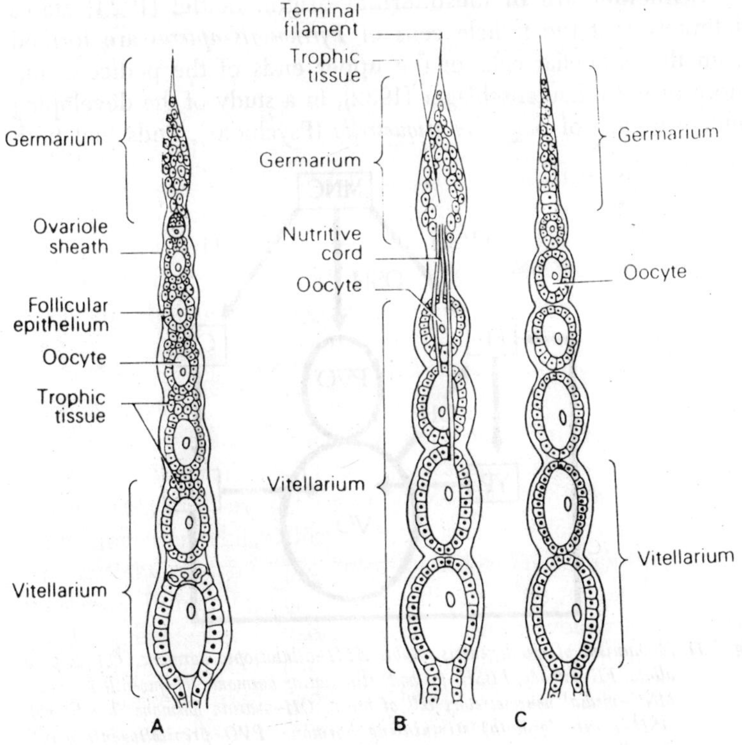

Fig. 2.10. Ovariole types. A–Polytrophic. B–Telotrophic. C–Panoistic.

pedicel are without doubt mesodermal parts of the ovariole derived from the primitive mesodermal covering of the germ cells. The tunica propria is probably a product or a remnant of the outer epithelial wall of the egg tube. The oogonia are direct descendants of the primary germ cells.

The oocytes and the usual trophocytes, or nurse cells that accompany the oocytes, are produced from the oogonia by mitotic division and nuclear changes. The principal question of origin, then, pertains to the derivations of the follicular cells. It is claimed by K. Schneider (1917) that in the moth *Deilephila* the ovarial follicle cells are proliferated anteriorly during the growth of the egg tube from the mass of the cells that closes the posterior end of the tube. The origin of the posterior cells Schneider did not discover, but he asserts that only the oocytes and the nurse cells are produced from the oogonia. There seems to be no question, however, that the generative cells of the follicle cells referred to by Schneider are of mesodermal origin. Seidel (1924) states definitely that the follicle cells of *Pyrrhocoris apterus* are formed from the epithelial cells of the upper ends of the pedicels, and more recently Lautenschlager (1932), in a study of the developing female organs of *Solenobia triquetrella* (Psychidae), finds that both

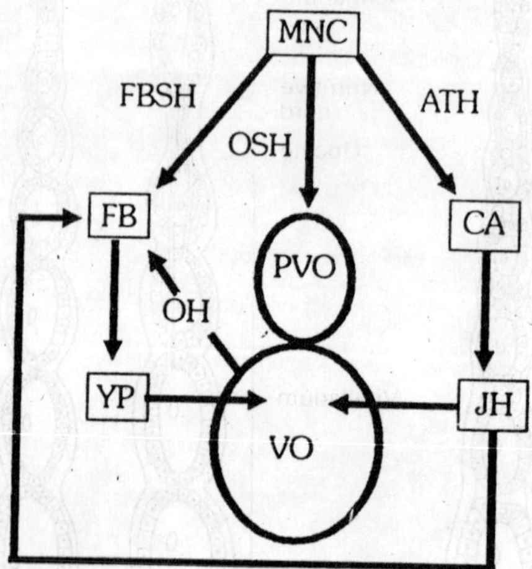

Fig. 2.11. Mechanism of vitellogenesis. Abbr. ATH–allatotropic hormone, CA–corpora allata, FB–fatbody, FBSH–fat body stimulating hormone, JH–juvenile hormone, MNC–medial neurosecretory cell of brain, OH–ovarian hormone (b-ecdysone), OSH–oocyte (growth) stimulating hormone, PVO–previtellogenic oocyte (penultimate), VO–vitellogenic oocyte (terminal), YP–yolk protein.

the follicle cells and the posterior masses of cells that give rise to the ovariole pedicels are derived from the mesodermal sheath of the primitive gonad.

It seems most probable, therefore, that the follicular egg tube, which, during the multiplication of the oocytes, is interpolated between the germarium and the pedicel of the ovariole, is formed by cells proliferated from the upper end of the pedicel and is hence of mesodermal origin. Earlier investigators believed that the follicel cells are descendants of the oogonia along with the oocytes and the usual trophocytes, though some claimed that they are derived from the mesodermal sheath of the gonad.

Formation of the Chorion and the Discharge of the Eggs

The mouth of each tube, as we have seen, is closed behind the last oocyte by a plug of the follicular epithelial cells, and the plug abuts against a transverse septum formed of the terminal wall of the ovariole pedicel. The last oocyte is thus completely enclosed in the follicular egg chamber and is prevented from escaping prematurely into the oviduct. When the egg is fully formed, the epithelium of the chamber begins a secretive activity producing a substance which is discharged upon the egg and there hardens to form the egg shell. This shell is the *chorion*. The substance of the chorion resembles in appearance the harder parts of the body wall cuticula, but it is invariably found to be nonchitinous. On its outer surface the chorion retains the marks of the cell that produced it in the form of a honeycomb pattern of fine ridges reproducing the outlines of the cells of the follicular wall. Only at the upper end of the egg is the chorion incomplete, there being left here a point not covered by the chorion deposit, which becomes the *micropyle* of the egg, an opening in the shell through which the spermatozoa gain entrance to the interior of the egg.

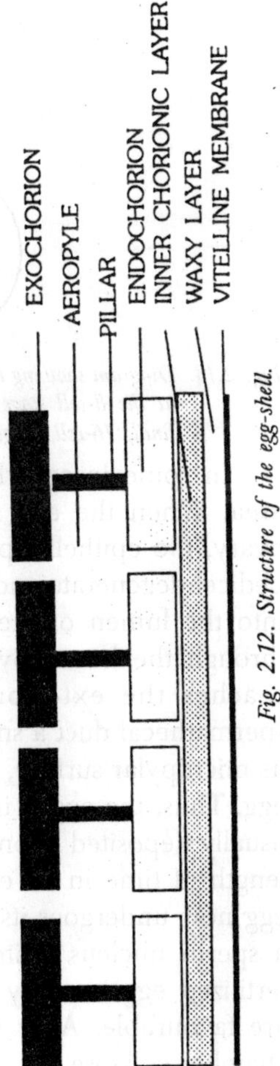

Fig. 2.12. Structure of the egg-shell.

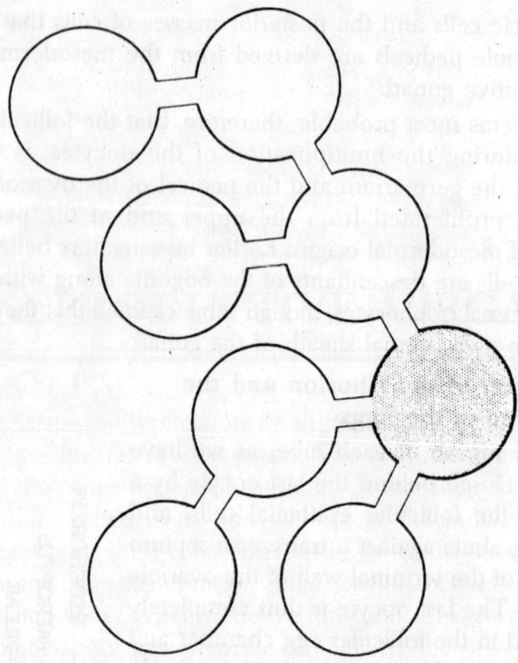

Fig. 2.13. Diagram showing the interconnections of the oocyte (stippled) and the trophocytes at the 8-cell stage in Drosophila. Each cell will divide again to produce the final, 16-cell, stage.

In some insects there are several apertures in the micropyle area. When the egg is finally ready to be discharged from the ovary, the epithelial plug behind it and the adjoining wall of the pedicel degenerate and open a passage through which the egg slips into the lumen of the pedicel and then goes through the calyx, through the lateral oviduct, into the median oviduct, and finally reaches the exterior. As the egg passes the mouth of the spermathecal duct a small mass of spermatozoa is discharged upon its micropylar surface, and some of the spermatozoa here enter the egg. Thus, the egg is inseminated just as it leaves the oviduct. It is usually deposited at once, but with some insects it is held a varying length of time in an external genital chamber of the female. The egg now undergoes its maturation divisions, and shortly thereafter a sperm nucleus unites with the nucleus of the ovum, and the fertilized egg is ready for development when external conditions are favourable. After an egg has left the follicle, the walls of its chamber collapse.

The epithelial cells, including the remanants of those that formed the closing plug, degenerate and are at last mostly dissolved and absorbed. The mass of degenerating cells in the lower end of an egg tube is sometimes called a *corpus luteum* in reference to its likeness to the degenerating Graafian follicle of a vertebrate ovary. With the disappearance of the posterior chamber in the egg tube, the next oocyte and its investing follicle assume a terminal position as the egg tube is lengthened by a growth in its cellular walls. The successive eggs thus probably do not literally pass down the egg tubes, as they are often said to do; more exactly, the tube shortens posteriorly by the degeneration of each emptied chamber and increases its length anteriorly by renewed growth to accommodate the newly forming oocytes.

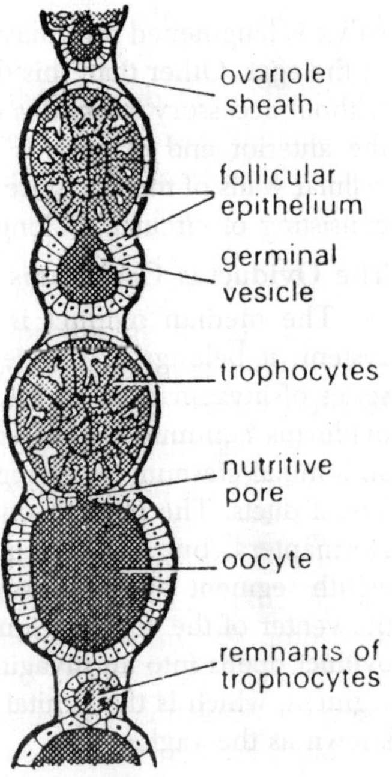

Fig. 2.14. Diagram of part of an ovariole of Bombus, a polytrophic ovariole in which the trophocytes are in a separate follicle from the oocyte.

The Lateral Oviducts

The duct that lead posteriorly from the ovaries are probably for the most part the primary mesodermal exit tubes of the gonads, though in some of the higher insects the mesodermal ducts are largely or entirely replaced by ectodermal tubes formed as branches of the median oviducts. In the early developmental stages of many insects the lateral oviducts are attached posteriorly to the body wall at the posterior margin of the seventh abdominal venter, but it is only in the Ephemerida that the lateral ducts have their permanent openings in this position. With all other insects the lateral oviducts discharge into a median invagination of the body wall, which becomes the oviductus communis. The anterior end of each lateral duct is generally somewhat expanded, forming a receptacle known as the calyx, into which open the pedicels of the ovarioles. When the ovarioles open serially into the oviduct the calyx is lengthened and may be greatly enlarged for the reception of the eggs. Other than this the oviducts are generally simple tubes without accessory structures of any kinds, though in the Acrididae the anterior end of each is prolonged into a tubular gland. The cellular walls of the ducts are usually covered by a muscular sheath consisting of circular or longitudinal fibers or both.

The Oviductus Communis

The median oviduct is not a part of the primitive genital system; it belongs to the secondary exit apparatus formed as a series of invaginations of the body wall. The first rudiment of the oviductus communis is an ectodermal pouch behind the seventh abdominal sternum receiving the two approximated mesodermal lateral ducts. The median ducts retains this primitive condition in Dermaptera, but in other insects it has been extended into the eighth segment by the closure of a groove continued from it on the venter of the eighth segment. Generally, the definitive median oviduct opens into an invagination of the body wall on the eighth segment, which is the genital chamber or a derivative of the latter known as the vagina.

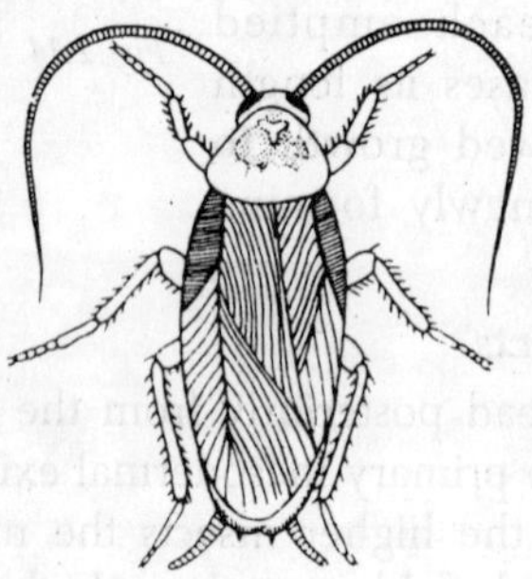

Fig. 2.15. Periplaneta. Ootheca formation. A–Abdomen of female showing the release of ootheca; B–A released ootheca.

The posterior opening of the median oviduct is the *female gonopore.* Primarily it is located on the rear part of the seventh abdominal segment; but, by the posterior extension of the oviduct, it occurs generally at the posterior end of the eighth segment, where usually it is concealed in the genital chamber or the vagina. The gonopore serves for the discharge of the eggs from the oviduct and is not a copulatory opening. It should be distinguished, therefore, from the *vulva,* which is the external opening of the genital chamber. The oviductus communis has a cuticular lining continuous with the cuticula of the body wall, and the entire epithelial tube is surrounded by a strong muscular sheath consisting of circular and longitudinal fibers. The length of the tube varies much in different insects, and its anterior end is sometimes bifurcate. In some insects branches of the median duct partially or entirely replace the mesodermal lateral ducts.

The Genital Chamber and its Derivatives

The median oviduct extended into the eighth segment does not ordinarily open directly to the exterior. Its aperture, the gonopore, is generally concealed in an inflection of the body wall behind the eighth sternum. The cavity thus formed in the *genital chamber.* The genital chamber receives the median oviduct and the duct of the spermatheca into its anterior end. It serves as a copulatory pouch during mating and is therefore properly termed the *bursa copulatrix.* Its external opening is the *vulva.* The genital chamber in its more primitive form is an open pocket of the body wall; but in many insects it becomes an internal pouch or takes the form of a tubular passage continuous with the median oviduct, in which case it is distinguished from the latter as the *vagina.* Since the vagina is a derivatives of the genital chamber, it opens primarily on the posterior part of the eighth abdominal segment. In many insects, however, as in some Cicadidae, Panorpidae, most Trichoptera, Lepidoptera, and Coleoptera, the vagina is continued through the ninth segment and has acquired an opening on this segment. In most such cases the primary anterior opening on the eighth segment is closed, and the posterior opening on the ninth segment becomes the functional vulva, serving both for copulation and for the discharge of the eggs.

In the majority of Lepidoptera, however, the anterior aperture is retained as a copulatory opening. The posterior vaginal opening, serving only for the discharge of the eggs. In the majority of Lepidoptera, however, the anterior aperture is retained as a copulatory opening. The posterior vaginal opening, serving only for the discharge of the eggs, may be distinguished in this case as the *oviporus.* The vulva of Lepidoptera having two genital openings leads into a passage connected with the vagina, which usually has a diverticulum serving as a copulatory pouch of the latter collectively represent the genital chamber of more generalized insects. The vagina is continuous with the median oviduct, and the spermatheca opens dorsally into its anterior end. A similar condition exists in certain species of Cicadidae. It should be observed that,

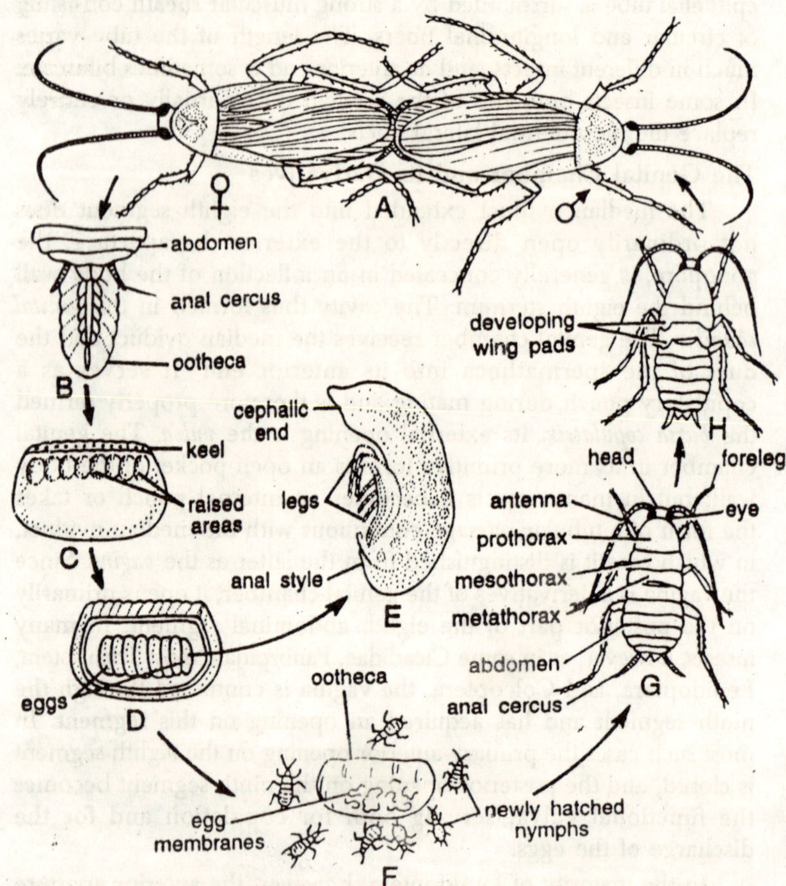

Fig. 2.16. Periplaneta. Life history. A–Copulation; B–Laying of ootheca; C–A single ootheca; D–Ootheca in section showing eggs; E–Early embryo; F–Hatching; G–Early nymph; H–Late nymph with wing pads.

in the continuous egg passage, the point of union between the true oviduct and the vagina is marked approximately by the opening of the spermatheca into the anterior end of the latter. Some confusion has arisen in the terminology of the female exit apparatus owing to a failure to distinguish between the true oviductus communis and the vagina, and because it has not been perceived that the vagina is a direct derivative of the genital chamber and not a continuation of the oviduct.

When the genital chamber is an open external pocket of the body wall, the facts are clear, and it should be recognized that an internal pouch or tube receiving both the oviduct and the spermatheca, though called the "vagina", is still the genital chamber or a part of it. Thus, in the Diptera, the saclike posterior part of the median egg passage, receiving the spermathecal ducts into its dorsal wall, and continuous anteriorly with the median oviduct, is the homologue of the open genital chamber of the Orthoptera. In the honey bee there is a shallow genital cavity concealed above the seventh abdominal sternum at the base of the sting, but from his external depression there open a large internal median pouch and two lateral pouches. The median pouch is functionally the vagina, but the fact that both the spermatheca and the oviduct discharge into it shows that it is a part of the genital chamber, as are also the lateral pouches. In the viviparous Diptera the anterior part of the genital chamber forms a pouch known as the *uterus*, into the anterior end of which open the oviduct, the ducts of the spermathecae, and the accessory glands.

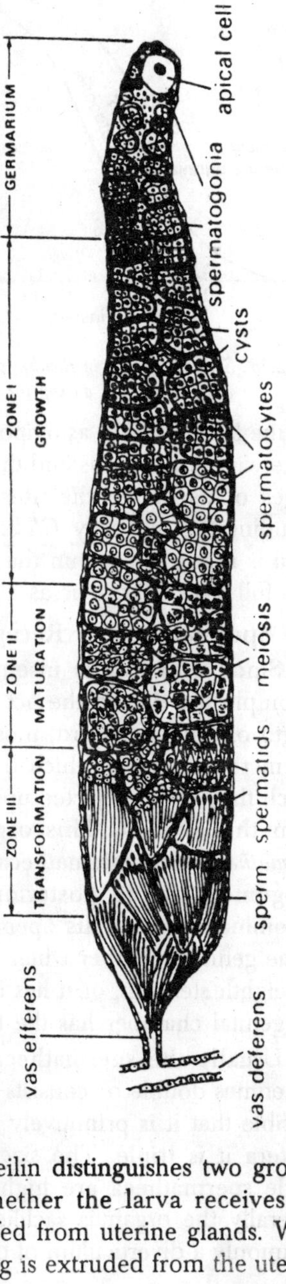

Fig. 2.17. Diagram of a testis follicle showing the stages of development of the sperm.

The egg is fertilized and in some species hatched, and the larva retained a varying length of time, even to maturity, within the uterus, where it may be fed from the secretion of the accessory glands. Keilin distinguishes two groups of viviparous flies according to whether the larva receives no nourishment from the mother or is fed from uterine glands. With some species of the first group the egg is extruded from the uterus,

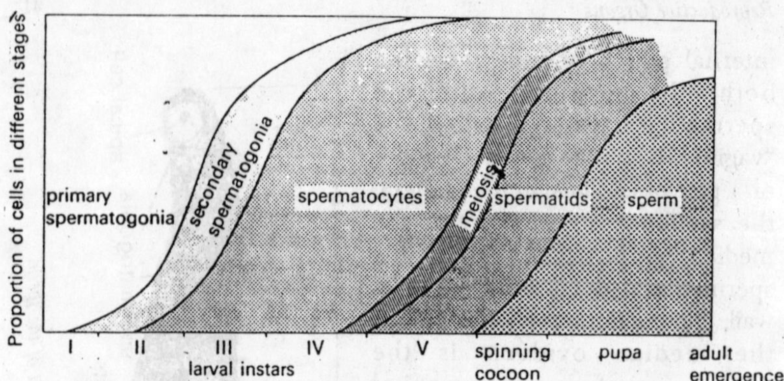

Fig. 2.18. Spermatogenesis in Bombyx showing the proportion of germ cells in each stage at different periods during the development of the insect.

but the larva hatches as oviposition occurs; in others hatching takes place within the uterus and the larva passes the first stage or several stages of its life in the uterine chamber. In the second group, including the testse fly *Glossina* and Hippoboscidae, in which the larva is nourished within the uterus, the young insect is deposited as a full-grown larva or as a pupa.

The Spermatheca, or Receptaculum Seminis

Since with insects insemination of the egg is not generally accomplished during the act of mating but takes place a varying length of time afterward, most female insects are provided with a sperm receptacle in which the spermatozoa are stored, and from which they can be ejected upon the eggs as the latter are extruded from the oviduct. This organ is the *receptaculum seminis,* or *spermatheca.* The spermatheca is primarily an invagination of the integument at the posterior end of the venter of the eighth abdominal segment. Its opening, therefore, comes to be enclosed in the genital chamber when a copulatory pouch is formed behind the eighth sternum, or it lies in the dorsal wall of the vagina when the genital chamber has the form of a vaginal tube.

Usually the spermatheca is a single organ; but since it is sometimes double or consists of two branches of unequal size, it is possible that it is primitively bifurcate or paired, thought in some Diptera it is triple. The size, shape, and structure of the usual single spermatheca are highly variable in different insects, but generally the organ is saclike in form with a slender duct. Very commonly a diverticulum of the duct forms a tubular spermathecal

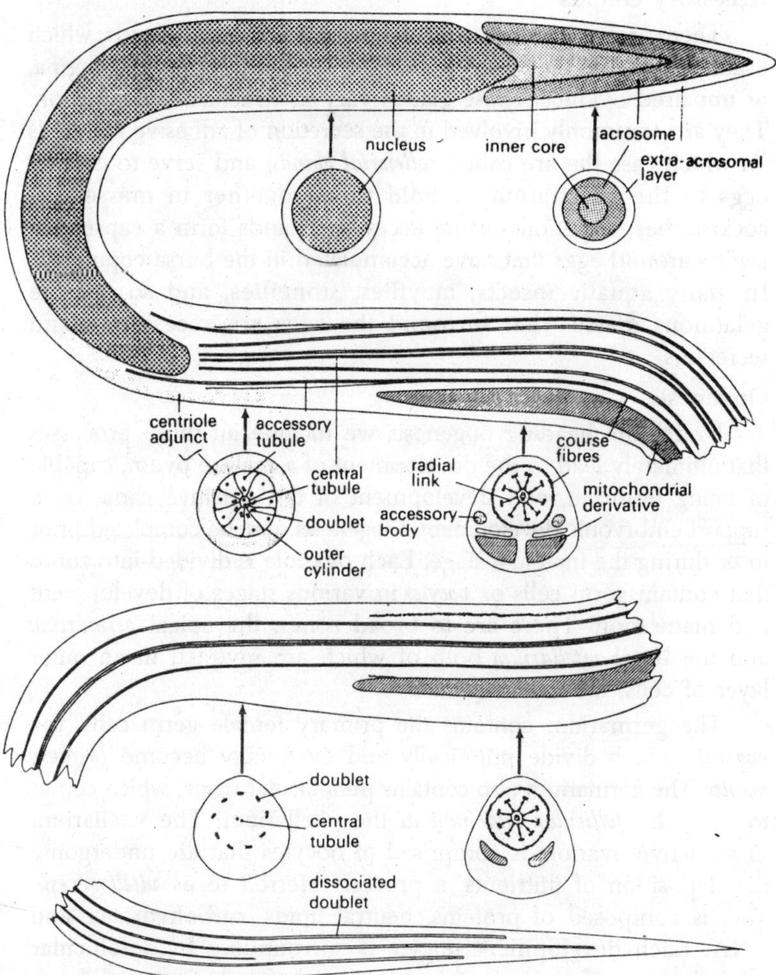

Fig. 2.19. Diagram showing the structure of a sperm in longitudinal section with representative transverse sections at the points shown. The anterior end shown in the upper half of the figure comprises only a small proportion of the total length.

gland, which secretes a fluid in which the sperm are discharged. On the outer surface of the duct there is a muscular sheath, and the muscle fibers are sometimes so arranged as to form a special pumping apparatus for ejecting the sperm, or a certain quantity of sperm-containing fluid, upon each egg as it issues form the oviduct into the genital chamber or vagina.

Accessory Glands

There are generally one or two pairs of accessory glands, which usually open into the apical portion of the bursa copulatrix, vagina, or unpaired oviduct. These glands vary in structure and function. They are commonly involved in the secretion of adhesive materials (in which case the are called *colleterial glands*) and serve to cement eggs to the substratum or hold them together in masses. In cockroaches, secretions of the accessory glands form a capsule or *ootheca* around eggs that have accumulated in the bursa copulatrix. In many aquatic insects, mayflies, stoneflies, and so on, the gelatinous masses that surround the eggs are accessory gland secretions.

Oogenesis

Under the heading oogenesis we include all those processes that ultimately lead to the development of a mature ovum, capable of being fertilized, and development of the nutritive capacity to support embryonic development. Oogenesis may be completed prior to or during the imaginal stage. Each ovariole is divided into zones that contain germ cells or *oocytes* in various stages of development and maturation. There are to broad zones, the apical *germarium* and the basal *vitellarium* both of which are invested in an outer layer of cells, the *ovariole sheath.*

The germarium contains the primary female germ cells, the *oogonia,* which divide mitotically and eventually become *primary oocytes.* The germarium also contains prefollicular tissue, which comes to form the *follicular eithelium* in the vitellarium. The vitellarium in an active ovariole is comprised of oocytes that are undergoing the deposition of nutrients, a process referred to as *vitellogenesis.* Yolk is composed of proteins, neutral lipids, carbohydrates, and RNA. Each development oocyte is surrounding by a follicular epithelium, and the oocyte and its associated epithelial layer comprise a *follicle.*

The oocytes in the vitellarium have progressively more yolk in an apical to basal sequence, the most mature basal oocyte being separated from the lumen of the pedicel by an epithelial plug, which ruptures when the oocyte is ready to leave the ovariole and proceed into the lateral oviduct. This process of exiting from the ovariole is called *ovulation.* Following ovulation, the follicular epithelial cells remain behind and eventually degenerate. There are three major types of ovarioles, based on the method by which yolk deposition occurs: *polytrophic, telotrophic,* and *panoistic.* Bonhag (1958), Mahowald (1972), and Telfer (1975) review ovarian structure and vitellogenesis.

Polytrophic ovarioles have nutritive cells, *nurse cells* or *trophocytes,* associated with each developing oocyte. Trophocytes are within the follicular epithelium, which surrounds each oocyte. Characteristically, the oocyte and trophocytes all originate from a single oogonium that undergoes a series of mitoses with incomplete cytokinesis. The results in the presence of cytoplasmic canals interconnecting the trophocytes and oocyte. A follicular plug exists between each oocyte with its accompanying trophocytes. Polytrophic ovarioles have been described in Neuroptera, Lepidoptera, some Coleoptera, Diptera and Hymenoptera.

Telotrophic ovarioles differ from polytrophic ones in that the trophocytes are not directly associated with each developing oocyte but are all located in the apical region of ;the ovariole and are connected to the various oocytes by means of "nutritive cords." Otherwise, the two types are similar, each oocyte being surrounded by a follicular epithelium and being separated from the others by follicular plugs. Telotrophic ovarioles are characteristic of Hemiptera and some Coleoptera. Polytrophic and telotrophic ovarioles are sometimes collectively referred to as *meroistic ovarioles.*

The third type of ovariole listed, panoistic, has no trophocytes. However, each developing oocytes is surrounded by the follicular epithelium, and follicular plugs exist between adjacent oocytes. Panoistic ovarioles are found in the apterygotes, Orthopter, Isoptera. Odonata, Plecoptera (stoneflies). Siphonaptera, and some Coleoptera.

In polytrophic and telotrophic ovarioles, the trophocytes furnish all or nearly all of the RNA contained in the mature egg. In the panoistic type, the oocyte nucleus produces all of the RNA in the mature egg. Much of the yolk protein originates in the fat body (as *vitellogenin*) and is transferred to the ovaries via the hemolymph. Yolk protein is called *vitelline.* There are few exceptional ovariole structures that do not fit into the above classification. For example, in ovarioles from the beetle *Steraspis speciosa,* the germarium and vitellarium are very short, there are no nutritive cords or trophocytes, and there is a long region of glandular tissue proximal to the vitellarium.

In the vast majority of insects oogenesis occurs in the last larval instar, the pupa, and the adult stages. However, in some species immature stages are capable of producing mature oocytes that may commence and even complete embryogenesis. This phenomenon, *pedogenesis*, has been described in a number of insect species, among which are *Micromalthus debilis* (Coleoptera) and cecidomyiid flies in the genera *Miastor* and *Oligarces.*

Eggs

Mature insect eggs are typically elongate and oval in longitudinal section, although some assume other forms. In most instances the largest portion of an egg is filled with *yolk* or *deutoplasm*, and the cytoplasm and nucleus occupy only a small portion. The yolk contains carbohydrates, protein, and lipid bodies, the protein bodies being the most abundant.

Cytoplasm is located around the nucleus (*nuclear cytoplasm*) and around the periphery of two membranes: the *vitelline membrane,* which apparently represents the cell membrane of the egg, and the *chorion*, or eggshell, which is secreted by the follicle cells. Several species lack a chorion (e.g., some viviparous species). When a chorion is present, it is nonchitinous, may be composed of two to several layers, varies considerably in thickness in different species, and may be smooth or sculptured in a variety of ways. Since insect eggs are usually deposited outside the parent female, the are subject to a drying atmosphere, attacks by parasites and predators, and other dangers.

The chorion serves the same functions as the cuticle. It serves as a protective coating (against physical damage, attack by parasites, etc.) and a barrier against water loss and is important relative to ventilation of the egg. Coatings secreted by the accessory glands that cover eggs may aid in water conservation in addition to serving to hold eggs together or glue them to the substratum. Some eggs laid in moist situations are capable of absorbing water from their surroundings. In insects in which the chorion is thin, ventilation may occur over the entire surface. In others the chorion may be lined with a porous, gas-filled layer that communicates with the outside of the egg by means of channels or *aeropyles.* Some eggs possess specialized structures that act as physical gills or plastrons, allowing them to obtain oxygen from water.

Hinton (1969) reviews and synthesizes the major literature pertinent to respiratory systems of insect eggs. Miller (1974a) also provides information on this topic. Spermatozoa gain entrance to an egg by means of one to several special channels or *micropyles,* which are perforations of the chorion and are located at various places on the eggs of different species. Hinton (1979) provides a comprehensive three-volume treatise on insect eggs. Furneaux and Mackay (1979) consider the structure, formation, and composition of the chorion and vitelline membrane.

Control of Oogenesis

It has been established for a number of species that the development of eggs–vitellogenesis, in particular–is controlled by a secretion of the corpora allata (or in the cyclorrhaphous Diptera, the region of the ring gland that is homologous with the corpora allata). Some of the kinds of experiments that have enabled investigators to arrive at this conclusion are

1. Histological observation of a correlation between ovarian activity and activity in the corpora allata.
2. Ovarian activity being attenuated by the microsurgical removal of the corpora allata (allatectomy) and restored by implantation.
3. Allatectomy, attenuating ovarian activity and hemolymph transfusion from a donor with active corpora allata restoring this activity.

The corpus allatum hormone may affect egg maturation by stimulating the incorporation of yolk into the oocyte and simultaneously regulating metabolism, particularly of proteins. In certain species (e.g., *Rhodnius* sp.) corpora allata from adults have been transplanted into larvae, and their secretion have had the same effect as the juvenile hormone. Conversely, larval corpora allata stimulate ovarian activity (gonadotropic effect) in adults. This has led many to hold the view that juvenile hormone and the gonadotrophic hormone may be one and the same. Recently, certain substances (e.g., farnesol and furnesyl methyl ether) have been found that mimic the effect of the corpus allatum hormone, acting as "juvenile" hormone and having a gonadotropic action (i.e., stimulating the ovaries). In some species of insects allatectomy does not prevent egg maturation, but there is evidence that the corpus allatum hormone may be present in the hemolymph from earlier stages.

In most species studied, secretions from the median neurosecretory cells in the pars intercerebralis in the brain are necessary for oogenesis. These secretions have been found to activate the corpora allata or stimulate protein synthesis (necessary for yolk formation) or produce a gonadotrophic hormone. Recall that the corpora cardiaca serve as neurohaemal organs storing and releasing the products of the neurosecretory cells in the brain. This allows the accumulation of secretory products in the corpora cardiaca for some time prior to their release from this organ. Thus, the action of the median neurosecretory cells is a function of the release of their products from the corpora cardiaca. Such release of material is probably under neural control. Other factors, mediated hormonally or via the nervous system, that have been found to influence the activity of the corpora allata in different species include mating, light (photoperiod), chemical stimulation (pheromones from males), various nutritional factors, and the presence or absence of eggs in the brood chamber. These factors may have a stimulatory or inhibitory effect on the corpora allata. In addition to secretions from the median neurosecretory cells and the corpora allata, ecdysone has been found to be involved in the control of oogenesis in female mosquitoes.

In response to ingestion of a blood meal, secretions from the median neurosecretory cells (i.e., *egg development neurosecretory hormone*) induce the ovaries to release ecdysone, which in turn triggers the synthesis of vitellogenin in the fat body. Juvenile hormone secreted by the corpora allata "activates" the fat body and ovaries. Several factors probably interact in the control of oogenesis in any given insect species, and the present level to the understanding does not permit broad generalizations. The reviews of Davey (1965), de Wilde and de Loof (1973b), Engelmann (1968), Riddiford and Truman (1978), and Telfer (1965) should be consulted for detailed information.

Male Reproductive Organs

The internal organs of reproduction in male insects having a single genital opening are in many respects similar to those of the female. In an adult insect the essential parts of the male reproductive system include a pair of *testes*, a pair of lateral ducts, the *vasa deferentia*, corresponding to the lateral oviducts of the female, and a median ectodermal exit tube, or *ductus ejaculatorius,* functionally comparable with the median oviduct of the female. Besides these constant parts there are generally present also accessory structures of a more variable nature. Frequently a section of each vas deferens, for example, is enlarged to serve as a sperm reservoir, or *vesicula seminalis*; or, again, a considerable length of the duct is thrown into a compact coil of irregular convolutions, forming an *epididymis.*

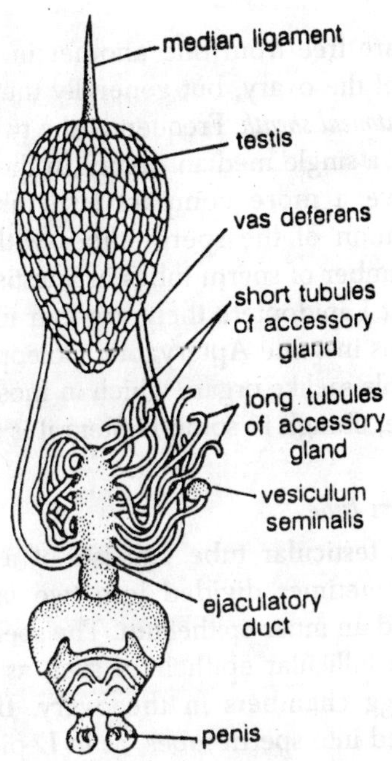

Fig. 2.20. Grasshopper. Male reproductive system.

Ectodermal *accessory glands* are commonly present in the form of pouches or blind tubes branching from the upper end of the ejaculatory duct. The external opening of the exit duct, or *male gonopore*, is generally situated on or within a median intromittent organ, the *penis* or *phallus*.

The Testes

Each testis consists typically of a group of short *sperm tubes.* The tube contain the male germ cells in successive stages of development, and other cells associated with the germ cells in various capacities. The sperm tubes in their origin and development correspond to the egg tubes of the ovaries, but they are usually called the testicular "follicles."

General structure of a testis

The testis in some of the more generalized insects closely resembles an ovary in that the sperm tubes arise serially from the distal part of the exit duct. Each tubule is attached to vas deferens by a small stalklike *ras efferens,* but the testicular tubes have no terminal filaments. In some insects, as in Apterygota and Plecoptera, the sperm tubes are free from one another in the adult stage, as are the ovarioles of the ovary, but generally they are all contained in an investing *peritoneal sheath.* Frequently the two testes of opposite sides are united in a single median organ. In the higher insects the testes usually have a more compact structure because of the incomplete separation of the sperm tubes within the peritoneal sheath (C). The number of sperm tubes in a testis varies in different insects; but in most Lepidoptera there are four tubules in each sex. In certain insects, as in some Apterygota, Coleoptera, and Diptera, each testis is a simple saclike organ, which in most cases is probably a single sperm tube, though in some Diptera it is said to be partially subdivided.

Structure of a sperm tube

The wall of a testicular tube consists of a cellular *epithelial sheath,* which is sometimes divided into two layers, forming an outer epithelium and an inner epithelium. The sperm tubes, however do not have a true follicular epithelium such as that which forms the walls of the egg chambers in the ovary. If the testis is but incompletely divided into sperm tubes, as in Lepidoptera, the septa between the compartments appear as folds of the epithelial sheath extending posteriorly toward the mouth of the duct, while the entire organ is invested in the peritoneal sheath. The walls of the testicular tubules probably serve as trophic intermediaries between the blood surrounding the gonads and the germ cells within them, as do the walls of the ovarioles.

The single long, coiled tube forming the testis of *Dytiscus,* according to Demandt (1912; Korschelt, 1924); is covered by two epithelial sheaths inside the peritoneal sheath. The thick outer epithelium consists of a spongy, granular plasma in which cell boundaries are not visible, but the cytoplasm is vacoulated by numerous small cavities, indicating that the outer epithelium has a secretory function and probably elaborates nutritive products discharged into the lumen of the tube. The outer epithelium is bounded externally by a basement membrane. The inner epithelium is a thin elastic layer having a fibrillated appearance and containing a large number of small nuclei. The single saclike testis of Diptera also is said to be surrounded by two envelopes, distinguished as the *tunica externa* and *tunica interna* by Keuchenius (1913). The outer tunic, as described by Lomen (1914) in the mosquito, is a thick connective tissue layer abundantly vacuolated by small spaces filled apparently with stored nutritive material. The inner tunic forms the lining of the testis.

In Lepidoptera, according to Ruckes (1919), the walls of the incompletely separated testicular compartments have the appearance of connective tissue and apparently serve for storage of reserve materials, including fat in some cases. In species with coloured testes this coat is the repository of the pigment granules. Within each sperm tube there are to be distinguished successive regions according to the state of development of the germ cells. The upper part containing the primary spermatogonia is the *germarium,* as in the ovary; beyond the germarium is a *zone of growth (I)* in which the spermatogonia enter a stage of multiplication and are usually encysted; next is the *maturation zone (II)* in which the maturation divisions take place; and lastly comes the *zone of transformation (III)* where the spermatocytes develop into spermatids and finally into mature spermatozoa. The entire process of spermatogenesis thus takes place regularly within the tubes of the testis.

The cellular elements within a testicular tube

A characteristic feature of the testicular tubes is the presence of a large cell or nucleated mass of protoplasm in the apex of

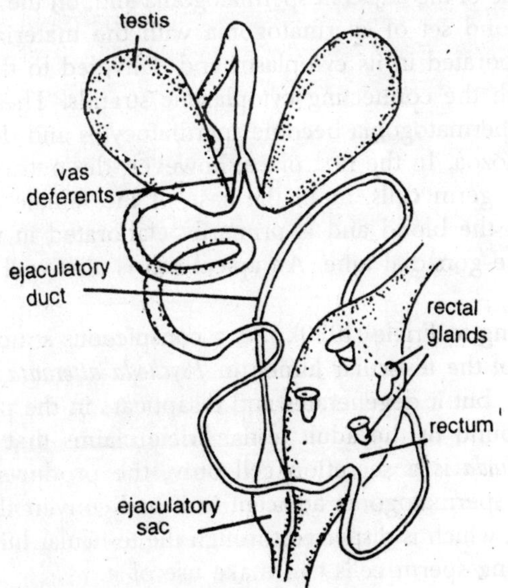

Fig. 2.21. Musca: Male reproductive organs.

germarium. This cell is known as the *Versonian cell,* or *apical cell.* Earlier investigators believed the apical to be the primary spermatogonium of the tube, from which by division all the other spermatogonia are produced. The majority of recent workers, however, regard the apical cell as a spermatogonium specialized as a trophocyte. The apical cell is particularly well developed in the Lepidoptera, where it consists of a large mass of cytoplasm containing a nucleus. Surrounding the apical cells are several concentric rows of spermatogonia, of which those nearest the apical cell are seen to be connected with the latter by protoplasmic strands containing dark granules that appear to originate tin the apical cell.

The spermatogonia of Lepidoptera, therefore, as pointed out by Grunberg (1903), must be nourished directly from the apical cell. The nutritive material utilized by the apical cell, it is claimed both by Grunberg and by Zick (1911), is derived from certain spermatogonia in the immediate neighbourhood of the apical cell which are dissolved and absorbed by the latter. According to this view, then, the apical cell of a testicular tube in the Lepidoptera is a spermatogonial nurse cell, which, on the one hand, dissolves and absorbs some of the adjacent spermatogonia and, on the other hand, feeds a second set of spermatogonia with the material from the first set elaborated in its cytoplasm and delivered to the recipient cells through the connecting cytoplasmic strands. These specially nourished spermatogonia become spermatocytes and develop into the spermatozoa. In the first place, however, the nutrient material of the male germ cells, as in the case of the female germ cells, comes from the blood and is primarily elaborated in the trophic sheath of the gonadial tube. An apical cell is also well developed in Diptera.

According to Friele (1930), it is a conspicuous structure in the upper end of the testicular lumen in *Psychoda alternata* during the larval stages, but it degenerates and disappears in the pupa and is not to be found in the adult testis. Friele claims that the apical cell of *Psychoda* is a secretion cell only, the products of which dissolve the spermatogonia adjacent to it and convert them into a fluid plasma, which is dispersed through the testicular lumen where the developing sperm cells can make use of it.

The spermatogonia that are destined to become spermatozoa undergo a series of divisions, but the cell produced from each

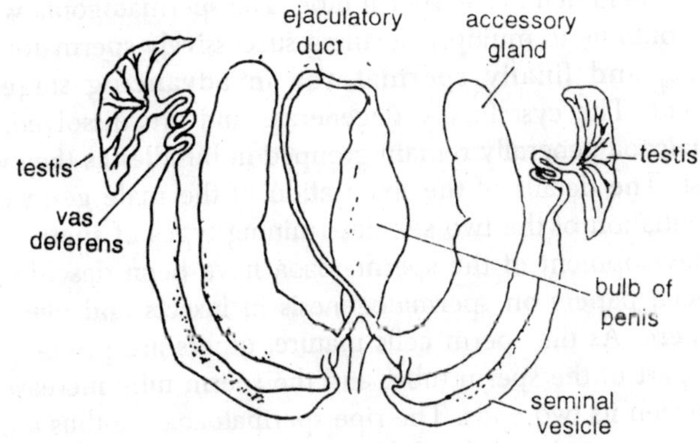

Fig. 2.22. Apis. Male reproductive organs.

primary sperm cell usually remain attached to one another centrally by protoplasmic strands and assume a radial position, giving a rosette pattern to the spherical mass is cross section. Each spermatogonial group in most insects soon becomes enclosed in a cellular envelope, known as a *sperm cyst.* The origin of the male cyst cells has been definitely observed in many insects, but, while some investigators have believed that these cells are derived from the sheath of the sperm tube, most writers regard them as products of the germ cells. According to Zick (1911), the differentiation between the secondary spermatogonia and the cyst cells in Lepidoptera is a matter of nutrition.

The primary spermatogonia immediately surrounding the apical cell, and which are directly nourished by the latter, he says, become the functional spermatogonia; those of the next rank, receiving little nourishment from the apical cell, become the cyst cells. Adjacent cells of these two groups attach to each other in pairs; the poorly nourished cell by division forms the cyst, the other produces the enclosed group of spermatogonia. On comparing the internal cellular organization of a testicular tube with that of an ovarial tube; it become evident that the cyst cells of the former correspond functionally at least to the follicel cells of the latter, though investigations seems to show that the ovarial follicle cells and the testicular cyst cells have different origins.

In the ovary the cystocytes produce a continuous epithelium lining the egg tube; in the testis they invest groups of spermatogonia separately. In some insects it has been observed that the testicular cysts are connected by strands of cells forming a network throughout the lumen of the sperm tube. The spermatogonia within each cyst continue to multiply, forming successively spermatocytes, spermatids, and finally spermatozoa in advancing stages of development. The cyst finally degenerate and are dissolved, but the spermatozoa generally remain grouped in bundles as they were in the cyst. The details of the maturation of the male germ cells, the differentiation of the two sex-determining types of sperm cells, and the development of the spermatozoa have been described in many special papers on spermatogenesis in insects and need not be given here. As the sperm cells mature, others are produced in the apical part of the sperm tube, and the sperm tube increases in length between its two ends. The ripe spermatozoa are thus always located at the posterior ends of the testicular tubes in proximity to the exit ducts.

The Vasa Deferentia

The ducts leading from the testes are usually simple tubes, each having a thick cellular epithelium limited by a basement membrane, outside which is a strong muscular coat of circular fibers. Frequently a part of each duct is much convolated, and the coils may form a definite epididymis. In some insects an enlargement of the duct in the shape of a dilation or diverticulum constitutes a vesicula seminalis for the storage of the mature sperm as the latter leave the testis. The spermatozoa are generally found closely massed in the vesicula, with their heads imbedded in the epithelial wall and their vibratile tails projecting like cilia into the lumen. The vasa deferentia are primarily of mesodermal origin, but they may be extended posteriorly by ingrowths from the ectoderm or more or less replaced by the latter. Thus, in Ephemerida, as shown by Wheeler (1983), the terminal parts of the male ducts may be lined with an ectodermal cuticula, and in the Diptera, according to Bruel (1897) and Friele (1930), the definitive ducts of testes are found entirely as lateral diverticula from the ectodermal ductus ejaculatorius. The lateral genital ducts of the male open separately to the exterior in Protura, Ephemerida, and some Dermaptera; in other insects they unite with the anterior end of a median ejaculatory duct.

Ducts

A tiny duct, the *vas efferens,* lead from each follicle to a common lateral duct, the *vas deferens*, and finally the vasa deferentia from each testis join to form the *ejaculatory duct,* which ends in the penis or aedeagus at the *gonopore.* The vasa deferentia are at least partly of mesodermal origin. In some insects (Protura, Ephemeroptera, and some Dermaptera) each vas deferens has its own opening to the exterior. Both the vas deferens and ejaculatory duct are invested with a layer of muscles, which are involved in the propulsion of semen. Each vas deferens may be formed into a series of convolutions forming an *epididymis* or have a dilated portion (i.e., *seminal vesicle*) in which the spermatozoa are stored in a quiescent state.

Accessory Glands

Various glandular structures are typically associated with the vasa deferentia or with the ejaculatory duct, although some insects (e.g., Apterygota and some Diptera) lack accessory glands. These accessory glands and their ducts are either of mesodermal origin (i.e., evaginations of the visa deferentia) or of ectodermal origin (i.e., evaginations of the ejaculatory duct). They usually occur as a single pair, but in some insects there may be several in a cluster (e.g., the mushroom body in male cockroaches). In some insects portions of the vasa deferentia or ejaculatory duct may also have glandular functions. Among the known functions of male accessory glands is the secretion of seminal fluid, activation of spermatozoa, and production of spermatophores.

In addition, accessory gland secretions may influence and inseminated female in a variety of ways including: stimulation of oviposition; acceleration of oocyte maturation; stimulation of contractions of the genital ducts that aid in sperm movement; and inhibition of subsequent insemination by formation of vaginal plugs or by exerting an effect on the female's behaviour. Hinton (1974) and Leopold (1976) discuss functions of male accessory gland secretions at length. There is evidence for hormonal control of seminal fluid production and growth of the accessory glands. Removal of the corpora allata from young males of several species retards accessory gland growth.

Spermatogenesis

The testicular follicles contain the *germ cells* and are hence the sites of the meiotic cell divisions that give rise to *spermatozoa,* the entire process being referred to as *spermatogenesis.* This process usually occurs during the last larval instar or pupal stage and in some species continues in the adult stage. Each follicle contains a large apical cell or complex of cells that apparently serve a trophic function, providing nutrients for the developing *spermatogonia.* Each follicle is divided apically to basally into zones that represent the different stages of spermatogenesis. Apically the *germarium* or *zone of spermatogonia* is comprised to the germ cells (spermatogonia) and somatic mesodermal cells. In the next region, several mitotic divisions, forming *primary spermatocytes* that become encysted in somatic cells.

The primary spermatocytes (diploid cells) undergo meiosis and produce haploid daughter cells in the next region, the *zone of maturation and reduction.* With the first and second meiotic division the primary spermatocytes become *secondary spermatocytes* and *spermatids,* respectively. In the basal *zone of transformation,* the secondary spermatids become transformed into flagellated spermatozoa. When the cysts in which they have been encaused throughout spermatogenesis rupture, the spermatozoa, enter the vas efferens and vas deferens and lodge in the seminal vesicles. Commonly, when the spermatozoa are released into the ducts, they remain in bundles (*spermatodesms*) held together by gelatinous material. The spermatozoa of most insects studied are filamentous with poorly developed "head" regions. The movement of spermatozoa within the male reproductive system is due not to their inherent motility but to contractions of the muscles associated with each vas deferens and the ejaculatory duct.

Further information on insect spermatozoa and spermatogenesis can be found in Baccetti (1972), King (1974), and Phillips (1970).

Control of Spermatogenesis

In cell culture experiments in humoral factor has been implicated in the differentiation of spermatocytes in giant silkworm moths (Lepidoptera, Saturniidae). Ecdysone may play a role in influencing the permeability of the testis walls to this humoral factor.

Spermatophores

In some insects the spermatozoa are produced and transferred to the female in specialized packets held together by proteinous secretions of the accessory glands. These packets are called *spermatophores* and often assume rather distinct forms. Spermatophores are common in the lower orders (e.g., the apterygotes) and rare or absent in some of the higher orders, such as Hymenoptera. Once spermatophore are emptied of spermatozoa, they may be digested

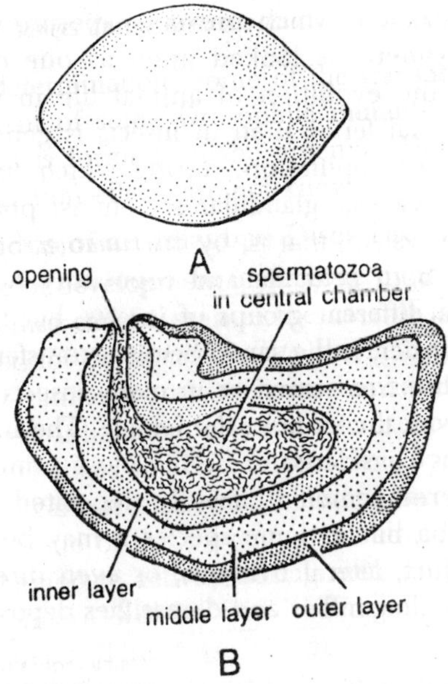

Fig. 2.23. P. americana. A spermatophore. A–Complete. B–In section.

and absorbed within the genital ducts if placed there during copulation or be eaten by the female in those species in which the male places the spermatophore on the sustrate.

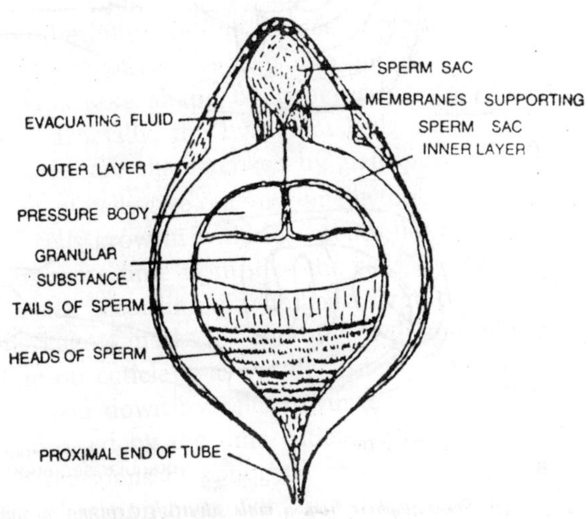

Fig. 2.24. Spermatophore of Acheta.

Seminal Transfer

Internal fertilization, which prevent exposure of gametes to the drying atmosphere, is looked upon as one of the several prerequisites for the evolution of animal life in the terrestrial environment. Internal fertilization in insects is generally brought about by the act of copulation, during which time the *semen* (spermatozoa plus various glandular secretions) produced in the male reproductive system, that is, by *insemination.* Suffice it to say at this point that both genitalia and copulatory behaviour vary greatly among the different groups of insects, but their common "goal", seminal transfer, is the same. Seminal transfer may involve the passage of either free semen or in many insects one or more spermatophores from the male to the female. The involvement of a spermatophore is considered to be the more primitive situation (Hinton, 1964). Free semen is usually deposited in the bursa copulatrix or vagina but in some species it may be deposited in the common oviduct, lateral oviducts, or even directly into the spermatheca. Male dragonflies and damselflies deposit semen in a

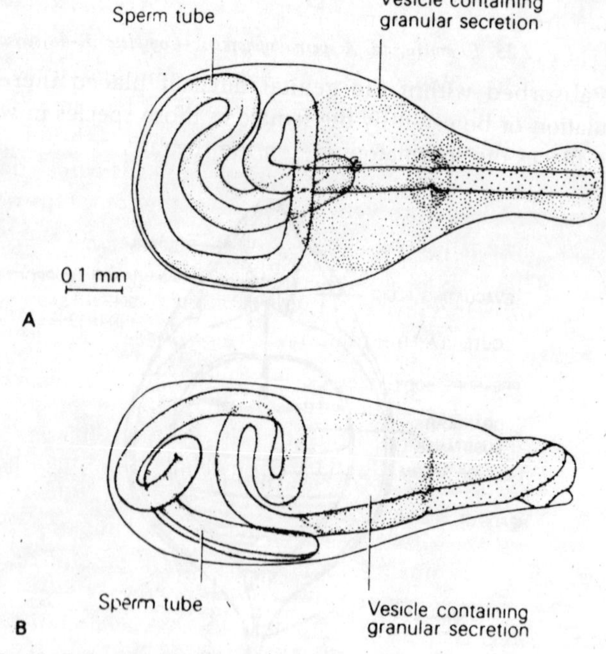

Fig. 2.25. Spermatophore from a male silverfish, Lepisma saccharina.

specialized organ on the venter of the second abdominal sternite. A portion of that organ is then placed in the female's vagina and seminal transfer is accomplished.

Many species in the superfamily Cimicoidea (e.g., bed bugs; family Cimicidae) have a rather bizarre method of seminal transfer. The males of these species inseminate the females by perforating the integument at a specialized site in their abdomen and ejaculating semen directly into the hemocoel. This method of semen introduction has been called *hemocoelic* or *traumatic insemination.* The seminal fluid and many of the spermatozoa are digested by the female, while some of the spermatozoa reach the ovaries. The adaptive value of this method of insemination is thought to be that it provides the female with additional nutritive substance (Hinton, 1964). Spermatophores are usually deposited by the male somewhere in the female reproductive system: the bursa copulatrix, vagina, or, rarely, the spermatheca. However, in the apterygotes the male deposits a spermatophore on the substrate and the female picks it up and deposits it within herself. This method of indirect sperm transfer without copulation also occur among the Diplura, Collembola, and in several noninsect arthropod group (Schaller, 1971), Gerber (1970) considers the methods of spermatophore formation in pterygotes.

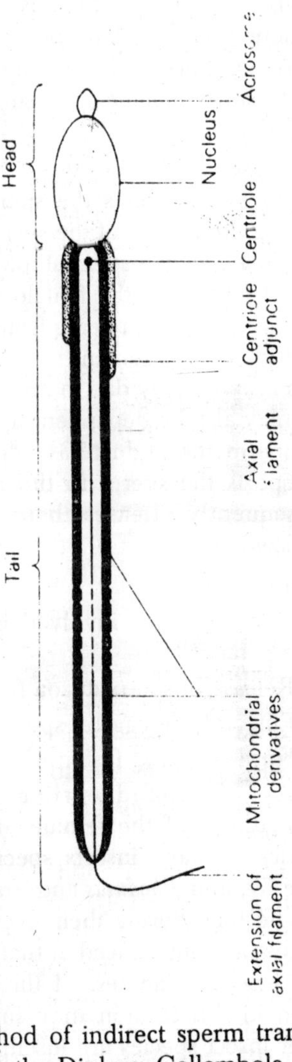

Fig. 2.26. ***Generalized insect spermatozoon.***

Once a spermatophore is in the female reproductive system, various mechanisms may account for the release of the spermatozoa. In most insect the spermatozoa are either forced out by pressure applied by the female or by the mechanical performation of the spermatophore. In the house cricket an osmotic-pressure mechanism is involved where there is a significant difference between the osmotic pressure of a block of protein or pressure body within the spermatophore and the osmotic pressure of the evacuating fluid, which surrounds the spermatophore at the time of ejaculation. The evacuation fluid is absorbed by the protein body, which swells up and forces the exit of the spermatozoa (Hinton, 1964).

In some insects the spermatophore may be digested away, causing the release of the spermatozoa. After the spermatozoa have been released, the spermatophore is in most cases probably digested and absorbed by the female and thus may have some nutritional significance. After release from the spermatophore or deposition in the form of free semen the spermatozoa move to the spermatheca. This is apparently due to muscular contractions of the female ducts but may also be dependent upon sperm motility. In the honey bee the spermathecal duct has a "pumping" structure, which seemingly transports the sperm to the receptacle following copulation and subsequently releases them when it is time for an egg to be fertilized.

Fertilization

The processes involved in fertilization may be divided into three parts.

1. Release of spermatozoa from the spermatheca.
2. Entry of the egg by spermatozoa.
3. Formation and fusion of the male and female pronuclei.

As mentioned earlier, spermatozoa are stored in the spermatheca of the female until it is time for fertilization. The females of many insects species (e.g., *Apis,* honey bee; *Glossina,* tsetse fly; and *Rhodnius* and *Triatoma,* conenose bugs) mate only at a single time during their lives with one or more males, and the spermatozoa introduced at that time are stored and used to fertilize their eggs for the rest of the reproductive period. Spermatozoa stored in this fashion may survive for several months or years. Other insect species are inseminated periodically throughout their reproductive lives, and in these species the storage of spermatozoa in the spermatheca may only for a short period of time. The mechanisms for release of spermatozoa form the spermatheca are not clearly understood.

Stimulation of sensory hairs by the passage of eggs in the oviducts, hemocoelic pressure forcing the axit of the sperm, and the inherent motility of spermatozoa that have been "activated" by some secretion have all been advanced as possible mechanisms in various species. Following ovulation, the egg is oriented in the reproductive passage in such a way that the micropylar region is in rough proximity of the site of sperm release.

The sperm migrate to the micropylar region of the egg, possible responding chemotactically, and one or more enter the egg through the micropyle. In the vast majority of insects more than one spermatozoon enters the egg, but usually only one fuses with the egg pronucleus. Excess sperm usually degenerate without disrupting the developing of the zygote. Shortly following the entry of sperm into the egg, the egg nucleus undergoes meiotic division, forming the female pronucleus. The spermatozoon that will fuse with this female pronucleus loses its tail, becoming the male pronucleus. The fusion of the two pronuclei forms the *zygote* and signals the commencement of morphogenesis.

Sex Determination and Parthenogenesis

Nearly all insects are bisexual; that is, male sex organs occur in one individual, and the female sex organs are in another individual. Several species in different groups are capable of reproducing *parthenogenetically* (i.e., they are able to produce individuals from unfertilized eggs). Insects in which both male the female sex organs occur in the same individual are said to be *hermaphroditic*. The cottony cushion scale (*Icerya purchasi*) and one or two relatives are the only insects in which true hermaphroditism has been established. Reproduction in these hermophroditic insects usually occurs by self-fertilization. Sex determination in bisexual insects is considered to depend upon a balance between genes for maleness and genes for femaleness. This balance is in most forms tipped in the direction of one sex or the other by a sex-chromosomes mechanism in which one sex prossesses two X (sex) chromosomes (i.e., the *homogametic sex*, XX) and the other a single X chromosome (XO) or a single X chromosome plus a smaller Y chromosome (XY) Individuals possessing the XO or XY configurations comprise the *heterogametic sex*.

In most insectan groups the males are hétrogametic and the females homogametic. However, the reverse is true in the Lepidoptera and Trichoptera. In some insects (e.g., Hymenoptera, Thysanoptera, and certain Homopterous insects) males develop from unfertilized eggs, females from fertilized eggs. White (1964) and

Bergerard (1972) discuss sex determination in insects. In some instances environmental factors have been shown to exert and influence on sex determination. For example, in the butterfly *Talaeporia* sp., more male than females are produced at high temperatures; the converse is true at low temperatures (Davey, 1965). Since extirpation of gonads or implantation of gonads into a previously castrated individual produces no effects, it is generally assumed that secondary sexual characters are not determined by humoral secretions associated with the gonads.

However, in the beetle *Lampyris noctiluca,* there is evidence that special secretory tissue associated with the testis rudiment controls the development of male structures, and implants of this tissue masculinize females. The appearance of this secretory tissue is determined by the presence of active neurodocrine cells in the brain (Naisse, 1966, 1969). Apparently every cell in the insect body is involved in sex determination, as is evident in the occasional occurrence of *gynadromorphs.* These individuals are literally sexual mosaics, some parts of the body possessing typically male traits and other parts typically female traits. This phenomenon is explained by differences in the sex chromosomes in the cells comprising the various tissues (i.e., some cells are "male" and others are "female"). These differences are known to occur by a number of mechanisms. One such mechanism is the loss of one X chromosome in the cleavage cells of a female (XX becoming XO, e.g., in *Drosophila,* vinegar or fruit flies). These cells thereafter give rise to male traits in whatever tissue they happen to form. Another mechanism occurs in the honey bee, in which the fusion nucleus and an extra sperm that has entered the egg both undergo cleavage. The cells that result from cleavage of the fusion nucleus (diploid) give rise to female traits, and those from the sperm nucleus (haploid) produce male traits.

3

COPULATORY ORGANS

Copulation is the mating of two individuals of different sexes and the organs specifically concerned with sexual mating and the deposition of the eggs are known collectively as the *external genitalia.* The copulatory organs pertain to both sexes, though they are particularly developed in the male; the female organs of oviposition are external genitalia in the sense that they are accessory to the reproductive function. The copulatory apparatus of the male includes primarily an organ for conveying the spermatozoa into a sperm receptacle of the female, and usually a group of associated structures adapted for grasping and holding the female. The recipient organ of the female is a copulatory pouch (genital chamber or vagina) or a spermathecal diverticulum of the latter. The principal clasping organs of the male are generally movable appendicular structures of the ninth segment, which serve as a pair of grappling hooks (harpagones), though accessory copulatory processes of various forms may occur on the same cases the cerci are transformed into grasping organs.

Coition, in most insects, is effected by a median intromittent organ located on the conjunctival membrane behind the ninth abdominal sternum, with which may be associated various accessory structures forming a group of phallic organs; but in some of the Apterygota and intromittent organ is absent, and in Odonata it is functionally replaced by a secondary copulatory structure on the anterior part of the abdomen. In certain lower pterygote insects there is a pair of intromittent organs. The external genitalia of the female, in addition of the copulatory pouch, consists of structural adaptations for the disposal of the eggs.

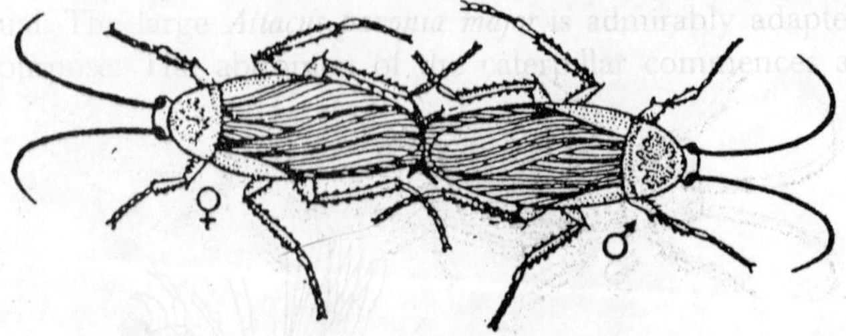

Fig. 3.1. P. americana. Copulation.

In the Thysanura (Machilidae and Lepidmatidae) and in many of the pterygote orders, the female is provided with a special egg-laying organ, known as the *ovipositor,* which appears to be formed of the appendages of the eighth and ninth abdominal segments. By means of the ovipositor the female is enabled to deposit her eggs in the ground, in the leaves, stems, and wood of plants, or into the bodies of other insects. With many insects, however, especially in these having the genital aperture from which the eggs are discharged located on the ninth abdominal segment, the ovipositor is reduced or absent, and in such cases the terminal segments of the abdomen are usually slender and tapering and capable of being protracted as a tube, from which the eggs are extruded and may be attached to smooth surfaces or concealed in crevices.

The Male Genitalia

The morphology of the male organs of copulation is not definitely known, notwithstanding the efforts that various investigators have given to the subject. In the following discussions, therefore, a minimum of attention will be given to theoretical views that do not appear to be in harmony with anatomical facts. Moreover, in order to avoid the nomenclatural confusion that has resulted from the lack of an understanding of the fundamental nature of the male organs, and in order to present simply the facts of structure, a terminology has been adopted that can be applied consistently to the major structural elements regardless of what may be the morphological relations of the latter. In each order, however, many special structures must be named individually, since it is clear that there are numerous modifications of the genital organs that have only a local significance. The primary mesodermal outlet tubes of the male genital system, as we learned in the last chapter, are attached during embryonic development in some

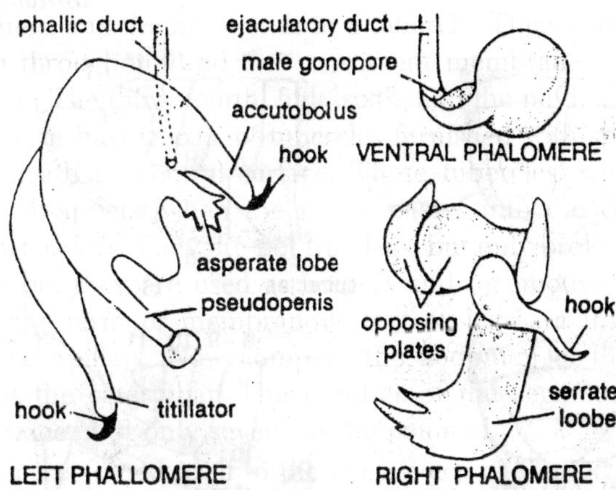

Fig. 3.2. Cockroach. Male gonapophyses.

orthopteroid insects to the ectoderm of the ventral wall of the tenth abdominal somite, or in some cases they terminate in ampullae located within the appendage rudiments of this segment.

In certain hexopods (Protura, Ephemerida, Dermaptera) the vasa deferentia retain their separate-openings in the adult, though each may terminate in an exit duct of ectodermal origin, and in such case the gonopores are bone on a pair of penes or on paired processes of a single organ. Since, however, the position of the intromittent organ in these several hexapod groups varies from the ninth to the eleventh segment, it seems doubtful that the paired adult structures represent rudimentary limbs, though we might conclude from the embryological evidence given above that the primitive male ducts opened on the bases of the appendages of the tenth abdominal segment. With the majority of adult insects the vasa deferentia open into the proximal parts of the ectodermal accessory glands and the latter then unite in a common outlet tube, the ductus ejaculatorius.

The ejaculatory duct opens usually on a median intromittent organ arising from the ventral conjunctival membrane between the ninth and tenth abdominal segments; this membrane is probably the posterior part of the venter of the ninth primary somite. It would appear, therefore, that the original orifices of the vasa deferentia in such cases have migrated forward to open in common with the accessory glands on the ninth abdominal venter, and that a median invagination of the body wall at this point has formed

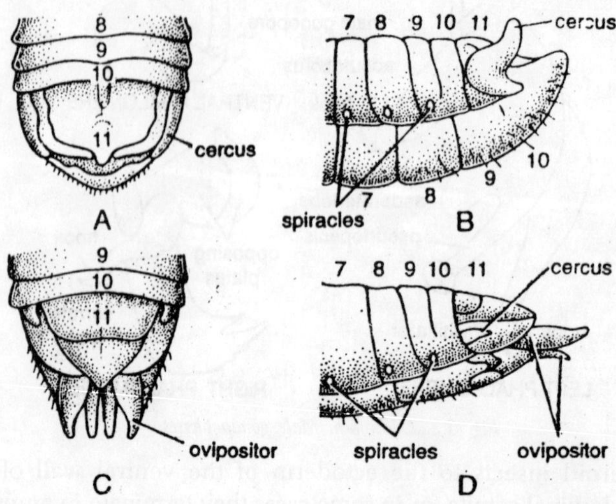

Fig. 3.3. Grasshopper. A–Male abdomen (dorsal view); B–Male abdomen (lateral view); C–Female abdomen (dorsal view); D–Female abdomen (Lateral view).

the common ejaculatory duct. This theoretical origin of the male exit apparatus, however, is not necessarily recapitulated in ontogenetic development, since in many insect the embryonic vasa deferentia have not been traced beyond the eighth or ninth abdominal somite, where they are said to unite directly with the ectodermal ductus ejaculatorious. The median gonopore, or external opening of the ejaculatory duct, may remain flush with the surface of the membrane in which it is situated; but with the majority of insects, it is situated on a tubular intromittent organ, the *median penis,* or *phallus.*

Male Genitalia of Hexapoda Having Paired Gonopores

Here are included the Proture, the Ephemerida, and the Dermaptera, though some of the last have secondary a single gonopore.

Protara

The males of Protura, as described by Berlese (1910) and by Prell (1913), have an elaborate bipartite intromittent organ eversible from between the sternal plates of the eleventh and twelfth abdominal segments, but they have no accessory lateral copulatory structures. The bifid distal part of the genital organ ends in a pair of long hollow processes on which the two genital ducts open separately through subterminal apertures. The position of the male

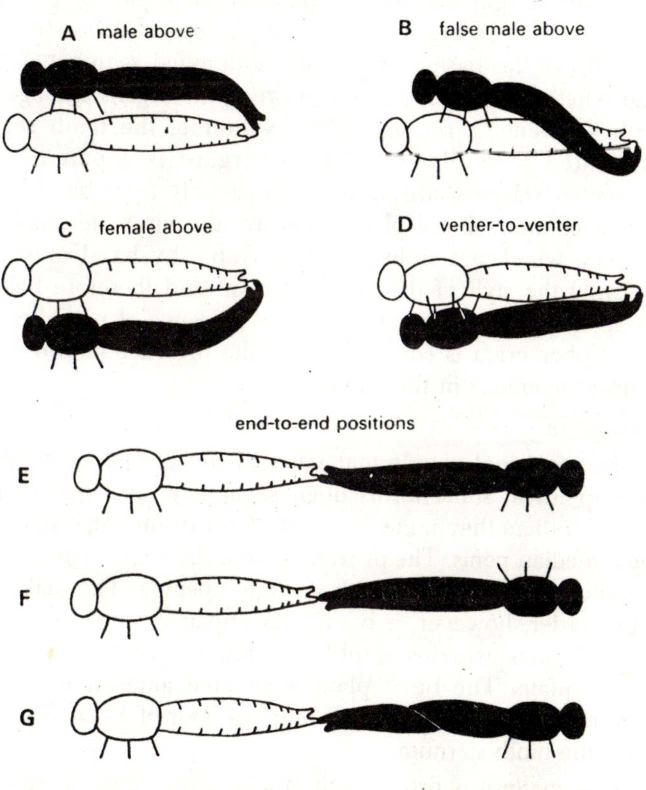

Fig. 3.4. Different positions assumed by the male and female during copulation, male black, female white. A–Male above (e.g. some Diptera). B–False male above (e.g. Acrididae). C–Female above (e.g. some Orthoptera). D–Venter-to-venter (e.g. Diptera, Culicidae–hypopygium inverted). E–End-to-end, male abdomen not twisted (e.g. some Hymenoptera). F–End-to-end, male inverted (e.g. some Tettigonioidea). G–End-to-end, male abdomen twisted (e.g. some Heteroptera).

organ in Protura would appear to preclude the possibility of any homology with the usual intromittent organ of insects, located on the ninth abdominal segments.

Ephemerida

The vasa deferentia of Ephemerida terminate in exit ducts of ectodermal origin that open separately on a pair of penes. The ephemerid penes are small, flattened, conical processes arising ventrally at the base of the tenth abdominal segment. It is not clear whether the membrane supporting the organs belongs to the ninth segment or represents the venter of the tenth segment. Associated with the intromittent organs is a pair of lateral segmented claspers, arising either separately from basal lobes or from a common basal plate borne by the ninth sternum. These claspers, which are individually movable by basal muscles are evidently the styli of the ninth segment and therefore belong to the appendages of this segment. The presence of paired penes in male Ephemerida is correlated with the presence of two separate oviducal openings in the female.

Dermaptera

The terminal ectodermal parts of the genital ducts of male Dermaptera in some forms open separately on a pair of penes, while in others they unite in a common exit duct that opens on a single median penis. The ovarial ducts of the female come together in a short median pouch with a single opening. The male organs in this order, however, is basally an unpaired structure, since the terminal parts, whether double or single, arise from a common median plate. The basal plate is situated anterior to the tenth abdominal sternum in the membranous floor of a genital chamber above the ninth sternum.

Proximally it is produced into a long apodemal process. Styli are absent in Dermaptera, and there are usually on other accessory structures associated with the intromittent organ. The bipartite type of intromittent organ is well shown in *Anisolabis maritima,* in which there are two elongate penes (*Pen, Pen*) arising from a common basal apodemal plate. Each penis ends in a distal lobe (a) and bears a strong lateral process, or "paramere" (b), arising at the base of the lobe. The left lobe is turned proximally in the usual condition. A strongly musculated ejaculatory duct penetrates each penis and opens into an eversible sac at the end of the terminal lobe.

The immature intromittent organ of *Anisolabis,* even in a full-grown nymph (B), has the form of four small simple lobes arising form a long thin apodemal plate. The double or deeply bipartite penis with two ejaculatory ducts opening separately to the exterior is characteristic, according to Walker (1922), of the superfamilies Protodermaptera and Paradermaptera. The unpaired type of organ occurs in Eudermaptera, where, as shown in *Forficula*, there is a single median penis with one terminal lobe (a) but provided with two lateral processes.

The vasa deferentia of *Forficula,* Walker says, unite in a single ejaculatory duct that opens on the median lobe, but there is present also a vestigial second duct with no external orifice, suggesting that one lobe of a primitively double penis has been suppressed. It is difficult to estimate the significance of the presence of paired penes in two such unrelated orders of the Pterygota as the Ephemerida and Dermaptera, when there is no suggestion of a double origin of the penis in any of the more primitive apterygote insects, nor any evidence that the single organ of other Pterygota is formed by the union of a pair of primitive penes. The essential differences in structure between the paired organs of Ephermerida and those of Dermaptera should not be overlooked.

The Male Genitalia of Thysanura

The genital equipment of male Thysanura is deserving of special attention because it presents in a simple form a structural complex of the type found in many pterygote insects, consisting of an unpaired median intromittent organ and of paired lateral accessories. In Machilidae and Lepismatidae the intromittent organ is a simple tubular median penis, or phallus, arising from the membrane behind the narrow membranous venter of the ninth abdominal segment. The organ is somewhat differentiated into a proximal part, or *phallobase,* and distal part, or *aedeagus.* In Diplura the penis is rudimentary. Closely associated with the penis are the appendages of the ninth segment, which are well developed in the Thysanua.

Each genital appendage, or gonopod, consists of a large, flat coxopodite and of a slender distal stylus movable by muscles arising in the coxopodite. In some species there arise from the mesal angles of the bases of the coxopodites of the ninth segment a pair of short gonapophyses, which closely embrace the penis. Certain species of *Machilis* have also a pair of smaller anterior gonapophyses arising at the corresponding angles of the coxopodites of the eighth segment. With such species, therefore, the appendages of the eighth and ninth abdominal segments of the male are identical in structure

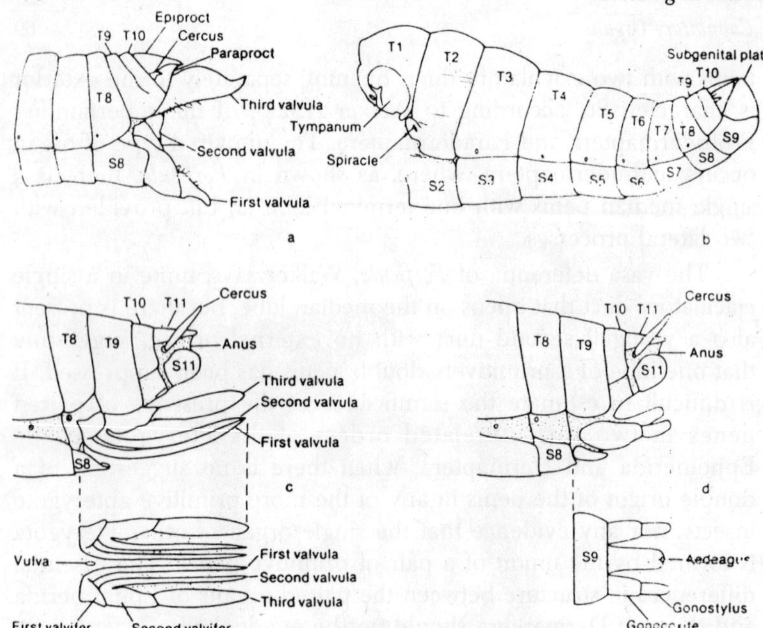

Fig. 3.5. Abdomen and external genitalia: (a) lateral view of female genitalia of Romalea microptera (Acrididae); (b) same of male abdomen and genitalia; (c) diagram showing general structure of female genitalia; (d) same of male genitalia.

with those of the same segments in the female, except for the greater length of the gonapophyses in the latter. Both pairs of gonapophyses, however, may be absent in the male. The ontogenetic organ of the median penis of Thysanura has not been carefully studied, but the simple structure of the adult organ suggests that the latter is merely a tubular outgrowth of the body wall around the mouth of the ejaculatory duct.

General Structure of the Male Genitalia of Pterygote Insects

The primary external genital organs or male pterygote insects are located medially on the venter of the ninth abdominal segment. This segment, therefore, is the *male genital segment,* or *gonosomite.* Accessory genital structures, however, may be present on the periphery of the genital segment or on the pregenital or postgenital segments. The genital parts, therefore, can be classed in two distinct groups of structures. Those of one group constitute the median intromittent apparatus of the ninth segment and may be designated the *phallic organs.* Those of the group are the peripheral accessory structures of the ninth or other segments and many be termed collectively the *periphallic organs.* The phallic organs are immediately concerned with the function of coition; they include the phallus and various accessory or supporting structures associated with the latter. The periphallic organs are movable or immovable lobes or processes that have for the most part a grasping or clasping role in the function of copulation.

The male genital segment

The genital segment of the male (ninth abdominal somite) may be a simple segmental annulus resembling those that precede it; but usually it is more or less modified, and it is sometimes greatly distorted, asymmetrical, turned upon its axis, or even inverted. Generally the tergum and the sternum of the genital segment are distinct plates. They are sometimes separated by lateral pleural plates bearing a pair of appendicular lobes, but more usually the pleural and sternal areas of the genital segment are united in a definitive pleurosternal plate, and frequently the entire segment is a continuously sclerotized annulus. The phallic organs arise from the conjunctival membrane behind the ninth sternum, but this membrane is usually invaginated within the ninth segment to form the *male genital chamber,* in which the phallic organs are ordinarily most concealed. The ninth sternum is thus, in most case, the *male subgenital plate* ("hypandium"), but often the external plate beneath the male genital apparatus is the eighth of the seventh sternum. The periphallic organs of the genital segment may include a pair of lateral movable claspers and various immovable lobes or processes arising from the tergum or the sternum.

In some of the more generalized pterygote insects a pair of styli is borne on the sternum or on coxosternal lobes of the ninth segment of the male, but it should be observed that typical styli nerve occur in conjunction with movable claspers (harpagones).

The phallic organs

The phallic organs of most insects other than Ephemerida and Dermaptera are the median genital outgrowths of the ninth segment, surrounding or containing the gonopore, that, as already defined, are immediately concerned with the function of coition. They take the form either of lobes, *phallomeres,* or of a median tubular penis, the *phallus,* and various accessory or supporting plates associated with the latter. In the Blattidae and Mantidae the phallic organs consist of three phallomeres arising close to the gonopore from the anterior wall of the genital chamber. These structures in the young nymph of *Blatta* are simple membranous, two of which are lateral and the other ventral with respect to the gonopore.

In the adults of both families, however, the phallomeres become greatly enlarged, somewhat altered in position, and take on highly irregular forms. Studies on the development of the male genitalia in Trichoptera, Lepidoptera, and Hymenoptera have shown that the tubular phallic organ of these insects in formed during larval development by the union of a pair of genital lobes that grow on at the sides of the gonopore (Zander, 1900, 1901, 1903; Singh Pruthi, 1924, 1925; Mehta, 1933). It is possible, therefore, that these larval phallic lobes of the higher insects are homologues of the lateral phallomeres of Mantidae and Blattidae.

According to Zander, the primitive phallic lobes divide each into a median lobe and a lateral lobe, the two median lobes uniting

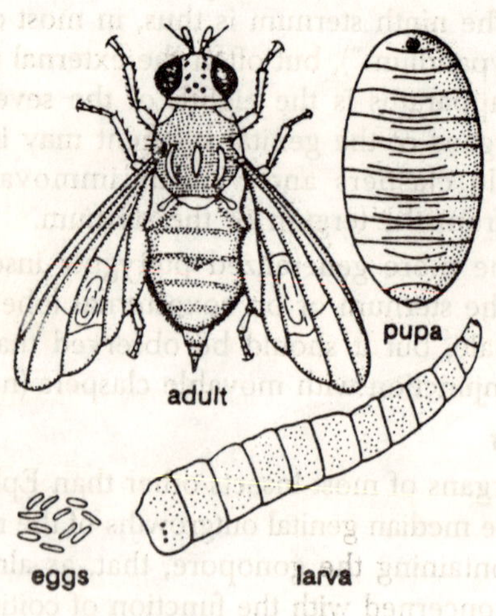

Fig. 3.6. Annual cycle of the apple-grain aphid, Rhopalosiphum padi.

to form the intromittent organ, while in Trichoptera and Lepidoptera, the lateral lobes move to the sides and become articulated to the margins of the annulus of the ninth segment, where they develop into the movable claspers of this segment. We might, therefore, regard the median lobes as gonapophyses of the gonopods, and the lateral lobes (valvae, or harpagones) as the styli. However, since it is claimed by Mehta that the lateral lobes in Lepidoptera arise separately from the median lobes, we cannot accept it as established that the gonopods of the male insect take any part in the formation of the intromittent organ, though there appears to be little doubt that they give rise to the styli or to the movable claspers of the genital segment. Some male Thysanura, as we have seen, have gonapophyses that are without doubt homologous with the processes of the female ovipositor, but there is here present also a well-developed median phallic organ between the gonapophyses of the ninth segment, and the gonapophyses may be entirely absent.

The lateral phallic lobes of nymphal blattids are far removed from the styli carried on the margin of the ninth sternum, and in the adult stage they are widely different in both form and

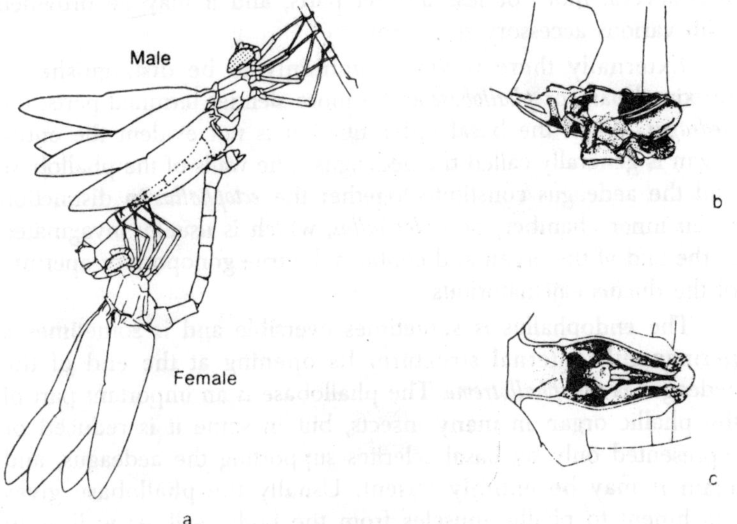

Fig. 3.7. Copulation in Odonata: (a) tandem pair of Macromia magnifica (Macromiidae) in "wheel position; (b), (c) secondary male genitalia of Macromia.

musculature from the gonapophyses of the female. The typical phallus is a conical or tubular structure containing the terminus of the ejaculatory duct. The organ, however, is highly variable in form and in the extent of its secondary modifications. Its musculature is simple or complex and so variable that in most cases it is impossible to trace any consistent scheme of homology in the various muscle patterns of the organ. The current view that male gonapophyses are involved in the formation of the phallus is disregarded in the present discussion because of the absence of positive ontogenetic evidence, and because of the entire lack of conformity in the phallic musculature.

It is assumed tentatively that the phallus is an independent median outgrowth of the body wall, independently musculated, and bound by no phylogenetic influences to conform with the structure of any ancestral appendage. On the other hand, the gonopods of the ninth segment appear undoubtedly to contribute the movable claspers to the male genital complex. A simple tubular phallus, similar to that of the Thysanura, is present in Plecoptera, though in the latter order the organ may have sclerotic plates in its walls, and it is ordinarily retracted into a pouch of the conjunctival membrane above the ninth sternum, from which it is exserted during copulation. With the majority of insects the phallus is differentiated into several more or less distinct parts, and it may be provided with various accessory structures.

Externally there is very commonly to be distinguished a proximal part, or *phallobase* and a more slender terminal parts, the *aedeagus*. When the basal differentiation is not evident the entire organ is generally called the aedeagus. The walls of the phallobase and the aedeagus constitute together the *ectophallus* in distinction to an inner chamber, or *endophallus,* which is usually invaginated at the end of the organ and contains the true gonopore or aperture of the ductus eajculatorious.

The endophallus is sometimes eversible and is sometimes a permanently internal structure. Its opening at the end of the aedeagus is the *phallotreme.* The phallobase is an important part of the phallic organ in many insects, but in some it is reduced or represented only by basal sclerites supporting the aedeagus, and again it may be entirely absent. Usually the phallobase gives attachment to phallic muscles from the body wall, as well as to muscles of the aedeagus, and may be provided with a *basal apodeme.*

Sclerites in its wall are termed the *basal plates* of the phallus. Very commonly the distal part of the phallobase forms a fold about the base of the aedeagus, and this fold is sometimes produced into a tubular sheath, the ***phallotheca,*** which partly or wholly encloses the aedeagus. In such cases, the aedeagus may be reduced or entirely suppressed; the phallic tube is then the theca, and its lining is the *endotheca.*

In some insects in which there is no evident phallobase, the aedeagus is more or less sunken into a *phallocrypt,* or pocket in the genital chamber wall, which possible represents the endotheca. The walls of the crypt may be membranous or variously sclerotized; the sclerotic part sometimes forms a ring or tube from which the aedeagus projects. Lobes or processes arising from the phallobase are of frequent occurrence.

Lateral basal lobes are particularly characteristic genital structures in the Coleoptera, where they are commonly called *parameres.* The term, however, is applied also to other processes of the genital armature and has been used by Walker (1922) and other writers as synonymous with "male gonapophyses," which are supposed to be components of the phallus. Since, however, as already shown, it appears doubtful that the gondpophyses are retained in the males of pterygote insects, the term "paramere" is here defined according to the usage by coleopterists as lateral process of the phallobase. Dorsal and ventral processes of the phallobase may then be named, respectively, *epimeres* or *hypomeres.* The aedeagus is usually tubular in form, though it assumes a great variety of shapes, and its walls are characteristically strongly sclerotized, except often for a terminal membranous part known as the *preputial sac,* or *vesica.* The organ is usually provided with its own muscles, and its base may be produced into one or more apodemal processes.

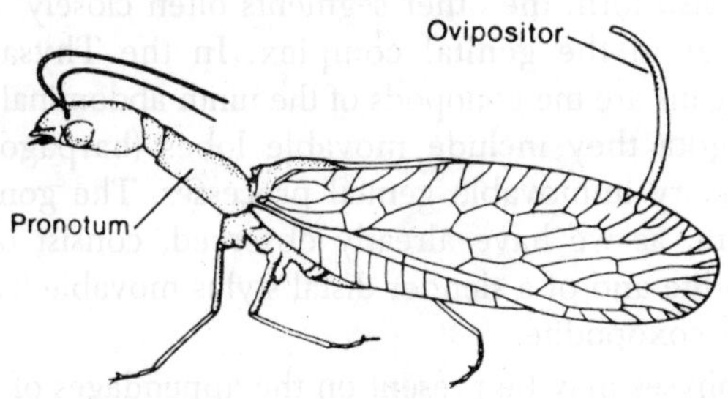

Fig. 3.8. Raphidioptera, Raphidiidae: adult of Agulla bracteata.

The distal extremity is frequently armed with spines, small plates, or slender processes often called *titillators.* Though the aedeagus is generally the conspicuous part of the phallic organ, it may be much reduced or even practically obliterated, and in such cases, as already mentioned, a tubular thecal extension from the phallobase may replace the aedeagus. The endophallus varies from a small invagination in the end of the aedeagus, containing the gonopore, to a long inner tube or an elaborate internal chamber of the phallus. In its tubular form the endophallus is often mistaken for the ejaculatory duct. When its walls are membranous, the endophallus is usually eversible and, when everted during copulation, it becomes the functional intromittent organ (sometimes called "the penis"), since it is the part of the phallus projected into the copulatory receptacle of the female. The length of the endophallus in such cases is probably correlated with the length of the genital tract of the female between the copulatory entrance and the mouth of the spermatheca.

The inner walls of the endophallus are often armed with spicules, spines, and plates, which become external with eversion. When the phallobase, the aedeagus, and the endophallus are each retracted, one with the other, the fully everted organ may become an extraordinarily long slender tube. On the other hand, when the endophallus is a permanently internal part of the phallus, it sometimes attains a high degree of development. An endophallus of this type is characteristic of the Acrididae, where it forms a complex, strongly musculated apparatus for discharging the spermatophores through the relatively small aedeagus.

The periphallic organs

The periphallic organs, in contrast to the median phallic organs, arise peripherally, generally from the annulus of the ninth abdominal segment but also form the other segments often closely associated with the latter in the genital complex. In the Thysanura the periphallic organs are the gonopods of the ninth abdominal segment; in the Pterygota they include movable lobes (harpagones) and various accessory immovable genital processes. The gonopods of the Thysanura, as we have already observed, consist of a large basal coxopodite and of a slender distal stylus movable by muscles arising in the coxopodite.

Gonapophyses may be present on the appendages of the ninth segment or absent rarely they occur on the appendages of the eighth segment. When present on the ninth segment they closely embrace the penis but do not form a part of the phallic organ. The movable periphallic genital lobes of pterygote insects are typically lateral appendages and, when present, always pertain to the annulus of the ninth abdominal segment. They are to be identified by the fact that they are individually provided with muscles inserted on their bases. In form they vary from slender processes to broad lobes or form small hooks to long falciform arms, and frequently they are armed with secondary outgrowths.

In general these *movable* periphallic organs serve as copulatory claspers, and for this reason they are here designated the *harpagones* (from a "grapplink hook"). They occur principally in Ephmerida, Hemiptera, Neuroptera. Mecoptera, Lepidoptera, and Diptera but are frequently absent in members of groups in which they are typically present. The harpagones always arise from some part of the lateral or ventral walls of the genital segment, nerve from the tergal region. In some insects they are borne on independent basal plates, which either are interpolated laterally between the tergum and sternum of the genital segment or are attached to the sternum. When such plates are present the muscles of the claspers take their origins upon them; otherwise the claspers are borne on the sternal (coxosternal) plate of the segment and their muscles arise on this plate.

The position and musculature of the harpagones at once suggest that these movable genital claspers represent the styli of the more generalized insects, and that the basal plates on which they are sometimes supported are the coxopodites of the gonopods. When the basal plates of the claspers are not individually evident, it is to be supposed that they have united with the sternum, producing the same condition as in some Thysanura and Orthoptera where the styli arise directly from a coxosternal plate of the genital segment. Movable genital lobes and typical styli, as we have observed, do not occur together in any insect, though both may be absent. The numerous fixed or merely flexible processes arising from the walls of the genital segment, or also from the other segments associated in the genital complex, are so variable in their occurrence that there can be little homology between those occuring in the different major groups of insects, though their presence and form are often highly characteristic of smaller groups and of species. Such structures are never specifically provided with muscles, though in rare cases they may be movable by segmental muscles attached at their bases.

The immovable genital processes are of such diverse forms that they can have no constant function; in general they appear to be adapted for grasping or holding various parts of the female apparatus, but only a close study of insects in copulation will reveal their exact uses in the genital mechanism.

Characteristics of the Male Genitalia in the Principal Pterygote Orders

It is impossible to given an adequate treatment of the many modifications in the copulatory apparatus of male insects within the space that may be allotted to the subject in a general text. The following descriptions, therefore, present only a sketch of the salient or distinctive features of the male genitalia in the principal orders, with suggestions as to how the various parts may be related to one another.

Odonata

The Odonata are of particular interest in a study of the male genitalia because of the development of a secondary copulatory organ on the anterior part of the abdomen. The true gonopore of the male is situated on a rudimentary penis of the ninth abdominal segment concealed beneath two small plates, which possibly represent the gonopods. A large postgenital plate appears to be a secondary sclerotization of the intersegmental area behind the genital organ. The functional intromittent organ of the Odonata is a secondary structure situated in a median depression, or genital fossa, on the ventral wall of the second abdominal segment. This organ is strongly sclerotized tubular structure composed of several segment-like parts movable upon each other. Various accessory lobes, differing much in different species, may arise from the surrounding walls of the first and second abdominal segments.

The copulatory organ contains a chamber open to the exterior which serves as sperm receptacle. Before copulation the male dragonfly transfers spermatozoa from the genital opening on the ninth segment to the receptacle of the intromittent organ by flexing the abdomen ventrally and forward until the two apertures are in contact. In copulation the male grasps the female with the cerci by the neck or the back of the thorax, or sometimes by the head, and the female brings the end of her abdomen forward beneath that of

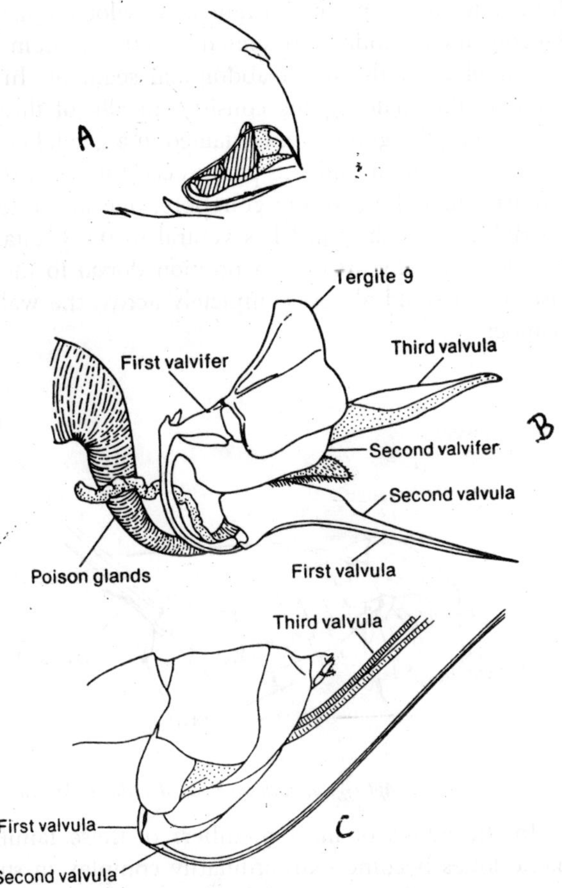

Fig. 3.9. Structural feature of Hymenoptera: (a) medial section through abdomen of Apis mellifera, showing sting in repose; (b) dissected sting of Apis mellifera; (c) end of abdomen and ovipositor of Megarhyssa atrata (Ichneumonidae).

the male of effect a union with the anterior genitalia of the latter. Details of the structure of the male organs in Anisoptera and Zygoptera have been described by Ingenitzsky (1894), Backhoff (1910), E. Schmidt (1916), and Kennedy (1917, 1922).

Orthoptera

The external genitalia of male Orthoptera are mostly phallic structures. Styli of the ninth segment are present in several families, but they take no part in the genital apparatus and depart but little from the typical stylus form. The coxopodites of the styli are plates distinct from the ninth sternum in Grylloblattidae, but otherwise the genital coxopodites are united with the sternum in the definitive sternal plate of the ninth abdominal segment. In Mantidae and Blattidae the male organs consist typically of three phallic lobes surrounding the gonopore, contained in a genital chamber between the ninth sternum and the paraprocts. Of the three phallomeres, two are situated above the gonopore, one to the left, the other to the right; while the third lies ventral to the genital opening. The right lobe usually assumes a position dorsal to the others and its base may extend almost completely across the wall of the genital chamber.

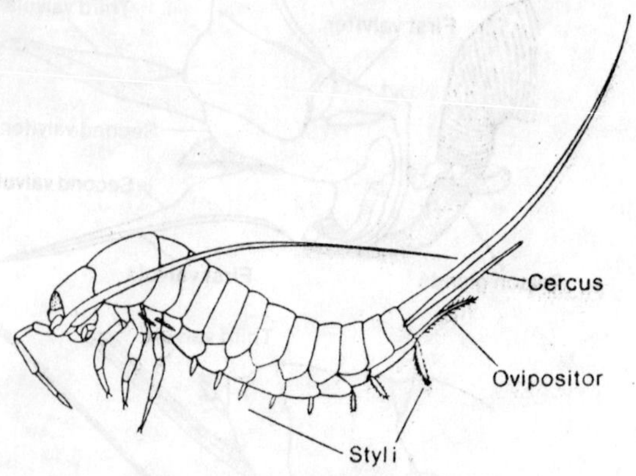

Fig. 3.10. Apterygota, Archeognatha, Machilidae: Machilis, female.

In the adults of most members of these families the lateral phallic lobes become extraordinarily complex in structure by the development of secondary lobes and processes of various forms. The ventral lobe is usually more simple, and, since the ejaculatory duct opens in a membranous fold at its base, it is often called the "penis." An intricate system of muscles arising in the ninth segment is inserted on the base of the phallomeres, and in addition there are numerous muscles within the lateral lobes inserted on their various secondary parts. In some of the roaches the phallic organs are simpler, and in certain forms, as in *Ectobia*, the lateral lobes are retracted into deep pouches of the genital chamber.

In the nymph of *Blatta* the high complex genital organs of the adult are represented by three simple lobes projecting from the anterior wall of the genital chamber, two of which lie immediately lateral of the gonopore; and the third below it. These three simple phallic structures appear to be merely outgrowths of the genital chamber wall in the immediate neighbourhood of the gonopore. The lateral lobes have been regarded as gonapophyses of the ninth segment, but in the nymph there is nothing to suggest that they have any relation to the gonopod bases incorporated in the ninth sternum. Among other Orthoptera the male genital structures are very different in the several families. The Phasmidae have a short compact intromittent organ, which possibly is formed by the union of primitive phallic lobes about the gonopore.

In Tettigoniidae and Gryllidae the lateral lobes are reduced and retracted, while the ventral lobe becomes enlarged and may be the only phallic structure ordinarily visible. In Acrididae the phallus is a large conical structure distinctly divided into a phallobase and aedeagus and contains a highly developed endophallus forming a sperm ejection pump into which opens the ejaculatory duct. A large, often complex sclerite, known as the *epiphallus*, or pseudosternite," lies dorsally at the base of the phallic organs in Tettigoniidae, Gryllidae, and Acrididae and forms and important element of the copulatory mechanism.

Coleoptera

The male genitalia in Coleoptera, as in Orthoptera, are phallic structures only, there being in general no accessory or periphallic armature on the genital segments. There are, therefore, no elements of the genital complex that can be referred directly to the gonopods; movable claspers (harpagones) of the ninth segment are always absent, and, in the male at least, styli are never present in any form. The ninth and tenth segments of the abdomen are usually much reduced and retracted into the eighth segment, and in some forms the eighth is concealed within the seventh. The phallic organs consist essentially of a tubular aedeagus and a variously developed phallobase usually provided with parameres.

The innumerable variations in the genital apparatus of male Coleoptera have been described by Sharp and Muir (1912), but the fundamental structure of the parts involved may be understood from a few typical examples. The external part of the abdomen consists usually, as illustrated in *Phyllophaga,* of the first eight segments. Within the eight segment is an invagination cavity into which are retracted the reduced ninth and tenth segments, but which is continued forward through the narrow annulus of the ninth segments as a large genital chamber containing the phallic organs.

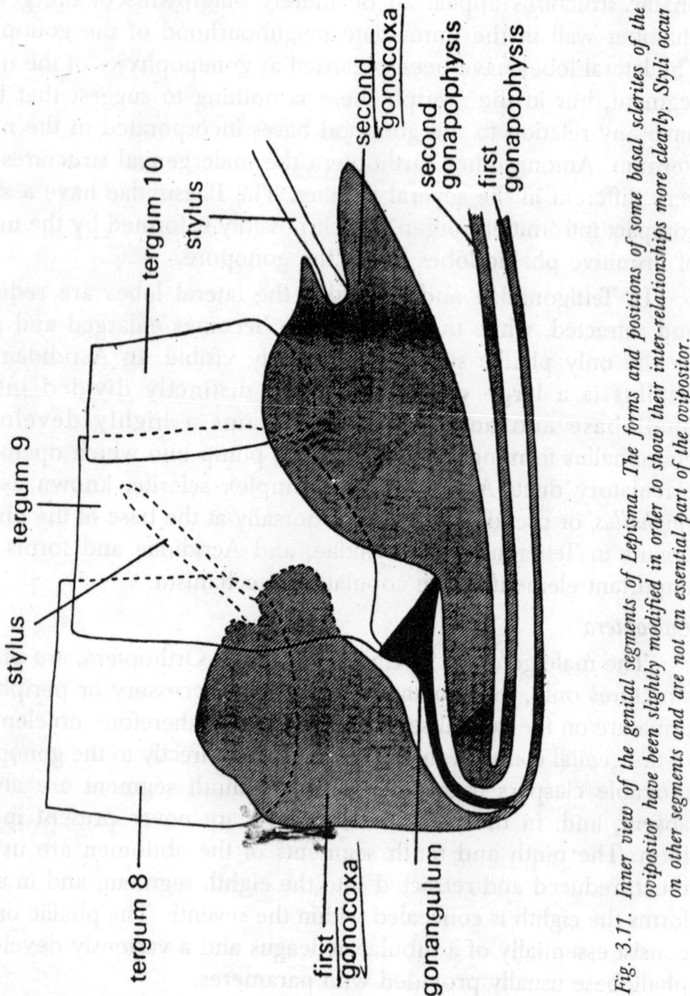

Fig. 3.11. Inner view of the genital segments of Lepisma. The forms and positions of some basal sclerites of the ovipositor have been slightly modified in order to show their inter-relationships more clearly. Styli occur on other segments and are not an essential part of the ovipositor.

The tenth segment appears as a small projection from the dorsal wall of the entrance chamber, bearing the anus.

The region of the ninth segment that encircles the mouth of the genital chamber may contain a complete, through narrow, sclerotic annulus, as in Carabidae, but generally only a sternal sclerite is present, and this, as in *Phyllophaga* and various other beetles, is reduced to a U-shaped or V-shaped bar, often provided with a ventral apodeme, which gives it a Y-shaped form. The phallic organs arise from the anterior wall of the genital chamber. The phallobase (*tegmen* of Sharp and Muir) is usually well defferentiated from the aedeagus but varies much in form and character; it maybe a membranous fold about the base of the aedeagus containing several basal plates, but often it is a sclerotic ring and sometimes forms a cylindrical theca completely investing the aedeagus. Generally, the phallobase bears a pair of parameres and in some cases a median dorsal lobe, or epimere; from its base an apodeme projects into the body cavity for muscle attachments.

The aedeagus is typically a sclerotic tube with a membranous distal part; from its base an apodeme projects into the body cavity beneath the apodeme of the phallobase. An endophallic chamber or tube is usually present and is generally eversible; when everted in coition it becomes the functional intromittent organ. In some beetles the eighth segment also is partly or entirely retracted, and, as in Oedemeridae, it may take the form of a sheath enclosing the ninth and tenth segments, the genital chamber, and the phallic organs, the last finally, consisting of a phallobase, an aedeagus, and an endophallus. In such cases, the genital apparatus assumes the complicated form of numerous folds successively ensheathing each other, all of which probably are protracted during the act of copulation.

Hemiptera

In the Hemiptera there are present, in addition of well-developed phallic organs, various periphallic structures having the form of lobes or processes arising from the eighth, ninth and tenth abdominal segments. Among these structures there is usually one pair movably articulated to some part of the ninth segment and individually provided with muscles. These movable claspers are thus to be identified as the herpagones, that is, as derivatives presumably of the styli of the gonopods of the ninth segment. The harpagones vary from small hooks to long slender processes or broad spatulate lobes, often of irregular shapes. In some cases they are absent. In Homoptera they arise from the floor of the genital chamber, where their bases are usually associated with one of the sclerites of the phallobase or supporting plates of the aedeagus.

In Heteroptera the harpagones are small but strongly musculated processes articulated to be sclerotic wall of the ninth segment inflected into the genital chamber. The principal segment involved in the genital modification of the hemipterous abdomen is the ninth, but in Heteroptera the eighth is often reduced and closely associated with the ninth. In most Homoptera the tenth and eleventh segments are distinct annuli, the tenth being sometimes provided with accessory genital processes in the form of lateral lobe. In Heteroptera the two postgenital segments apparently are united in a tubular proctiger. The ninth segment in Homoptera often bears, in addition to the harpagones, accessory periphallic structures having the form of short processes or of long arms or broad lobes. Such structures may be flexible at their bases and they have been mistaken for the harpagones, but they are to be distinguished from the latter by the fact that they are never provided with muscles.

The phallus of the Hemiptera comprises in most cases a phallobase and an aedeagus, though either one or the other may be suppressed. The aedeagus in its simplest development is a tubular structure, but more usually it takes on an irregular shape, which by exaggeration may produce bizarre forms, often with curious terminal outgrowth. It is usually provided with apodemal processes for muscle attachment. The phallobase is variously developed. It may consist merely of one or two basal plates in the wall of the genital chamber supporting the aedeagus and giving attachment to phallic muscles. In the cicada the single large basal plate of the phallus, is articulated upon fulcral arms of the sternal margin of the ninth segment. In some cases the phallobase bears parameral processes, or again it may be produced into a thecal sheath more or less investing the aedeagus and subject to many variations in form. When the theca is well developed the aedeagus is sometimes greatly reduced, as in Fulgoridae and Cicadellidae, except for terminal processes that may protrude through the theca.

In the Cicadidae it is evident that the aedeagus has been almost entirely suppressed, and that the long tubular intromittent organ is the theca. The inner tube of the organ is therefore the endotheca. The only remnant of the aedeagus in the cicada is an apodemal process (i) attached to the inner extremity of the endotheca where the ejaculatory duct opens into the latter. The homologies of the various genital structures of male Hemiptera can be determined only by a very close study of the relations of the parts to one another.

Mecoptera

In the Mecoptera the periphallic genital claspers of the ninth abdominal segment are distinctly two segmental. The basal segments appear to be the coxopodites, the distal segments the true harpagones. Each segment is individually movable by muscles. In *Panorpa* the organs are strongly developed, but the harpagones extend but little beyond the ninth segment; in *Merope tuber* they are long slender, weakly sclerotized appendages, each of which is bifid terminally and bears a suckerlike disk on its inner surface. The abdomen of the typical scorpionflies ends with a recurved bulbous structure formed of the ninth segment and its appendages, which contains the phallic organs and mostly conceals the tubular

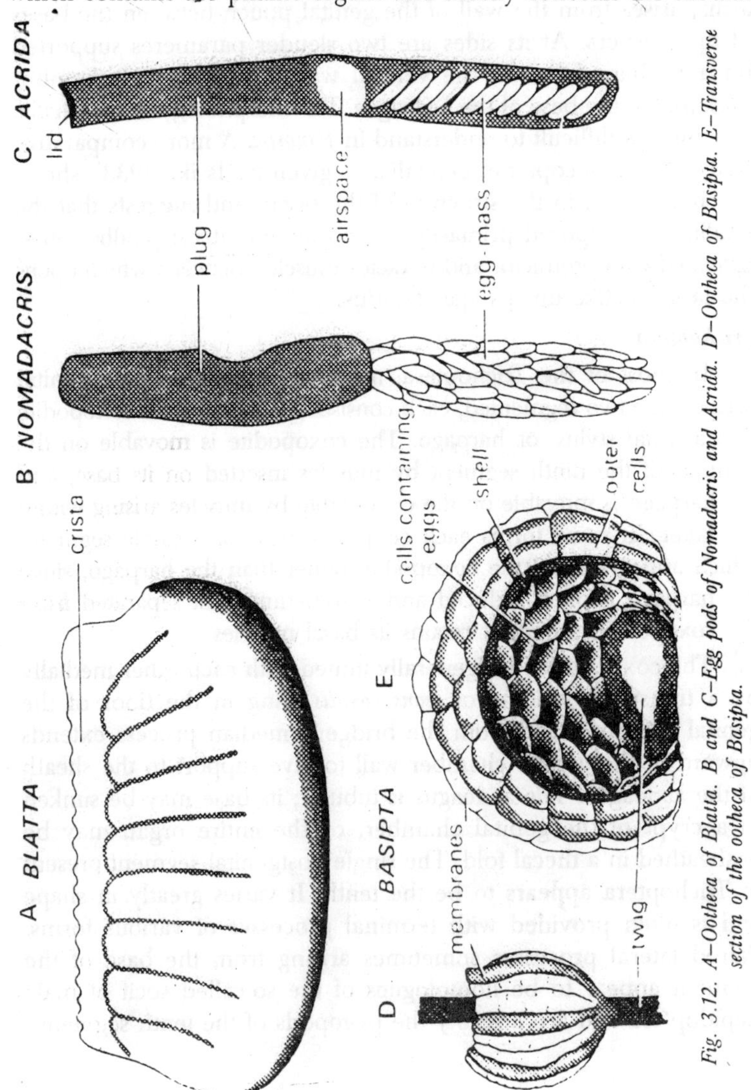

Fig. 3.12. A–Ootheca of Blatta. B and C–Egg pods of Nomadacris and Acrida. D–Ootheca of Basipta. E–Transverse section of the ootheca of Basipta.

proctiger. In *Panorpa* the genital segment is a continously sclerotized annulus, the tergal region of which is greatly prolonged posteriorly, and the short sternal region produced into two long lobes. The large lateral coxopodites are broad oval lobes articulated basally to the annulus of the ninth segment. Distally each bears a strong, hooked harpago, which is individually movable by antagonistic muscles arising in the coxopodite.

The aedeagus, a flat irregular structure with a pair of distal arms, arises from the wall of the genital pouch between the bases of the claspers. At its sides are two slender parameres supported on a U-shaped bar in the ventral wall of the genital chamber proximal to the base of the aedeagus. The morphology of the phallic structures is difficult to understand in *Panorpa.* A more comparative study of the mecopteran genitalia, as given by Issiki (1933), shows much variation in the structure of the organ and suggests that the phallus is composed primarily of two united lateral phallic lobes, provided with protractor and retractor muscles, between which opens the large saclike ductus ejaculatorius.

Trichoptera

In most of the Trichoptera, as in the Mecoptera, the genital claspers are two segmented, each consisting of a proximal coxopodite and a distal stylus, or harpago. The coxopodite is movable on the annulus of the ninth segment by muscles inserted on its base, and the harpago is movable on the coxopodite by muscles arising within the latter. In some forms each clasper consists of a single segment, which appears to be the coxopodite rather than the harpago, since the harpago is often reduced and is sometimes not separated from the coxopodite, though it retains its basal muscles.

The coxopodites are generally united with each other medially by a transverse bridge, or *pons coxalis* lying in the floor of the genital chamber, and from the bridge a median process extends upward in the genital chamber wall to give support to the sheath of the aedeagus. The aedeagus is tubular; its base may be sunken in a crypt of the genital chamber, or the entire organ may be ensheathed in a thecal fold. The single postgenital segment present in Trichoptera appears to be the tenth. It varies greatly in shape and is often provided with terminal processes of various forms. Paired lateral processes sometimes arising from the base of the segment appear to be homologues of the so-called socii of male Lepidoptera and are possibly the pygopods of the tenth segment.

Lepidoptera

The genital complex of male Lepidoptera includes the eighth, ninth and tenth abdominal segments. The eighth segment forms at least a protractile base for the copulatory apparatus, and in some cases, it bears accessory genital lobes. The ninth segment may be a simple sclerotic ring, but usually it is irregular in form, with

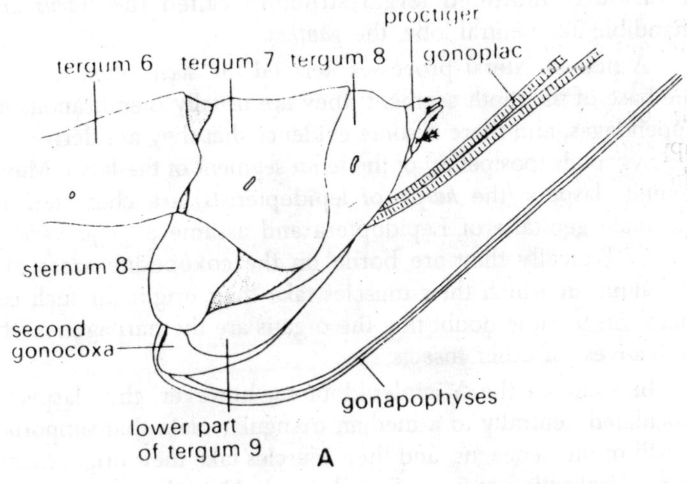

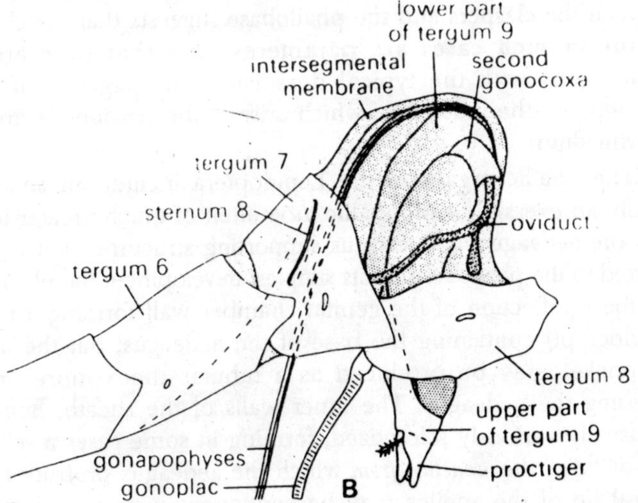

Fig. 3.13. Basal part of the ovipositor of Megarhyssa (Hymenoptera). A–At rest. B–Positions of abdominal sclerites and ovipositor during oviposition. Membranes stipped.

distinct tergal and coxosternal areas of sclerotization. The tergum is the *tegumen* of lepidopterists, and the coxosternal are the *vinculum.* The sternal region of the vinculum is often extended forward in a deep inflection known as the *saccus,* the membranous ventral wall of which is usually adnate with the sternal surface. The tenth segment may be a simple membranous tube, but usually it presents a variously modified tergal structure called the *uncus* and a mandible-like ventral lobe, the *ganthos.*

A pair of lateral processes, termed the *socii,* often arise from the base of the tenth segment; they are usually membranous hairy appendages, and there is some evidence that they are derived from the pygopods (postpedes) of the tenth segment of the larva. Movable genital claspers (the *harpes* of lepidopterists) are characteristic of the male genitalia of Lepidoptera and assume a great variety of forms. Typically they are borne on the coxopodite areas of the vinculum, in which their muscles take their origin. In such cases, there can be little doubt that the organs are the harpagones (stylus derivatives) of other insects.

In some of the Microlepidoptera, however, the claspers are articulated ventrally to a median triangular plate that supports the sheath of the aedeagus, and their muscles take their origin medially on a sclerotization of the aedeagal sheath. Here the intimate relation between the claspers and the phallobase suggests that the clasping organs in such cases are parameres, and that they are not homologous with the typical dorsolateral harpagones of other Lepidoptera, the muscles of which arise in the coxopodite areas of the vinculum.

The phallic organs of the Lepidoptera include an aedeagus, usually an eversible endophallic tube often of much greater length than the aedeagus, and various supporting structures that may be referred to the phallobase. In its simplest development the phallobase is a mere inflection of the genital chamber wall forming a pocket (phallocrypt) containing the base of the aedeagus; but the lips of the pocket may be produced as a tubular theca more or less enclosing the aedeagus. The inner walls of the sheath, however, are usually variously sclerotized, forming in some cases a sclerotic ring or tube, the *annelus, from* which the aedeagus protrudes. The ventral lip of the anellus may be continuous with a median arm from a supporting basal plate in the ventral wall of the genital chamber.

Hymenoptera

The male genitalia of Hymenoptera appear to be phallic structures only, periphallic appendicular organs being absent in most cases. The phallus is usually a large, highly complex structure arising from the wall of the genital chamber above the ninth sternum. It consists of a central aedeagus, often provided with lateral or terminal processes, and of a large two segmented phallobase bearing various lobes and processes surrounding the aedeagus. The proximal segment, or basal ring ("cardo"), of the phallobase opens form the body cavity by a large foramen that gives passage to the ejaculatory duct. The distal segment bears usually one or two pairs of movable lobes provided individually with muscles. Of these the more lateral ventral pair may be termed *parameres*, since they are at least analogous with the parameres of Coleoptera, though the dorsal lobes are accessory structures of the same nature.

Hymenopterists generally call the ventral lobe on each side the *volsella*, and the dorsal lobe the *squama*. Some, however, regard the basal parts of the genital organ as being formed of the united coxopodites of the ninth segment, and therefore regard the appendicular lobes as stylus derivatives. The aedeagus is a relatively simple structure. As represented in *Pteronida* it is mostly membranous but contains two lateral plates produced proximally as apodemes on which the aedeagal muscles are attached. A median dorsal groove leads into a proximal aperture from which a membranous endophallus is eversible.

In the Apidae the aedeagus commonly bears a pair of proximal lateral processes, and often a dorsal lobe. In the honey bee the entire phallic organ is much simplified; the basal structures so characteristic of other Hymenoptera are absent, the organ appears to consists of the aedeagus with a highly developed eversible endophallus. In the ontogenetic development of the Hymenoptera, as shown by Zander (1900), the entire group of phallic structures proceeds from two primary genital lobes at the sides of the gonopore. First, each lobe divides into two, then those of the resulting median pair unite to form the aedeagus, while those of the outer pair become the lateral lobes of the phallobase. The basal ring is said to be differentiated as a circular fold of the wall of the genital chamber. Only one postgenital segment is present in Hymenoptera, which, judging form the larva, is the tenth abdominal segment. In some of the lower families it bears a pair of cercuslike appendages, which, since they occur on the tenth segment, are perhaps to be identified with the socii to Trichoptera and Lepidoptera.

Diptera

The male genitalia of Diptera show a great proclivity toward the development of secondary lobes and processes, both phallic and periphallic. In the more generalized families the genital segments have a tendency to form a terminal enlargement (hypopygium) of the abdomen; in higher families the distinction between the genital and visceral regions of the abdomen becomes accentuated by a reduction of the sixth and seventh segments and a close association between the eight, ninth and tenth segments to form a genital complex, which becomes mostly concealed within the fifth segment. Asymmetry is of frequent occurrence in the genital region, and the ninth segment is sometimes partly revolved upon its axis of completely inverted. The harpagones are well developed in lower Diptera; they are often bilobed and sometimes are bipartite. In some of the Tipulidae the genital coxopodites are distinct pleural plates on the sides of the ninth segment, and distinct pleural plates on the sides of the ninth segment, and they may be partly exerted from between the tergum and sternum as in Mecoptera and Trichoptera; but more commonly the coxopodites unite with the sternum or entirely lose their identity in the continuously sclerotized annulus of the ninth segment. The venter of the genital segment may be entirely membranous, but usually it contains a sternal plate.

The greatly specialized type of abdomen characteristic of higher Diptera is well exemplified in *Pollenia rudis.* The visceral part of the abdomen consists of segments *I-V,* but the first segment is usually observed by reduction and union with the second. In *Pollenia* segment VI appears to be obliterated, and segment VII contains only a small tergal plate. Segment VIII has a well-developed tergum, but its sternum is reduced to a narrow sclerotic band, which is incomplete on the right side. Segment IX presents externally a small tergal plate behind and below the eighth tergum; the sternum of the ninth segment, however, ordinarily projects forward and upward from the lower angles of the tergum in the dorsal wall of a large pouch with membranous wall invaginated within the eighth sternum. This pouch contains the phallic organs. Seen from below, the ninth sternum is a broad plate having its posterior angles produced as two arms in the membranous wall beneath the ninth tergum, from which a pair of lateral bars extends to the tenth segment. Two median plates arises from the posterior margin of the ninth sternum and support a pair of free lobes that embrace the best of the aedeagus. Movable claspers that can be identified with the harpagones, or styli of the gonopods, are absent in muscoid Diptera, but the ninth tergum commonly bears on its lower posterior angles a pair of long lobes, which may be flexible at their bases but are not provided with muscles.

The small flat membranous tenth segment, in which the anus appears as a median slit, also in usually provided with a pair of lateral lobes associated with those of the ninth tergum. On the bases of these lobes of the tenth segment is inserted a pair of muscles from the ninth tergum. The phallic organs of Diptera consist principally of a variously developed aedeagus, though supporting basal structures also may be present. The aedeagus in its simpler form varies from a short tapering process to a long slender tube usually curved or coiled. The typical muscoid aedeagus is a large irregular structure with basal, lateral, and ventral lobes or processes.

The phallobase is represented by a low thecal fold surrounding the base of the aedeagus, in the walls of which are two small plates supporting a large basal apodeme for muscle attachments. Ordinarily the aedeagus is turned forward in the phallic pouch above the eighth sternum. The ductus ejaculatorius is provided with an ejaculatory bulb, a syringe-like structure with a thick muscular sheath supported on a flat central apodeme arising form one side of the cuticular lining of the duct. Distal to the bulb the duct enters the base of the phallus.

The Female Genitalia

The primitive individual openings of the lateral ducts of the female genital system, situated on the seventh abdominal segment, are retained in modern adult insects only in the Ephemerida, the females of this order having a pair of gonopores located in the conjunctival membrane behind the seventh sternum, from which the eggs issue in two distinct masses. In Dermaptera the lateral oviducts unite in a very short median oviductus communis opening immediately behind the seventh abdominal sternum. In all other insects the median egg passage is extended posteriorly, and the exit aperture is located either on the eighth abdominal segment or on the ninth. When the genital opening is established on the eighth segment, there is generally associated with it an organ formed of appendicular parts of the eighth and ninth segments serving for the deposition of the eggs. This organ is the *ovipositor.* An ovipositor having a uniform basic plan of structure is of wide occurrence among pterygote insects, and an organ of the same type though of more primitive structure occurs in the Thysanura.

Hence, there is little reason to doubt that the common ancestors of the Pterygota and Thysanura were equipped with an egg-laying organ from which the modern ovipositor has been evolved. The ovipositor of present-day insects is nearly always rudimentary or suppressed in insects having the egg exit on the ninth segment, and it is often reduced or absent in forms having he genital opening on the eighth segment. Insects in which the ovipositor is absent or nerve fully developed include Collembola, Diplura, most Odonata, Ephemerida, Plecoptera, Mallophaga, Anoplura, the tubuliferous Thysanoptera, Coleoptera, most Neuroptera, Mecoptera, Trichoptera, Lepidoptera and Diptera.

Insects lacking an ovipositor, or in which the organ is not functionally developed, may have no special provision for placing the eggs; but with many of them the posterior segments of the abdomen are so modified that they can be protracted in the form of a slender telescope tube having the opening of the egg passage near its distal end. A substitutional "ovipositor" of this kind is characteristic of the tubuliferous Thysanoptera, the Mecoptera, the Lepidoptera, the Coleoptera, and the Diptera. Most insects of these groups insert their eggs into crevices or attach them to smooth surfaces by a cementing substance discharged from the collecterial glands. In some cases, however, the end of the abdomen forms a piercing or cutting organ, or it may have a simple modification for manipulating or placing the eggs.

The abdomen of the fruit flies (Trypetidae, Lonchaeidae), the distal part of which is narrowed and tapering or sometimes greatly elongate, terminates in a sharp point that enables the insects to pierce the skin or rind of fruit in which they deposit their eggs. In most Lepidoptera the female abdomen is provided with two terminal lobes at the side of the egg exit, which serve to grasp the issuing eggs, or which, when spread out flat, form a disc for pressing the eggs against the surface on which they are attached by the secretion of the cement glands.

The Ovipositor of Thysanura

The ovipositor of the Thysanura presents in a simple form structural elements from which, there can be little question, the more highly perfected ovipositor of pterygote insects has been evolved. More than this, its basal parts show an exact serial identity with ventral plates of the pregenital segments that are almost certainly rudiments of abdominal limbs. The under surface of the pregenital segments of the Machilidae, have each three plates, namely, a small triangular median plate, which is the true sternum, and two large lateral plates, which appear to be the coxopodites of the otherwise rudimentary limbs.

The coxopodites of each pair are united medially behind the sternal plate, and each bears a slender appendage known as a stylus, which is movable by muscle arising in the supporting basal plate. In the genital segments of female Thysanura the coxopodites are free plates or flat lobes, and sternal plates are absent. Each coxopodite bears distally a stylus, but in addition it has a long slender genital process, or *gonapophysis,* arising from the mesal angle of its base, the four of which are closely associated to form the saft of the ovipositor. Since the gonapophyses are never represented on the pregenital segments they would appear to be special developments of the gonopods. If so, judging from their position, they are of the nature of coxal endites. Each is provided with a short muscle arising in the supporting coxopodite.

The principal movements of the gonapophyses, however, are probably brought about by muscles of the coxopodites, which arise on the corresponding terga of the genital segments. In some cases, the second gonapophyses are fused at their bases, and in other they may be united throughout their length, but the first gonapophyses are always entirely free from each other and are apparently capable of independent movement. The two pairs of gonopods are particularly alike in Machilidae; but in *Thermobia,* and probably in other Lepismatidae, the coxopodites of the second pair are each divided into an anterior plate bearing the gonapophysis, and a posterior plate supporting the stylus. The tergal muscles of the coxopodite are inserted on the anterior sclerite. This division of the second coxopodite occurring in the Lepismatidae is most interesting since it suggests the similar condition that is characteristic of most pterygote insects.

General Structure of the Ovipositor of Pterygote Insects

The ovipositor of pterygote insects, in its typical form, consists of a *shaft* and a *basal apparatus* and usually includes a pair of accessory lobes. The shaft, in most insects, is composed of two pairs of closely appressed elongate processes, the *first* and the *second valvulae,* the first valvulae being usually ventral, the second dorsal. The second valvulae are often united in a single median dorsal piece. The basal apparatus consists essentially of two pairs of lobes or plates, the *first* and the *second valvifers,* which support the ovipositor shaft by the bases of the valvulae. The proximal parts of the valvulae, by which the latter are attached to their respective valvifers, are distinguished as the *rami of the valvulae.*

The accessory lobes, or *third valvulae,* are borne on the posterior ends of the second valvifers. The usually ensheath the distal part of the shaft, but in Orthoptera they form a third pair of blades in the shaft. The ventral body wall between the bases of the second valvifers sometimes contains sclerites, which are termed *intervalvulae.* The principal muscles of the ovipositor are inserted on the valvifers. The dorsal muscles of the first valvifers take their origin on the tergum of the eighth abdominal segment, those of the second valvifers on the tergum of ninth segment. It is thus evident that the first valvifers pertain to the eighth segment and the second valvifers to the ninth segment. The second valvifers usually retain a close connection with the tergum of the ninth segment, but the first valvifers are often more or less dissociated from the eighth segment. An exception of the general structure and musculature of the ovipositor occurs only in the Acrididae and Tridactylidae. When the separated elements of the pterygote ovipositor are compared with those of the thysanuran ovipositor, it is to be seen at once that there is almost an exact correspondence between the parts of the organ in the two cases.

It becomes evident, therefore, that the first and second valvifers of the pterygote ovipositor are derived from the coxopodites of the gonopods respectively of the eighth and ninth abdominal segments, and that the first and second valvulae are respectively the first and second gonapophyses. The third valvulae of the Pterygota, which are carried by the second valvifers, are also derivations of the second coxopodites, but they are not the styli. True styli are present on the second coxopodites, but they are not the styli. True styli are present on the second coxopodites of a few pterygote insects, as in nymphs of Blattidae and in the adults of some Odonata. We may now recall that in *Thermobia* each coxopodite of the ninth segment is divided into a proximal plate carrying the gonapophysis and a distal stylus-bearing plate Styli of the coxopodites of the eighth segment are never present in Pterygota. The development of the pterygote ovipositor in entirely in accord with the homology of its parts suggested by the, structure of the adult organ.

The first valvulae are median outgrowth of the coxal areas of the eighth abdominal segment, the latter developing into the first valvifers. The primary genital processes of the ninth segment are the third valvulae, which are distal outgrowths of the coxopodites of the second gonopods, which become differentiated proximally into the second valvifers, while mesal (endite) processes become the second valvulae. Styli never appear on the coxopodites of the eighth segment, and those of the ninth segment, if present an immature stages, are lot with the transformation to the adult, except in some Odonata. While the theory of the origin of the parts of the ovipositor form the primitive limbs of the genital segments may thus seem to be well substantiated by ontogenetic development, due consideration should be given to the evident fact that all outgrowths of the body wall must necessarily look much alike in their early stages of growth, whether they represent primary or secondary organs. Similarly in such structures, therefore, is no proof of identity. The principal groups of pterygote insects in which the ovipositor is well developed are the Orthoptera, Hemiptera, Thysanoptera, and Hymenoptera.

The Ovipositor of Orthoptera

The ovipositor of Orthoptera is in some respects more generalized than that of other insects, while, on the other hand, it has one unique feature and certain structure specializations that adapt it to the particular mechanism developed this order. The orthopteroid ovipositor is generalized in that the coxoprodite of the ninth segment are not anatomically separated into second valvulae and third valvulae, though each is distinctly differentiated into a valvifer region and a distal valvular.

Also there are present in some families anterior and intervalvular sclerites in the venter of the ninth segment between second valvifers, and muscles from the ninth tergum and from the valvifers that do not generally occur in insects are inserted on these sclerites. The unique feature of the orthopteroid ovipositor is the including of the third valvulae as a pair of blades in the shaft of the organ. When the three pairs of valvulae, therefore, are all developed, the shaft consists of three pairs of valvular lobes or (A, B). In such cases, the first valvulae are ventral, the valvulae dorsal, and the second valvulae median and usually concealed between the others. The first valvulae may be movably both the second and the third valvulae by interlocking groove ridges. The second valvulae, however, are in some cases reduced or rudimentary, as in Acrididae and Gryllidae, and the shaft then consists of only two pairs of blades, which, it should be noted, are the first and third valvulae, and not the first and second as in the usual four-valve ovipositor of other insects. The particular mechanical features characteristic of the orthopteroid ovipositor are in the relations of the valvifers to the terga of the genital segments, and in the interrelation of the valvifers to each other.

The first valvifer is never closely attached to the eighth tergum, and it may be displaced so far posteriorly that it appears to belong to the ninth segment. By its posterior angle it articulates with the second valvifer, and it may also have a strong articulation with the ninth tergum. Internally the two valvifers of *Gryllus* osculate by special articular processes–a feature of much importance in the mechanism of the gryllid ovipositor.

The second valvifer has no direct articulation with the ninth tergum, its fulerum of movement being the articulation with the first valvifer. It is interesting to note that this same anatomical relation of the valvifers to each other and to the ninth tergum is characteristic of Hymenoptera also, though the structure and mechanism are not exactly the same in the two orders. The definitive first valvifer of Gryllidae is a composite structure formed of the true first valvifers and a small plate derived from the coxopodite of the ninth segment. The musculature of the ovipositor, as illustrated in *Gryllus*, is somewhat more complex than in insects of other orders owing to the presence of the muscles inserted on the intervalvular sclerites. Each first valvifer has a muscle from the seventh sternum and a large muscle from the eighth tergum, both inserted on an anterior apodemal arm of the sclerite. In its musculature, therefore, the first valvifer asserts its relation with the eighth abdominal segment regardless of its mechanical connections. Each second valvifer has a pair of large antagonistic muscles arising on the ninth tergum. This pair of muscles recurs in nearly all insects provided with an ovipositor.

The pair of muscles recurs in nearly all insects provided with an ovipositor. The muscles of the intervalvular sclerites include a pair of tergosternal muscles of the ninth segment, and two pairs of the muscles between the posterior intervalvula and the first and second valvifers, respectively. The motion of the valvifers produced by the tergal muscles inserted on them results in an alternate back-and-forth movement of the valvulae on each other. The articulations of the first valvifers with the ninth tergum, and the articulation of the two valvifers on each sides with each other constitute a mechanism of such a reversely to the other pair, with the result that the corresponding valvulae have opposite movement.

The Phasmidae, Mantidae, Blattidae, Grylloblattidae, and Tettigoniidae retain the three pairs of valvulae in the ovipositor, but otherwise the ovipositor has essentially the same structure in these families as in Gryllidae, though it is reduced and more or less modified in the first four and attains it highest mechanical perfection in Gryllidae. In the Acrididoidea and Tridactylidae, however, the organ departs widely in structure, musculature, and mechanism, not only form the ovipositor of other Orthoptera, but form that of all other insect, since the four terminal processes work by a divergent motion instead of sliding upon each other.

The Ovipositor of Hemiptera

The morphology of the pterygot ovipositor is perhaps best shown in the Hemiptera, because here the valvifers more nearly than in the other orders retain their proper segmental connections. In most of the Heteroptera the ovipositor reduced on rudimentary, but in Homoptera, except parasitic forms, is generally well developed. The eighth abdominal sternum is reduced or practically obliterated in all Hemiptera, and the subgenital plate is the seventh sternum. The shaft of the ovipositor in Homoptera, as illustrated in the cicac, issues at the base of the ninth segment between the eighth tergum and the seventh sternum. When dissected the part belonging to the two genital segments are easily distinguished. The first valvifers, implanted in the membranous ventrolateral part of the eighth segment (A), are smaller triangular plates carrying the first valvulae. Each articulates posteriorly with the ninth tergum, and small plate on the inner surface continuous with the dorsal margin of the first valvula, is directly fused with a lobe of the ninth tergum. This feature is produced into two rami, the outer of which is attached to the valvifer, while the inner one is fused with the ninth tergum.

The cicada a short muscle extends between the outer and inner plates the first valvifer. Other muscles of the first valvifer comprise a muscle from the seventh sternum and two muscles from the eighth tergum, all inserted on an apodeme of the dorsal margin of the valvifer. The second valvifers of the cicada are elongate plates mostly concealed within the projecting lower parts of the ninth tergum. At their anterior ends the second valvifers are directly continuous with the bases of the second valvulae, and third valvulae arise at their posterior ends. Each second valvulae is articulated at a point near the middle of its dorsal margin with condyle on the ventral margin of the ninth tergum. Two large muscles arising on the ninth tergum are inserted on each second valvifer respectively anterior and posterior to the articular fulcrum.

The second valvifers of the Hemiptera, therefore, rock directly on the ninth tergum, and not on the first valvifers as in Orthoptera and Hymenoptera. There are no other muscles in the ninth segment connected with the ovipositor, since intervalvular sclerites and their muscles, such as those of the cricket are absent in Hemiptera.

The shaft of the cicadid ovipositor consists of only three distinct parts, since the second valvulae are united with each other to form a strong median rod, to the sides of which the first valvulae are attached by the usual ridge-and-groove device. The lateral position of the first valvulae with respect to the second valvulae is characteristic of Homoptera, but more commonly the valvulae have the form of flattened blades. When the ovipositor is not in use it is ensheathed between the concave inner surfaces of the third valvulae.

The Ovipositor of Hymenoptera

The ovipositor of Hymenoptera, in its general form and in the composition of its shaft, resembles the ovipositor of Hemiptera more closely than that of Orthoptera, but it has one special character, namely, the articulation of the second valvifers with the first valvifers and not with the ninth tergum, that is a highly developed feature in the mechanism of the ovipositor of Gryllidae. The basic structure of the hymenoptera ovipositor is well shown in the Tenthredinidae, though the organ here does not have the form typical of the ovipositor of clistogastrous Hymenoptera, in which it is usually long and slender.

The shaft of the tenthredinid ovipositor, as illustrated in *Pteronidea ribesii*, is short and broad with an acute apex and strong lateral ridges. It is composed of the first and second valvulae, and is ordinarily ensheathed between the broad third valvulae. The basal part of the ovipositor consists of the first and second valvifers lying beneath the long lower margin of the ninth abdominal tergum. The first valvifer is a small triangular plate articulated by its dorsal angle with the ninth tergum and by its posterior ventral angle with the second valvifer. Anteriorly the first valvifer is continuous with the narrow ramus of the first valvulae. The second valvifer is a relatively large, elongate plate with the ramus of the second valvula attached to its anterior extremity, and the third valvula forming a broad lobe at its posterior end. The second valvifer has no articular connection with the ninth segment, its point of movement being the articulation with the first valvifer. The two second valvulae are united to each other by a median membrane. It should be observed that in the Tenthredinidae the terga of the eighth and ninth abdominal segments are normally developed and fully exposed plates. There are, however, as in all Hymenoptera, no sternal plates in these segments, the subgenital plate being the seventh sternum.

In the higher hymenopterous families the eighth and ninth segments are retracted into the seventh, and the terga of these segments are progressively reduced, until, in the bees, they consist only of two pairs of lateral sclerites associated with the base of the sting. In the more typical form of the hymenopterous ovipositor the shaft is long and slender, and the ensheathing third valvulae are correspondingly lengthened, but the general structure of the organ is in no way essentially different from that of the Tenthredinidae. An extreme development of the slender type of shaft occurs in some of the Ichneumonidae, as in *Megarhyssa* and related genera. During oviposition by such species the terminal part of the abdomen is turned downward, exposing the wide conjunctival membrane between the seventh and eighth terga. The base of the ovipositor is now dorsal and is contained in a large pouch formed by inflection of the membranous ventral part of the body wall including the seventh sternum.

The ovipositor shaft protrudes from the pouch at right angles to the length of the abdomen and is ensheathed distally in the slender third valvular; but, as the shaft penetrates the wood in which the female lays her eggs, the sheath valves separate from the shaft and may curve upward at the side of the body, as shown in various families illustrations. In the stinging Hymenoptera the ovipositor loses its egg-lying function and is converted into a poison-injecting instrument, but even here there is little change in the structure of the organ except in the development of valvular lobes on the first valvulae for driving the position liquid through the shaft. The stinging apparatus of the bees includes not only the usual parts of the ovipositor but also the lateral sclerites (quadrate plates) of the ninth tergum and the lateral spiracular plates of the eighth tergum. The united second valvulae form an inverted trough, the enlarged proximal part of which is the bulblike swelling of the shaft, and the tapering distal part the median *stylet*.

The slender first valvulae, or *lancets* (*Let*), slide on the lower margins of the bulb and stylet. Proximal to the bulb the rami of the valvifers. The reservoir of the poison gland discharges directly into the base of the bulb; the tubular left accessory gland opens ventrally between the rami of the valvulae. When the sting is in repose it is concealed in a sting chamber within the seventh abdominal segment, where it is ensheathed between the third valvulae. To be used effectively, the sting must be protracted from the sting chamber and deflected at right angles to the basal plates.

4

Development

Embryonic development begins with the first mitotic division of the zygote nucleus and terminates at hatching. Not surprisingly, in view of their diversity of form, function, and life history, insects exhibit a variety of embryonic development patterns, though, as more knowledge is obtained, certain evolutionary trends are becoming apparent. Eggs of most species contain a considerable amount of yolk. In exopterygote eggs there is such a preponderance of yolk that the egg cytoplasm is readily obvious only when it forms a small island surrounding the nucleus. In eggs of endopterygotes, the yolk-cytoplasm ratio is much lower than that of exopterygotes and, as a result, the cytoplasm can be seen as a conspicuous network connecting the central island with a layer of periplams lying beneath the vitelline membrane. This trend toward reduction in the relative amount of yolk in the egg, carried to an extreme in certain parasitic Hymenoptera and viviparous Diptera (Cecidomyiidae), whose eggs are yolkless and receive nutrients from their surroundings, has some important consequences.

Broadly speaking, the eggs of endopterygotes are smaller (size measured in relation to the body size of the laying insect) and develop more rapidly than those of exopterygotes. The increased quantity of cytoplasm leads to the more rapid formation of more and larger cells at the yolk surface which, in turn, facilitates the formation of a large embryonic area from which development of endopterygotes is streamlined and simplified. There has been as Anderson (1972) puts it, "reduction or elimination of ancestral irrelevancies," which when taken to an extreme, seen in the

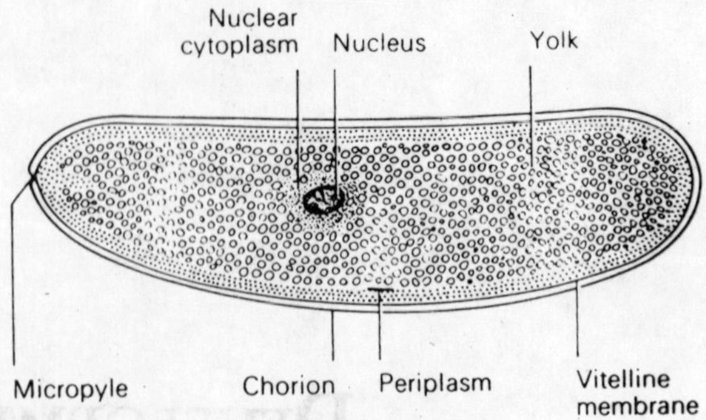

Fig. 4.1. Sagittal section of a typical insect egg.

apocritan Hymenoptera and cylorrhaph Diptera, results in the formation of a structurally simple larva which hatches within a short time of egg laying. However, superimposed on this process of short-circuiting may be developmental specializations associated with an increasing dissimilarity of juvenile and adult habits.

Cleavage and Blastoderm Formation

As it moves toward the center of an egg after fusion, the zygote nucleus begins to divide mitotically. The first division occurs at a predetermined site, the cleavage center located in the future head region, which cannot be recognized morphologically but which appears to become activated either when sperm enter an egg or when an egg is laid. Early divisions are synchronous, and as nuclei are formed and migrate through the yolk toward the periplam, each becomes surrounded by an island of cytoplasm. Each nucleus and its surrounding cytoplasm are known as a cleavage energid. In eggs of endopterygotes and possible exopterygotes, but not those of apterygotes, the energids remain inter-connected by means of fine cytoplasmic bridges. The rate at which nuclei migrate to the yolk surface and the method of colonization are variable.

In eggs of some species nuclei appear in the periplasm as early as the 64-energid state (after 6 divisions); in others, nuclei are not seen in the periplasm until the 1024-energid stage. In eggs of most endopterygotes and in those of paleopteran and hemipteroid exopterygotes, the periplasm is invaded uniformly by the energids. However, in eggs of orthopteroid insects the periplasm at the posterior pole of the egg receives energids first, after which there is progressive colonization of the more anterior regions. In eggs of most insects not all cleavage energids migrate to the periphery

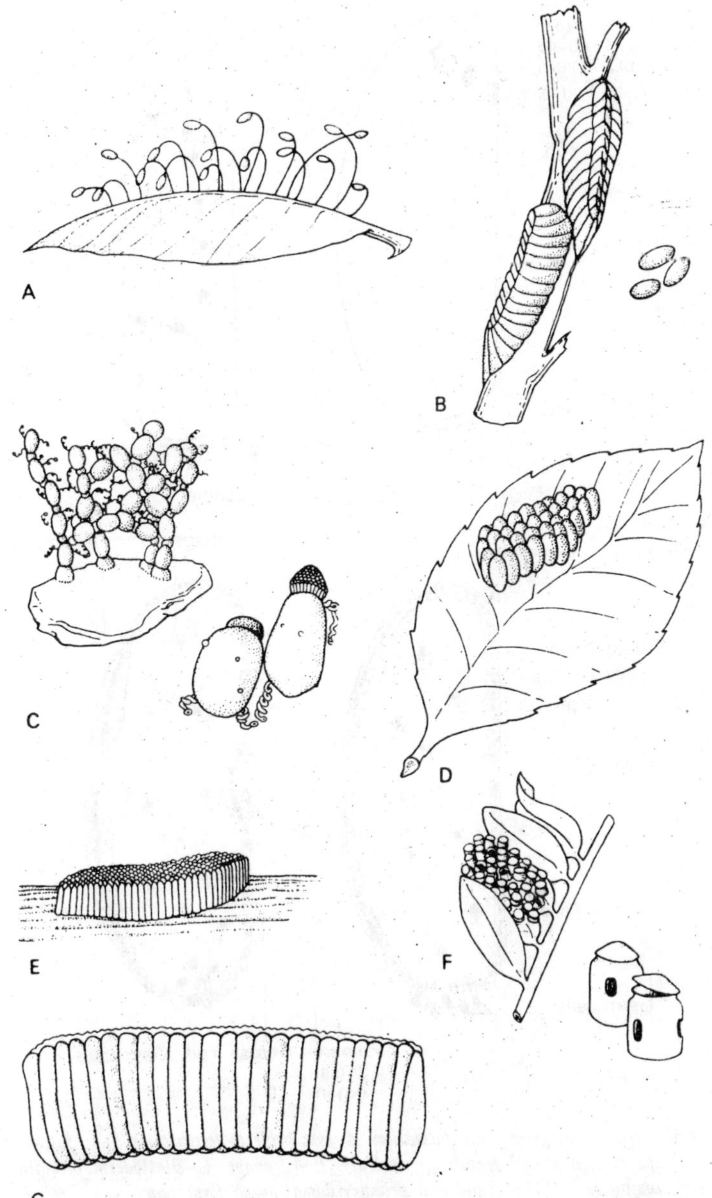

Fig. 4.2. Representative insect eggs. A–California green lacewing, Chrysopa californica. B–Egg cases (oothecae) and eggs of a praying mantid. C–Eggs and egg masses of the mayfly, Ephemerella rotunda. D–Eggs of the convergent ladybird beetle. E–Egg raft of the mosquito Culex pipiens. F–Eggs and egg masses of the stink bug, Pentatoma lignata. G–Ootheca of the German cockroach, Blattella germanica.

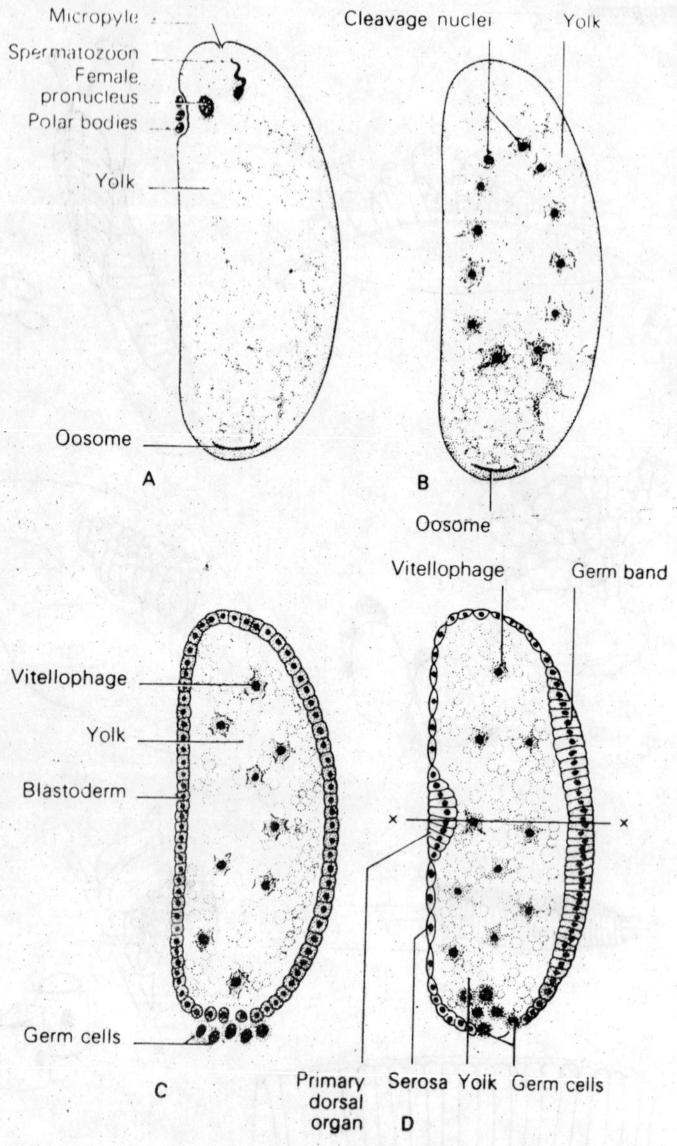

Fig. 4.3. Fertilization–germ band formation. A–Just prior to fertilization (i.e., fusion of spermatozoon and female pronucleus). B–Cleavage. C–Blastoderm formation complete. D–Germ band and primary dorsal organ formation.

but continue to divide within the yolk to form primary vitellophages, phagocytic cells whose function is to digest the yolk.

In eggs of Lepidoptera, Diptera and some Orthopteroid insects, however, all the energids migrate to the periplasm and only later do some of their progeny move back into the yolk as secondary vitellophages. Secondary vitellophages are also produced in eggs of other insects to supplement the number of primary vitellophages. So-called tertiary vitellophages are produced in eggs of some cyclorrhaph Diptera and apocritan Hymenoptera from the anterior and posterior midgut rudiments. After their arrival at the periplasm, the energids continue to divide, often synchronously, until the nuclei become closely packed (the syncytial blastoderm stage), after which cell membranes form (the uniform blastoderm stage).

Formation and Growth of Germ Band

The next stage is blastoderm differentiation, giving rise to the embryonic primordium (an area of closely packed columnar cells from which the future embryo forms) and the extraembryonic ectoderm from which the extraembryonic membranes later differentiate. Differentiation is controlled by two centers, as the classic experiments of the German embryologist Seidel first demonstrated. As energids move toward the posterior end of the egg they "interact" with a so-called "activation center" and differentiation subsequently occurs. Seidel's experiments showed that neither an energid nor the activation center alone could stimulate differentiation. It is presumed that the center is caused to release an as yet unidentified chemical which diffuses anteriorly. This diffusion is seen morphologically as a clearing and slight contraction of the yolk. As the chemical reaches the future prothoracic region of the embryo (the "differentiation center"), the blastoderm in this region gives a sharp twitch and becomes slightly invaginated. Blastoderm cells aggregate within this invagination and differentiate into the embryonic primordium. (Later in embryogenesis, other processes, for example, mesoderm formation and segmentation, begin at the differentiation center and spread anteriorly and posteriorly from it.) As a result of the differing amounts of yolk that exopterygote and endopterygote eggs contain, important differences occur in the formation of the embryonic primordium.

In exopterygote eggs where there is initially little cytoplasm, the embryonic primordium is normally relatively small, and its formation depends on the aggregation and, to some extent, proliferation of cells. In these eggs it usually occupies a posterior midventral position. In contrast, in endopterygote eggs with their

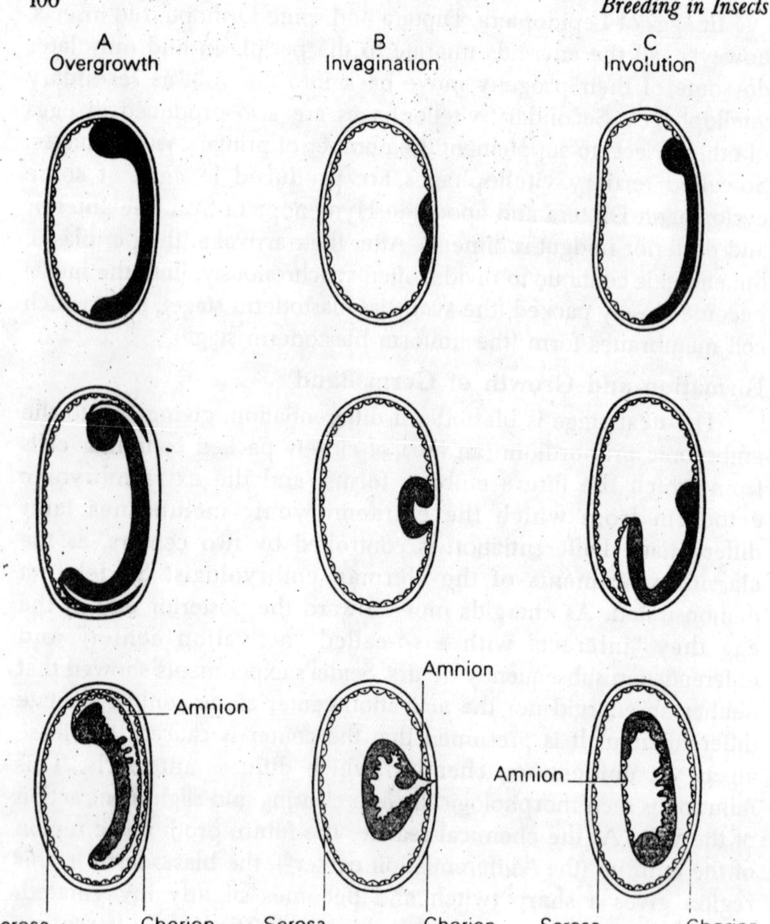

Fig. 4.4. Formation of extraembryonic membranes (diagrammatic).

greater quantity of cytoplasm, the primordium forms as a broad monolayer of columnar cells that occupies much of the ventral surface of the yolk. In other words, the primordium in endopterygote eggs does not require to undergo much increase in size, as is necessary in eggs of exopterygotes, so that tissue differentiation can occur directly and embryonic growth more rapidly. At its extreme, seen in eggs of some Diptera and Hymenoptera, the primordium occupies both ventral and lateral areas of the egg, with the extraembryonic ectoderm covering only the dorsal surface. The shape of the primordium is variable, though in most insects the anterior region is expanded laterally as a pair of head lobes

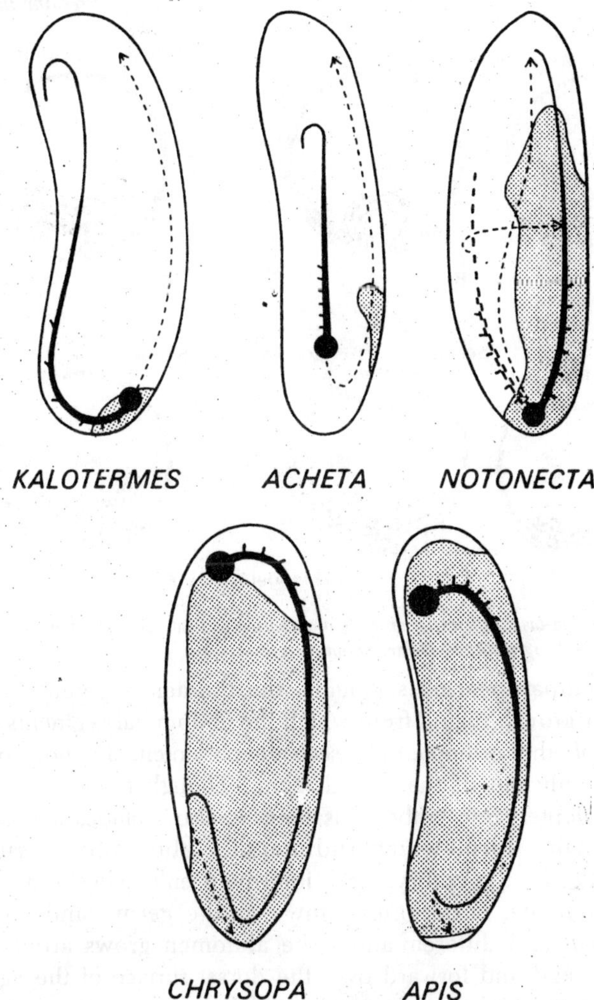

Fig. 4.5. Diagrams showing the extent of the germ band (shaded), the position of the embryo prior to katatrepsis (solid black, showing position of protocephalon, gnathal and thoracic appendages), and the movement of the embryo at katatrepsis (arrow). In all cases the embryo finally comes to lie on the ventral surface of the egg (to the right) with its head at the anterior (upper) pole.

(=protocephalon), behind which is a region of variable length, the protocorm (postentennal region).

In eggs of Paleoptera, hemiteroid insects, and some orthopteroid species, the protocorn is semilong and at its formation includes

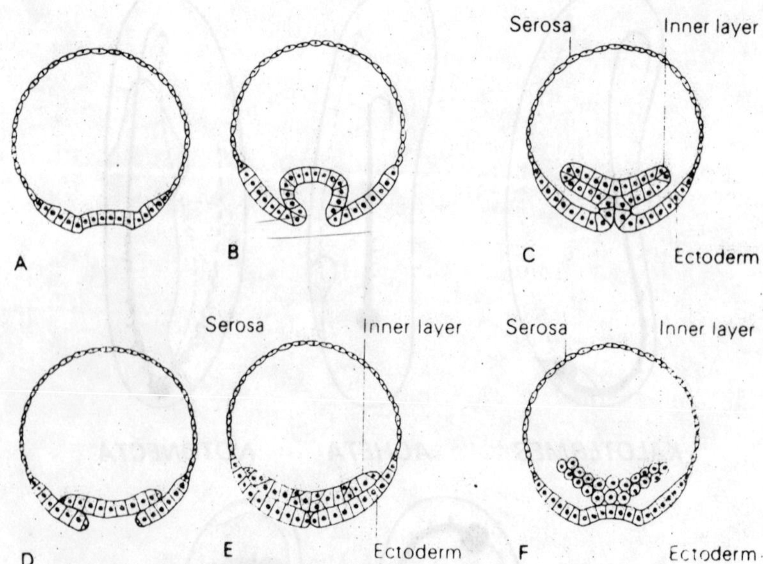

Fig. 4.6. Germ-layer differentiation (yolk and chorion omitted). A-C–Invagination. D and E–Overgrowth. F–Delamination.

the mouthpart-bearing segments, the thoracic segments, and a posterior growth region from which the abdominal segments arise. In eggs of other orthopteroid insects the postantennal region consists initially only out of the growth zone. Though the protocorn in most endopterygote embryos is long, is also includes a posterior growth zone form which rudimentary abdominal segments proliferate. As the embryonic primordium enlongates and begins to differentiate, it becomes known as the germ band. During elongation and differentiation, the abdomen grows around the posterior end and forward over the dorsal surface of the egg. In eggs of some higher endopterygotes (Hymenoptera-Apocrita and Diptera-Cyclorrhapha), there is no posterior growth zone and the abdominal segments arise directly from the primordium. It is during the differentiation and elongation of the germ band that the primordial germ cells first become noticeable in most endopterygote eggs, though in those of some Coleoptera they are distinguishable even as the syncytial blastoderm is forming. They are largish, rounded cells in a distinct group at the posterior pole of the yolk, and accordingly are referred to as pole cells. In eggs the Dermaptera, Psocoptera, Thysanoptera, and Homoptera also, the germ cells differentiate early at the posterior end of the primordium. In those of most exopterygotes, however, they are not apparent until gastrulation or somite formation has occurred. As the germ band elongates and becomes broader, segmentation and limb-bud formation are apparent externally and are accompanied internally by mesoderm and somite formation.

Fig. 4.7. Sagittal section of an embryo (diagrammatic).

Growth of the germ band may occur either on the surface of the yolk (superficial growth) as seen in eggs. Dictyoptera, Dermaptera, Isoptera, some other orthopteroid, insects and all endopterygotes, or by immersion into the yolk (immersed growth) as occurs in eggs of Paleoptera, most Orthoptera, and hemipteroid insects Immersion of the germ band (anatrepsis) forms the first of a series of embryonic movements, collectively known as blastokinesis. The reverse movement (katatrepsis), which brings the embryo back to the surface of the yolk, occur later. Anatrepsis has developed secondarily (i.e., superficial growth is the more primitive method) and convergently among those exopterygotes in which it occurs. Its functional significance is, however, not yet clear (Anderson, 1972a).

Gastrulation, Somite Formation and Segmentation

As the embryonic primordium begins to increase in length, its midventral cells sink inward to form a transient, longitudinal gastral groove. The invaginated cells soon separate from the outer layer which closes to obliterate the groove. It is from the anterior and posterior points of closure of the gastral groove that the stomodeum and proctodeum, respectively, develop. The outer layer can now be distinguished as the embryonic ectoderm. The invaginated cells, which proliferate and spread laterally from the mesoderm except adjacent to the development stomodeum and proctodeum where they become the anterior and posterior midgut rudiments, respectively. The mesodermal cells become concentrated into paired longitudinal tract which soon separate into segmental blocks, leaving

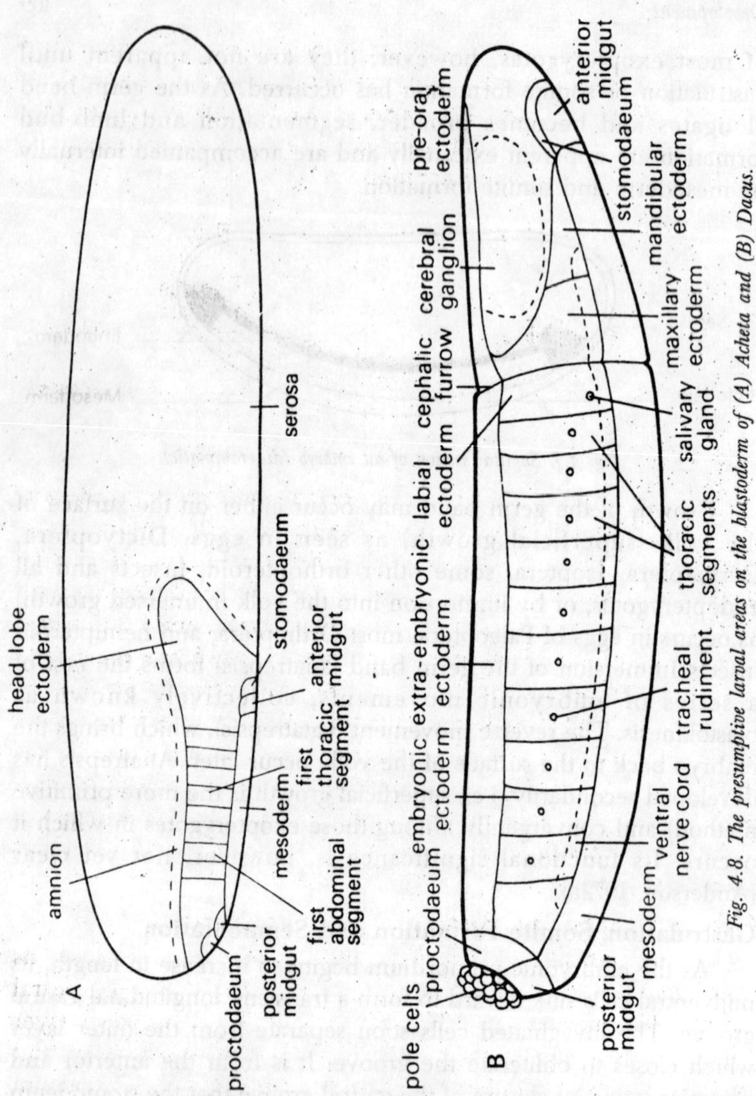

Fig. 4.8. The presumptive larval areas on the blastoderm of (A) Acheta and (B) Dacus.

only a thin longitudinal strip, the median mesoderm, from which hemocytes later differentiate. From these segmental blocks, paired hollow somites usually arise. Somite formation is initiated and occurs more or less simultaneously in the gnathal and thoracic segments, spreading anteriorly and posteriorly after gastrulation takes place.

Formation of the coelom (the cavity within a somite) may occur in one or two ways, by internal splitting of a somite or by median folding of the lateral part of each somite. In embryo of a given species, one or both methods may be seen in different segments. For example, internal splitting of the somites occurs in all segments of embryos of Phasmida, most hemipteroid insects, and most endopterygotes, and in the abdominal segments of *Locusta* embryos.

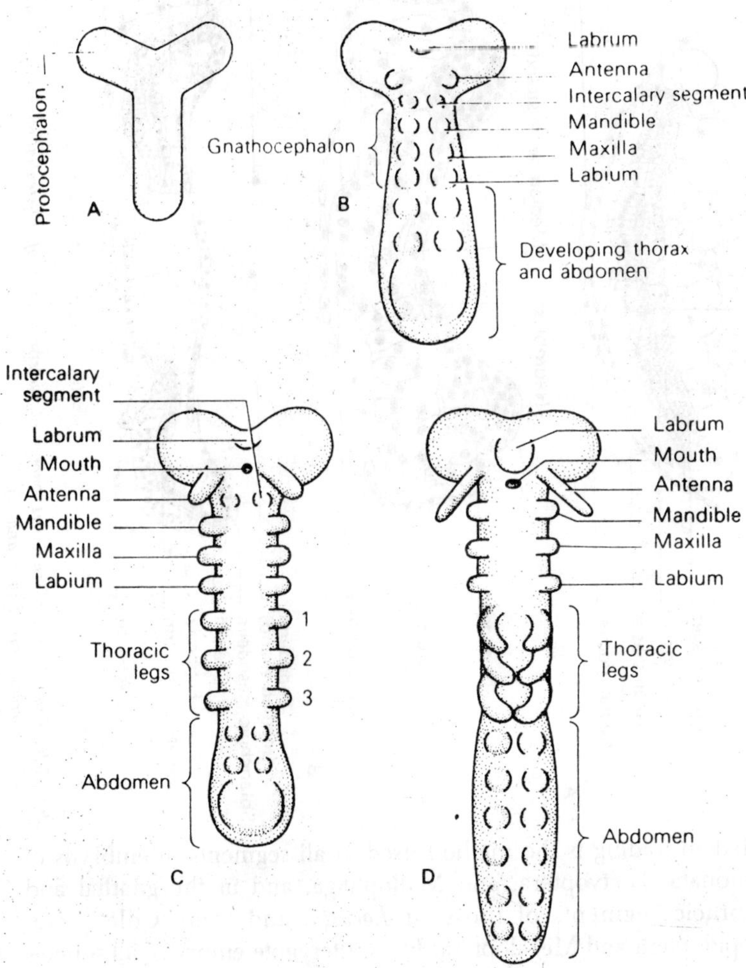

Fig. 4.9. Segmentation and appendage formation. A–Prior to segmentation. B and C–Successive stage. D–Segmentation complete.

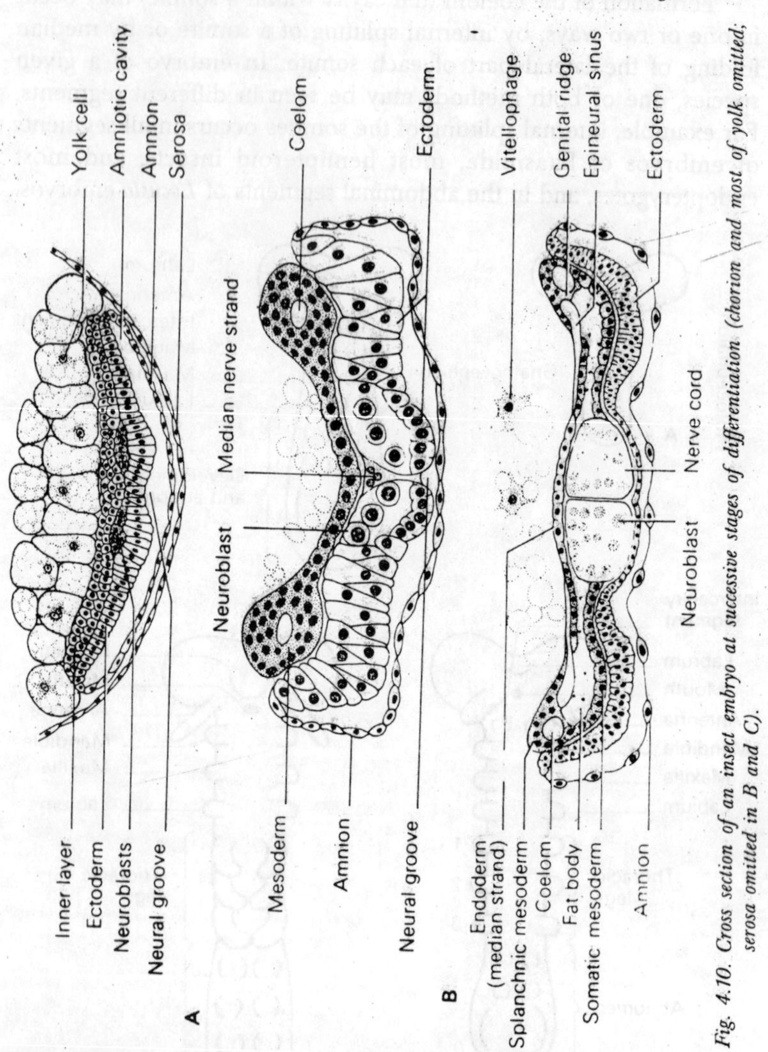

Fig. 4.10. Cross section of an insect embryo at successive stages of differentiation (chorion and most of yolk omitted; serosa omitted in B and C).

Median folding is the method used in all segments in embryos of Odonata, Dictyoptera, and Mallophaga, and in the gnathal and thoracic segments of those of *Locusta,* and some Coleoptera, Lepidoptera and Megaloptera. In exopterygote embryos, all somites usually develop a central cavity, though this may be only temporary. Among endopterygotes, members of more primitive orders retain a full complement of somites in their embryos and the latter usually develop a coelom. In embryo of some species,

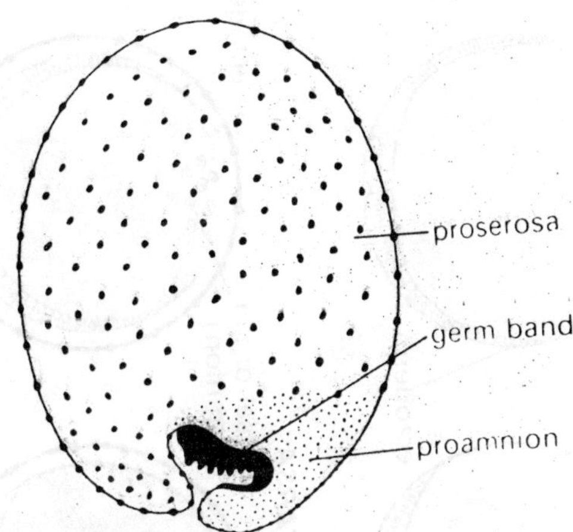

Fig. 4.11. Early stages of the invagination of the germ band in Machilis.

however, cavities may not form, and somite formation may be suppressed in the head segments. In embryos of Diptera and Hymenoptera, no distinct head somites appear, and in those of some Cyclorrhapha and Apocrita, somite formation is entirely suppressed, so that mesodermal derivatives are produced directly from a single midventral mass.

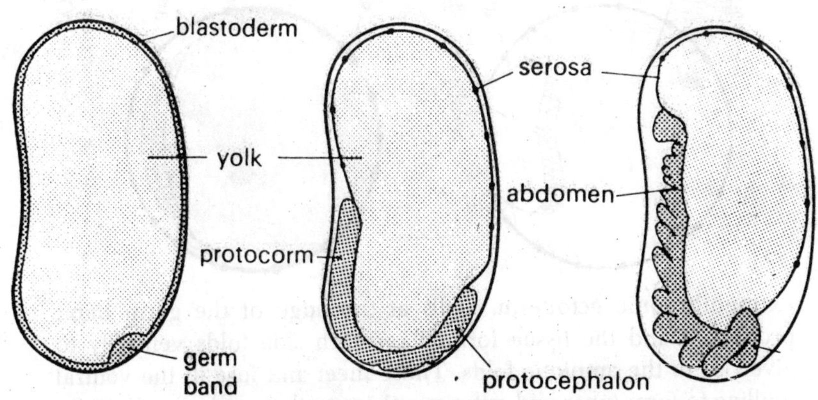

Fig. 4.12. Early stages in the development of the embryo of Zootermopsis, an example of an embryo development superficially on the yolk. The amnion, not shown, covers the outer surface of the embryo.

Formation of Extraembryonic Membranes

Simultaneously with gastrulation and somite formation, two extraembryonic membranes, the amino and serosa, develop from

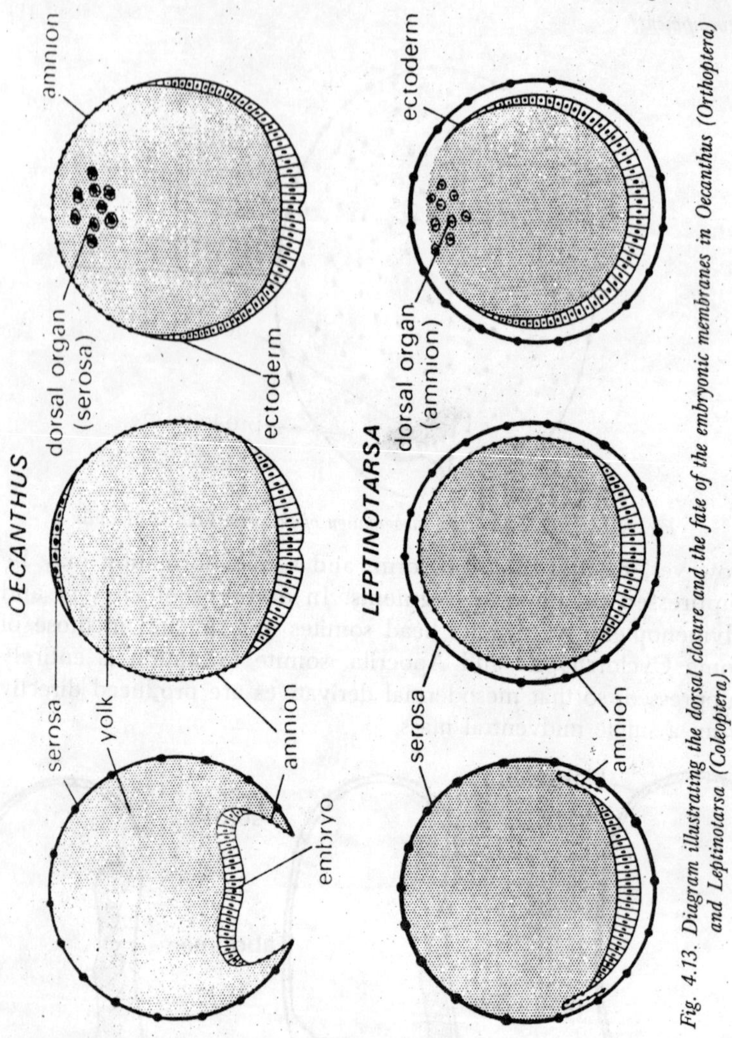

Fig. 4.13. Diagram illustrating the dorsal closure and the fate of the embryonic membranes in Oecanthus (Orthoptera) and Leptinotarsa (Coleoptera).

extraembryonic ectoderm. Cells at the edge of the germ band proliferate and the tissue formed on each side folds ventrally to give rise to the amniotic folds. These meet and fuse in the ventral midline to form inner and outer membranes, the amnion and serosa, respectively, the former enclosing a central fluid-filled amniotic cavity. Many authors have suggested that such a cavity would provide space in which an embryo could grow and also prevent physical damage. Anderson (1972a) considers, however, that these

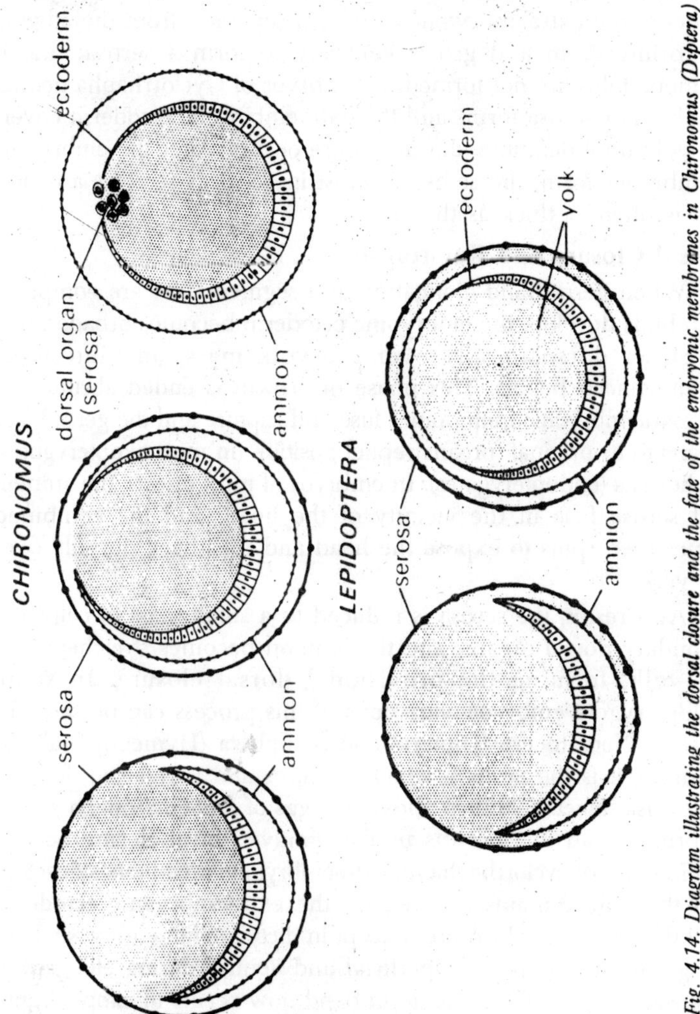

Fig. 4.14. Diagram illustrating the dorsal closure and the fate of the embryonic membranes in Chironomus (Diptera) and a lepidopteran.

functions are redundant and that the cavity must have an as yet unidentified function. Another possibility is that the amnion and its cavity are used to store wastes which are thus kept separate from the yolk.

The general method of amnion and serosa formation outlined above is found in all insect embryos (with some modification where immersion of the germ band into the yolk occurs) except those of Cyclorrhapha and Apocrita, in which, it will be recalled, the embryonic membranes are greatly reduced or lost. In embryos of Apocrita the extraembryonic ectoderm separates from the edge of the primordium and grows ventrally to form a serosa; that is amniotic folds are not formed. In embryos of Cyclorrhapha neither amnion nor serosa forms and the extraembryonic ectoderm covers the yolk until definitive dorsal closure occurs. After the embryonic membranes form, the serosa in most insect eggs secretes a cuticle that is often as thick as the chorion.

Dorsal Closure and Katatrepsis

When germ band elongation and segmentation are complete, limb buds develop, the embryonic ectoderm becomes broader, that is, grows dorsolaterally over the yolk mass, and internally organogenesis begins. This phase of growth is ended abruptly as the extraembryonic membranes fuse and rupture and the germ band reverts to its original (preanatreptic) position (in most exopterygotes) or shortens (endopterygotes). In embryos of most insects, the amnion and serosa fuse in the vicinity of the head, and the combined tissues then splits to expose the head and rolls back dorsally over the yolk.

As a result, the serosa is reduced to a small mass of cells, the secondary dorsal organs, and the amnion becomes stretched over the yolk, forming the provisional dorsal closure. In some endopterygote embryos, variations of this process can be seen. In those of Nematocera (Diptera) and Symphyta (Hymenoptera), for example, it is the amnion which ruptures and is reduced, leaving the serosa intact. As noted above, in eggs of Apocrita only a serosa is formed, and this persists until definitive dorsal closure occurs, and in those of cyclorrhapha no extraembryonic membranes develop, and the yolk remains covered by the extraembryonic ectoderm until definitive dorsal closure. Except in dictyopteran embryos where the germ band remains superficial and ventral during elongation, extension movement of the germ band now occurs in exopterygote eggs which serves (1) to bring an immersed germ band back to the surface of the yolk, and (2) to restore the germ band to its preanatreptic orientation, that is, on the ventral surface of the yolk with the head end facing the anterior pole of the egg. This movement, the reverse of anatrepsis, is known as katatrepsis.

At the beginning of provisional dorsal closure the germ band of most endopterygotes is quite long so that, although its anterior end is ventral, its posterior component passes round the posterior tip of the yolk and forward along the dorsal side. During closure, the germ band shortens and broadens rapidly so that its posterior end now comes to lie near the posterior end of the egg. Definitive dorsal closure, that is, the enclosing of the yolk within the embryo, then occurs. It is achieved in all insect embryos by a lateral growth of the embryonic ectoderm which gradually replaces the amnion or, rarely, the serosa.

Tissue and Organ Development

Appendages

Paired segmental evagination of the embryonic ectoderm appear on the thoracic, antennal, and gnathal segments while the abdominal part of the germ band is still forming. Their subsequent growth results from proliferation of the ectoderm as a single layer of cells and of mesodermal cells within. The cephalic and thoracic limbs ultimately differentiate into their specific form, except in eggs of secondary apodous species where they soon shorten or become reduced to epideal thickenings. In embryos of Cyclorrhapha, the thoracic appendages never develop beyond the epidermal thickening stage. In Paleoptera and orthopteroid insects; 11 pairs of abdominal appendages evaginate before provisional dorsal closure.

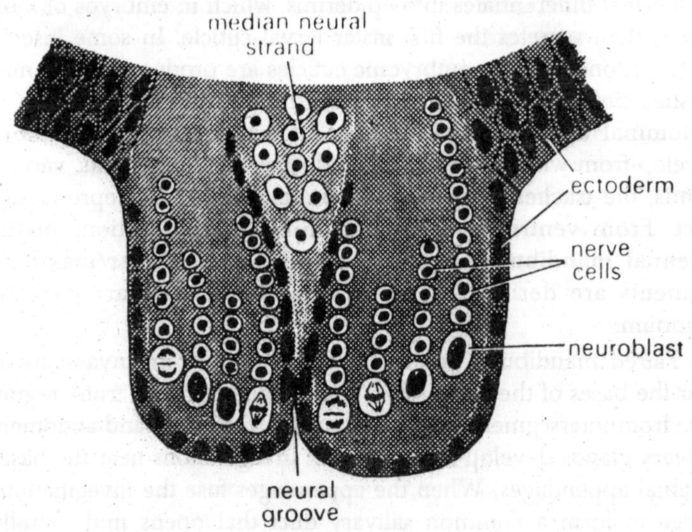

Fig. 4.15. Section through the developing ventral nerve cord showing the neuroblasts with their daughter cells.

In most hemipteroid embryos, no sign of abdominal limbs is evident, though in those of Hemiptera and Thysanoptera appendages develop on the first and last abdominal segments. Ten pairs of abdominal evaginations develop in most endopterygote embryos. The fat of the abdominal appendages varies, and some or all of them may disappear before embryonic development is completed. The first (most anterior) pair disappears after blastokinesis in embryos of Plaeoptera and some orthopteroid insects, but remains as glandular pleuropodia in those of Dictyoptera, Phasmida, Orthoptera, Hemiptera, and some Coleoptera and Lepidoptera. The function of the pleuropodia is uncertain, though some authors have suggested that in orthopteran embryos they secrete an enzyme which brings about dissolution of the serosal cuticle. The pleuropodia are resorbed or discarded before hatching. The appendages of the second though seventh abdominal segments are resorbed except in some endopterygotes where they persist as larval prolegs. Pairs 8 to 10 may differentiate into the external genitalia or disappear, while the last pair either persists as cerci or disappear.

Integument and Ectodermal Invaginations

Soon after definitive dorsal closure, the outer embryonic ectodermal differentiates into epidermis, which in embryos of most insects then secretes the first instar larval cuticle. In some insects, however, one or more embryonic cuticles are produced which may be shed before or at hatching. Concurrently with the formation of abdominal appendages a number of ectodermal invaginations develop from which differentiate endoskeletal components, various glands, the tracheal system, and certain parts of the reproductive tract. From ventrolateral invaginations at the junctions of the antennal mandibular segments and the mandibular/maxillary segments are derived the anterior and posterior arms of the tentorium.

Paired mandibular apodemes differentiate from invaginations near the bases of the mandibles. The apodemes of the trunk region arise from intersegmental invaginations in the thorax and abdomen. Salivary glands develop from a pair of invaginations near the bases of labial appendages. When the appendages fuse the invaginations merge to form a common salivary duct that opens midventrally on the hypopharynx. The corpora allata develop from a pair of ventrolateal invaginations at the junction of the mandibular/ maxillary segments. Initially, they exist as hollow vesicles, though these fill in as they move dorally to their final position adjacent to the stomodeum.

The most glands also originate as paired ventral ectodermal invaginations, usually on the prothoracic segment. Other invaginations on the head may give rise to specialized exocrine glands on the mandibles or maxillae. Elements of the tracheal system can be seen first as a paired lateral invaginations on each segment from the second thoracic to the eighth (ninth in a few Thysanura) abdominal. However, not all of these invaginations develop completely into tracheae. Those which do, bifurcate and anastomose with branches from adjacent segments and from their opposite partner of the same segment. The cells differentiate as tracheal epithelium and then secrete a cuticular lining. After cuticle secretion but before hatching gas is secreted into the tracheal system. Some of the invaginated ectodermal cells differentiate into oenocytes. These may remain closely associated with the tracheal system, form definite clusters in specific body regions, or become embedded as single cells in the fat body.

Central Nervous System

Soon after somite formation has commenced, specialized ectodermal cell on each side of the midventral line, the neuroblasts, begin to proliferate, resulting in the formation of paired longitudinal neural ridges separated by a neural grooves. As proliferation occurs, the cells move slightly inward so that they become separated from the ectoderm. They then begin to divide vertically, unequally and repeatedly, the small daughter cells eventually developing into ganglion cells. With the onset of segmentation, neuroblasts in the intersegmental regions become less active, so that paired segmental swellings, the future ganglia, now become apparent. As the ganglion cell take on the apperance of neurons, their cells bodies become arranged peripherally around the central axons (neuropile). Subsequent growth of the axons leads to formation of longitudinal connectives and transverse commissures.

The neurilemma is also formed from ganglion cells. As embryogenesis continues, fusion of ganglia occurs in the head region to form the brain and subesophageal ganglion and at the posterior end of the abdomen where ganglia from segments 8 to 11 form a composite structure. In embryos of species belonging to different order of Insecta, varying degrees of fusion of other ganglia may subsequently occur.

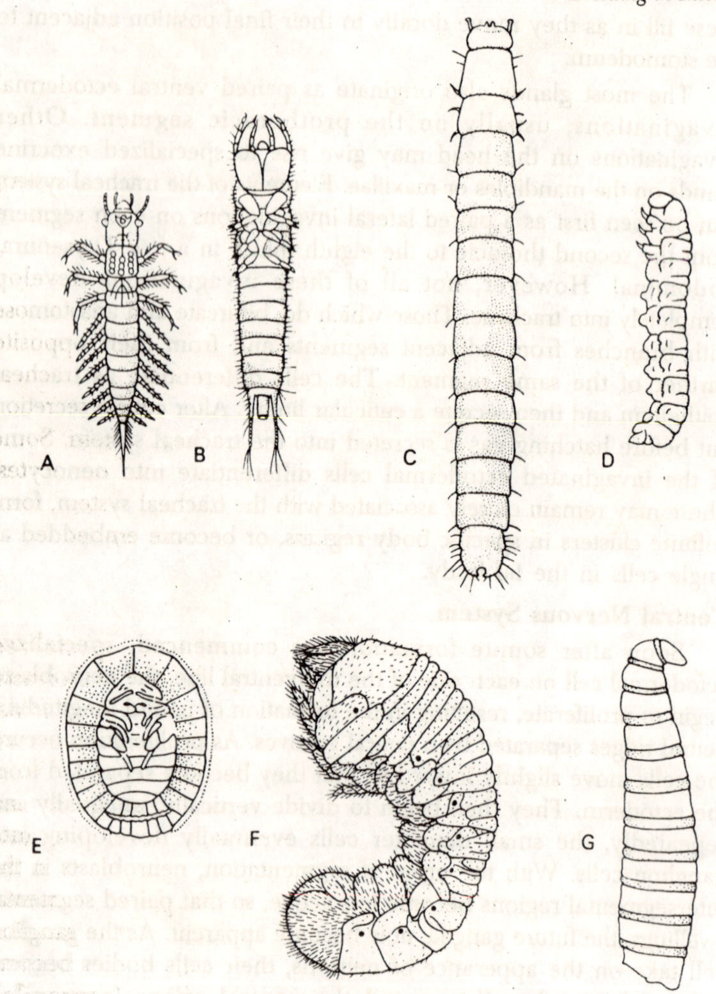

Fig. 4.16. Larval types. A–Campodeiform, alderfly, Sialis sp. (Neuroptera; Sialidae). B–Carabiform, ground beetle, Harpalus sp. (Coleoptera; Carabidae). C–Elateriform, click beetle (Coleoptera; Elateridae). D–Eruciform, clear-winged moth (Lepidoptera; Aegeriidae). E–Platyform, aquatic beetle. Eubrianax edwardsi (Coleoptera; Dascillidae). F–Scarabaeiform, branch and twig borer, Polycaon confertus (Coleoptera; Bostrichidae). G–Vermiform, flesh fly, Sarchophaga sp. (Diptera; Sarcophagidae).

Gut and Derivatives

As noted earlier, the stomodeum and proctodeum arises at the anterior and posterior ends of the gastral groove, respectively. Both develop as hollow invaginated tubes, the stomodeum slightly earlier than the proctodeum, and as they differentiate into the subdivisions of the foregut and hindgut, respectively, various associated structures arise. On the roof of the stomodeum, evaginations or thickenings give rise to the frontal ganglion, hypocerebral ganglion, intrinsic cells of the corpora cardiaca, and ingluvial ganglia.

At its distal end, the proctodeum develops pouches which are the rudiments of the Malpighian tubules. (In many insects additional tubules develop during larval life.) The anterior and posterior midgut rudiments which appeared at gastrulation begin to proliferate at about the time of provisional dorsal closure to form the midgut. Each rudiment proliferates a pair of strands which grow cauded or cephalad, respectively, between the nerve cord and yolk. After strands from each rudiment meet in the middle of the embryo, they laterally and dorsally so as to eventually enclose the yolk mass. The cells then differentiate as midgut epithelium. Prior to this, in embryos of some species, the vitellophages from a temporary "yolk sac" around the yolk mass.

Circulatory System, Muscle and Fat Body

The heart, aorta, musculature, fat body, lining of the hemocoel, and some components of the reproductive system are derived from the somites and median mesoderm formed after gastrulation. As noted earlier, the median mesoderm gives rise to hemocytes. In most embryos, each somite becomes hollow and forms three interconnected chambers, the anterior, posterior, and ventrolateral pouches. The latter grows into the adjacent ectodermal limb but and breaks up to form intrinsic limb muscles. The splanchnic walls (i.e., those facing the yolk) of the two remaining pouches spread round the gut, forming the gut musculature, some fat body, and part of the reproductive system.

The somatic walls (those facing the ectoderm) of the anterior and posterior pouches give rise to extrinsic limb muscles, dorsal and ventral longitudinal muscles, and more fat body. Thus, in insects as in other arthropods, the breaking up of the somite walls into discrete tissues means that there is not true coelom. Rather, the latter merges with the epineural sinus (the space between the dorsal surface of the embryo and the yolk) and is correctly called a mixocoel (hemocoel). From mesodermal cells at the dorsal junction of the somatic and splanchnic walls of the labial to the tenth abdominal somites, a sheet of cardioblasts develops. As the mesoderm grows around the gut, the sheets on each side become apposed to form the heart. Other somatic mesoderm cells adjacent to the cardioblasts differentiate as alary muscles, pericardial septum, and pericardial cells. The aorta develops from the median wall of the antennal somites which become apposed and grow posteriorly to meet the heart.

Reproductive System

The reproductive system includes both mesodermal and ectodermal components. In female exopteryotes, the vagina and spermatheca develop after hatching as midventral ectodermal invaginations of the seventh of eighth abdominal segment. In males, the ejaculatory duct and ectadenes (ectodermal accessory glands) are formed from a similar midventral invagination of the ectodermal of the ninth or tenth abdominal segment. The paired genital ducts and mesoderms (mesodermal accessory glands) arise in exopterygotes from mesoderm of the splanchnic walls of certain abdominal somites which first thickens the hollows out to form coelomoducts. Some of these soon disappear, but those of the seventh and eighth somites (in female) or ninth and tenth somites (in males) enlarge to form the ducts and or accessory gland components.

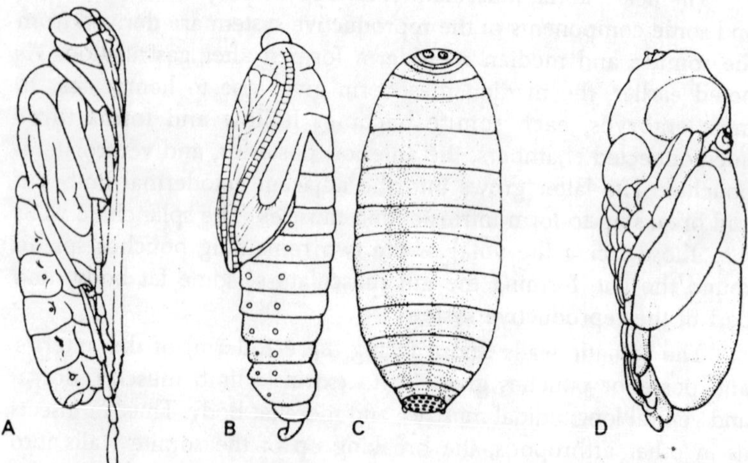

Fig. 4.17. Pupal types. A–Exarate, ichneumon wasp. B–Obtect, moth. C–Coarctate, house fly, D–Puparial case removed, exposing exarate house fly pupa.

In endopterygotes, the genital ducts are formed during postembryonic development. In *Drosophila* and other cyclorrhaph Diptera the reproductive system (excluding the gonads) develops form a single or pair of imaginal discs during metamorphosis. Development of the gonads varies, though two related trends can be seen namely, earlier segregation of the primordial germ cells and restriction of these cells to fewer abdominal segments. In the most primitive arrangement, seen in some thysanuran and orthopteran embryos, the germ cells do not become distinguishable until they appear in the splanchnic walls of several abdominal somites. In *Locusta* embryos, for example they are found initially in the somites of abdominal segments 2 through 10, though they remain only in segment 3 through 6. Eventually they fuse longitudinally to form a compact gonad on each side.

Such a segmental arrangement is presumably primitive as it is seen also in adult Annelida, Onychophora, Myriapoda, and noninsectan apterygotes (Anderson, 1972a). In embryos of other orthopteroid insects (Dictyoptera, Phasmida, and Embioptera) and some hemipteroid insects (Heteroptera) the germ cells become apparent early in gastrulation. Nevertheless, they still become associated with the splanchnic mesoderm of several anterior abdominal segments. In embryos of Dermaptera, Psocoptera, Thysanoptera, Homoptera, and endopterygotes, the germ cells differentiate as the blastoderm forms. After somite formation, they migrate to the third and fourth abdominal segments in exopterygote embryos or fifth and sixth abdominal segments in endopterygote embryos where they divide into left and right halves and become surrounded by splanchnic mesoderm.

Special Forms of Embryonic Development

The great majority of insects species are bisexual and females lay eggs that contain a considerable amount of yolk. However, in some species, males may be rare and females may produce viable offspring from unfertilized eggs (parthenogenesis). In another form of sexual reproduction, polyembryony, which is characteristic of some parasitic Hymenoptera and Strepsiptera, several embryo develop from one fertilized egg. In other insects, fertilized eggs may be retained within the female reproductive tract for variable periods of time so that a young insect may hatch from the egg almost as soon as or even before the latter is laid (viviparity). In few species paedogenesis may occur where mature larvae are able to produce, parthenogenetically and usually viviparously, a further generation of young.

Parthenogenesis

The ability of unfertilized eggs to develop is common to many insect species, and in some is the normal mode of reproduction under certain conditions. In all insects except Lepidoptera and Trichoptera, the female is the homogametic sex (that is, having two X sex chromosomes) and the male, heterogametic (XY or XO). Unfertilized eggs, therefore, will contain only X chromosomes. However, whether they contain one or two such chromosomes and, therefore, the sex of parthenogenetic offspring, depends on the behaviour of the chromosomes during meiosis in the oocyte nucleus (Suomalainen, 1962; White, 1973). In some species no meiotic division occurs during oogenesis. Therefore, offspring have the same genetic make-up as the mother; that is, they are diploid and female (ameiotic parthenogenesis).

In meiotic parthenogenesis, the typical reduction division is followed by nuclear fusion so that a diploid chromosome complement is retained. Again, therefore, the offspring are female. Haploid parthenogenesis, where the oocyte nucleus undergoes meiosis which is not followed by nuclear fusion, is of relatively rare occurrence, though typical of Hymenoptera. Thsanoptera, and some Homoptera and Coleoptera, and results in the production of males. In Hymenoptera, haploid parthenogenesis is facultative; that is, a female determines whether or not an egg will be fertilized. In the honeybee, for example, a queen normally lays fertilized eggs which develop into workers (diploid females). However, under certain conditions, for example, when the hive is crowded, and the workers construct larger than normal drone cells on the honeycomb, she will lay unfertilized eggs from which haploid males develop, as a preliminary to swarming. Parthenogenesis, producing in most species female offspring may confer two advantages.

In a species whose population density may be (temporarily) low, the ability of an isolated female to reproduce parthenogenetically may ensure survival of here genotype until the population density increases and males are again likely to be encountered. More often, however, parthenogenesis is employed as a mechanism that provides a rapid mode of reproduction, to enable a species to take full advantage of temporarily ideal conditions. Thus, a parthenogenetic female, who does to require to locate or be located by a male, can devote her time and energy to egg production. Further, all her offspring are female, so that her maximum reproductive potential can be realized.

The disadvantage of parthenogenesis is that the genotype of successive generations remains more or less constant so that adaptation of a species to changing environmental conditions is very slow. To counteract this, many species alternate one or more parthenogenetic generations with a normal sexual generation. Aphids, for example, reproduce for most of the year by ameiotic parthenogenesis. However, toward fall (and affected by changing environmental conditions) there occurs, during maturation of some oocytes, a separation of the two X chromosomes, one of which migrates to thc polar body and is destroyed. From such eggs (with an XO constitution) males will develop. As spermatogenesis occurs in these individuals, spermatocytes containing either one X chromosome or no X chromosome are produced. However, the latter do not mature, so that only sperm with an X chromosome result. Therefore, the overwintering eggs produced as a result of mating will have an XX sex chromosome complement and give rise the following spring only to females.

Polyembryony

Polyembryony, the development of several embryos from one egg, is known to be a normal occurrence in about 30 species of parasitic Hymenoptera (mostly Encyrtidae, Platygasteridae, and Branconidae) and one species of Strepsiptera (Ivanova-Kasas, 1972). In these insects it is always associated with either parasitism or viviparity and is presumed to have evolved in conjunction with the abundance of food offered by these two modes of life. Characteristically, the eggs of polyembryonic species are minute and devoid of yolk. Because they depend on the external (host or maternal) source of nutrients the chorion, which is initially thin and permeable, soon disappears. Further, in Hymenoptera, the serosa becomes modified for the uptake of nutrients and is known as a "trophamnion." Both the number of embryos formed and the point in development at which they become discernible vary.

In *Platygaster hiemalis,* a parasite of Hessian fly larvae, for example, at the 4-cell stage, the cells may separate into two groups so that twin embryos are formed. In contrast, in the chalcid *Litomastic truncatellus,* which parasitizes larvae of the moth genus *Plusia,* formation of embryos does not begin until the 220-225 blastomere stage. At this stage, certain of the blastomeres become spindle-shaped and fuse to form a syncytial sheath which divides the remaining blastomeres into groups, the primary embryonic masses. In due course, secondary, tertiary, etc., embryonic masses form so that the final number of potential embryos may exceed 1000. Eventually, the polygerm (the total embryonic mass within the trophamnion) disintegrates, and each embryo develops into a larva. The larvae feed within a host until all the soft parts are consumed and then pupate. At this point the host is nothing more than a cuticular bag of parasites.

Viviparity

Viviparity, the retention of developing offspring within the maternal genital tract, is found in a range of complexity within the Insecta. It is seen in different forms in species from several orders, but among Diptera the entire range of variation may occur. In its simplest form (ovoviviparity) the eggs retain their full complement of yolk for nourishment of the embryo.

The eggs may be retained within the mother for a variable period of time but usually are laid just before they hatch. Such an arrangement is seen in many Tachnidae where the first instar larvae actually escape from the chorion during oviposition. Associated with retention of the eggs, which presumably affords them greater protection, is a trend toward production of fewer of them. As fewer eggs are produced each can acquire more yolk so that larvae can hatch at more advanced stages of development. For example, many Sarcophagidae (flesh flies) produce only 40-80 eggs but are larviparous; that is larvae hatch from the eggs while the latter are still within the reproductive tract. At the extreme, the number of eggs that mature simultaneously is reduced to one, as, for example, in *Hylemya strigosa* (Anthomyiidae) and *Termitoxenia* sp. (Phoridae). In *Hylemya* the larva which emerges from a newly laid eggs molts immediately to the second instar; in *Termitoxenia*, whose egg is relatively larger, it is a third instar larva that emerges from an egg and it pupates within a few minutes. In truly viviparous species, developing offspring obtain their food from the mother.

Accordingly, the structures of the maternal reproductive system and eggs are modified to facilitate this exchange. As in ovoviviparity, the trend is toward reduction of the number of embryos being developed simultaneously. Some aphids, Psocoptera, and Dermaptera (*Hemimerus*) exhibit pseudoplacental viviparity (Hegan, 1951). Eggs of these insects contain little or no yolk and lack a chorion. They develop within the ovariole, where the follicle cells supposedly supply at least some nourishment to the embryo.

(In species with meroistic ovarioles, the nurse cells are also important).

In *Hemimerus,* for example, follicle cells adjacent to the anterior and posterior ends of an oocyte proliferate and become connected with the embryonic membrane forming pseudoplacentae. Later, the follicle cells degenerate because, it is assumed, they are supplying nutrients to the developing embryo. In *Glossina* spp. and pupiparous Diptera adenotrophic viviparity occurs. In this arrangement, an egg is normal, that is, contains yolk and possesses a chorion, yet is retained within the expanded bursa, the so-called uterus. Embryonic development, is, therefore, correctly described as ovoviviparous. However, after hatching, the larva remains within the uterus and feeds on secretions (uterine milk) of the enormous accessory glands which ramify though the abdomen. One larva at a time develops and pupation occurs shortly after birth. In hemocoelic viviparity, practiced by Strepsiptera and some paedogenetic Cecidomyiidae (Diptera), oocytes are released from the ovarioles into the maternal hemocoel.

In Strepsiptera fertilization occurs within the maternal body cavity, and, during embryonic development, nutrients are absorbed directly from the hemolymph. After hatching, the larvae escape from the female's body via the genital pores. In some cecidomyiids, oocytes develop parthenogenitically. Initially, development occurs within the ovarioles, but the larvae on hatching escape into the hemocoel. The larvae remain within the mother and feed on her tissues until just prior to pupation when they exit via the body wall.

Paedogenesis

Though not actually a form of embryonic development, paedogenesis, that is, precocious sexual maturation of juvenile stages, is conveniently mentioned here. Paedogenesis is usually associated with both parthenogenesis and viviparity, and, probably, is best exemplified in certain Cecidomyiidae, though it is known to occur also in some Chironomidae (Diptera). Coleoptera, and Hemiptera. In some cecidomyiids, the oocytes develop viviparously in the hemocoel of the last larval or pupal instar. In some chironomids, embryonic development begins in female pupae. In the viviparous hemipteran *Hesperoctenes* female larvae which have been inseminated hemocoelically develop embryos within their ovarioles. In an extreme situation seen in some aphids, development of young may begin in the mother while she herself is still in her own mother's reproductive system.

Factors Affecting Embryonic Development

Temperature is probably the single most important environmental variable affecting embryonic development. For eggs of most species, there are upper and lower temperature limits, outside which development is greatly retarded or completely inhibited. Within these limits, however, an inverse but linear relationship exists between temperature and time required to complete development; that is, the total heat requirement (temperature above minimum required x duration of exposure to this temperature) is constant for a given species. This heat requirement is typically measured in degree-days. Outside these developmental limits, yet within the limits of viability, an egg may survive but does not develop. Under these conditions, it is said to be quiescent and in this state may survive for a considerable length of time. In a quiescent state, an egg is always ready to take advantage of favourable conditions, even if only temporary, to continue its development. However, quiescence is a relatively sensitive developmental state; that is, outside certain temperature limits, an egg will be killed. For many species, therefore, which exist in habitats exposed to climatic extremes, especially of temperature but also of precipitation, a more resistant state of developmental arrest, diapause, has evolved to permit their survival. Diapause is discussed at greater length, though in the present context it is worth nothing that diapause may occur at different stages of embryonic development and with variable strength in different species. In all instances, however, it is characterized by a cessation of morphogenesis and a considerable lowering of the metabolic rate. Also, the water content of an egg is often low at this time.

In *Bombyx,* diapuse, which begins in overwintering eggs almost as soon as they are laid, is extremely strong; that is, even when eggs are experimentally maintained at 15-20°C from the time of laying they will not develop. Development begins only after they have been exposed to a temperature of about 0°C for several months. In eggs of the damselfly, *Lestes congener,* diapause is also strong but does not commence until after anatrepsis. Diapause in eggs of the grasshopper *Melanoplus differentialis* also occurs after anatrepsis but is weak. Should the temperature to which eggs are exposed be maintained at summer levels (around 25°C), some of the eggs will develop directly, though more slowly than those which undergo chilling. In eggs of some insects, for example contain mosquitoes (*Aedes* spp.) and the damselfly *Lestes disjunctus,* embryonic development is almost completed before diapause is initiated. Water is another important requirement and in eggs of many species must be acquired from the external environment before embryonic development can begin. When it is available to an egg in insufficient quantity, the embryo becomes quiescent or remains in diapause (though this was not inducted by the lack of moisture).

Some species can obtain sufficient water from moisture in the air. For example eggs of the beetle *Sitona,* when kept at 20°C and 100% relative humidity, hatch in 10.5 days; at the same temperature, but only 62% relative humidity, development takes twice as long. In other species contact of the egg with liquid water is necessary for continued development. Such is the case in the damselfly eggs mentioned above which pass the winter in snow covered, dried-out *Scripus* stems and do not continue their development until the stems become waterlogged following the spring thaw.

Advanced embryonic Development

When the larva is fully developed within the egg it escapes by rupturing the egg membranes and sometimes it has some special device which assists this process. In the course of hatching or immediately afterwards many insects shed an embryonic cuticle. Once it has hatched the larva begins to feed and grow, but since the cuticle will only stretch to a limited extent growth is punctuated by a series of moults. The number of moults which occurs is variable, but is generally less in more advanced insects.

In general, weight increases progressively, but linear measurements may increase in a series of steps corresponding with the moults, or more or less continuously if the cuticle is membranous as it is in many larvae. It is common for different parts of the body to grow at different rates so that simple mathematical relationships often do not hold. Growth of the epidermis and internal organs may entail an increase in cell size or an increase in cell number. Growth form larva to adult usually involves some degree of metamorphosis. In many insects the larval form is tied to that of the adult by morphogenetic considerations, but in others a pupal instar interposed between the last larval instar and the adult has permitted a great divergence of forms occurs. Sometimes the larva changes its habits during the life history and there is a corresponding change of form, a phenomenon known as heteromorphosis.

HATCHING

Escape From the Egg

Hatching stimuli

The fully developed larva within the egg escapes by rupturing the vitelline membrane, the serosal cuticle when it is present, and the chorion. The stimuli which promote hatching are largely unknown and in many cases insects appear to hatch whenever they are ready to do so. Even in these instances, however, it is possible that some external stimulus influences hatching. There is a suggestion, for instance, that the larvae of *Swchistocerca* hatch mainly at about dawn and those of *Epitheca* (Odonata) at about sunset (Corbet, 1962). In a few cases specific hatching stimuli are known. These vary, but are relevant to the insect concerned. The eggs of some *Lestes* (Odonata) species hatch when they are wetted provided the temperature is above a certain level.

Aedes eggs hatch when immersed in deoxyenated water, the lower the oxygen tension the greater the percentage hatching, but the responsiveness varies with age. The larvae are most sensitive soon after development is complete and will then hatch even in aerated water, but if they are not wetted for some time they will only hatch at very low oxygen tensions. The low oxygen tension

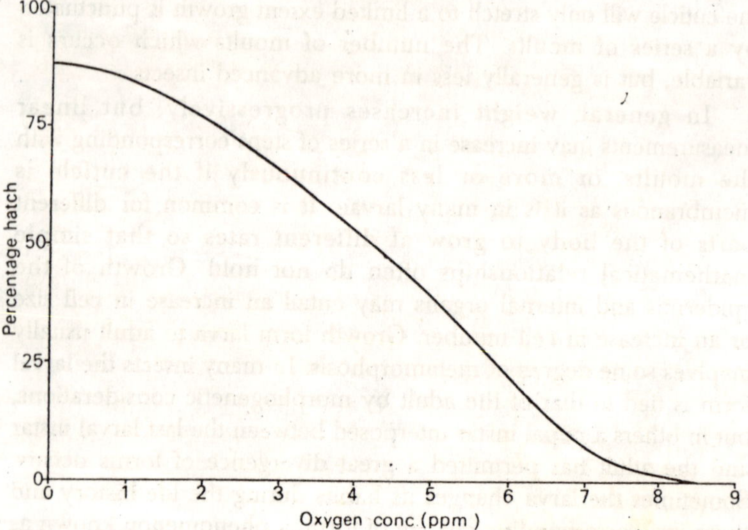

Fig. 4.18. The percentage of eggs of Aedes which hatch in water containing different concentrations of dissolved oxygen after incubation at 90-100% relative humidity.

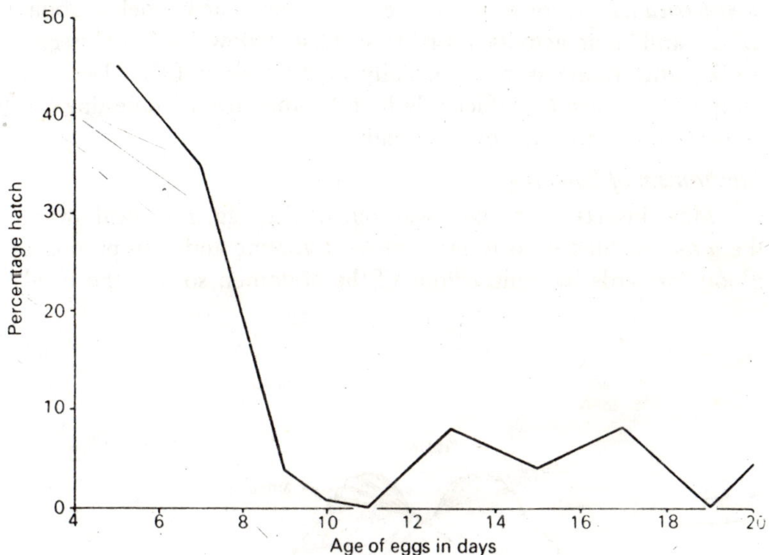

Fig. 4.19. The percentage of Aedes eggs of different ages which hatch under sub-optimal conditions, that is in aerated water. Only newly developed larvae hatch under these conditions, larvae which have remained some time in the egg require anaerobic conditions to induce them to hatch.

is perceived by a sensory centre in the head or thorax and maximum sensitivity coincides with a period of maximum activity of the central nervous system as indicated by the concentration of acetylcholine. Low oxygen tension has completely the opposite effect on hatching of *Agabus* larvae which only occurs in oxygenated water (Jackson, 1958). Amongst terrestrial insects the eggs of *Dermatobia* (Diptera) are stimulated to hatch by the warmth of the host, while in grasshoppers the mechanical disturbance produced by one larva hatching activates other, unhatched larvae in the egg pod so that they all hatch within a short time (Uvarov, 1966).

Suitable temperatures are necessary for all insect eggs to hatch and there is a threshold temperature below which hatching does not occur. This temperature varies in different insects, but is about 8°C for *Cimex,* 13°C for *Oncopeltus* and 20°C for *Schistocerca*. It is independent of the threshold temperature for full embryonic development which may be either higher, as in *Cimex,* or lower, as in *Schistocerca* (about 15°C). The failure to hatch at low temperatures may be related to the inactivity of the larva. Newly emerged

Schistocerca larvae, for instance, are not normally active below about 17°C., and their activity remains sluggish below 24°C. (Hussein, 1937), and *Cimex* is not normally active below 11°C. Further, temperature must be sufficiently high for the enzyme digesting the serosal cuticle to function efficiently.

Mechanism of hatching

Most insects force their way out of the egg, first swallowing the amniotic fluid so as to increase their volume and then pumping blood forwards by contractions of the abdomen so that the head

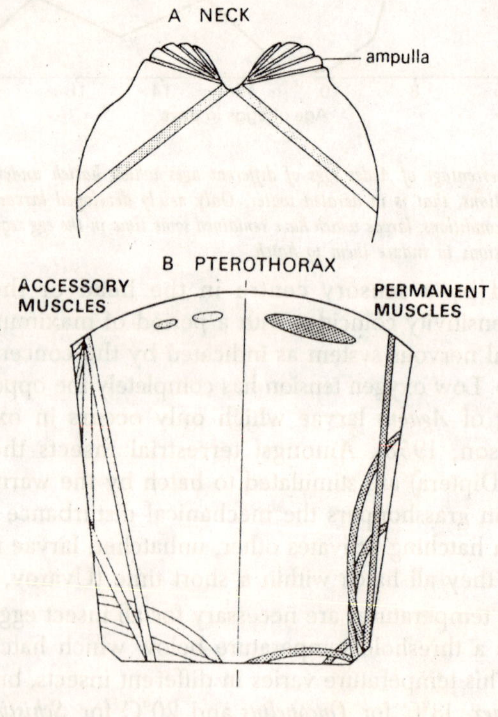

Fig. 4.20. Diagrammatic cross-section through (A) the neck and (B) the pterothorax of a newly hatched first instar larva of Schistocerca showing the accessory muscles.

exerts pressure against the shell. In some cases, the insects may increase their volume by swallowing air which diffuses through the shell or which enters following the initial rupture. In *Acheta* there are special muscles which assist the pumping movements and which degenerate after hatching. Similar muscles which are functional only at the time of a moult are known in other insects. The chorion may split in a more or less irregular manner depending on where the pressure is exerted; in *Agabus* a longitudinal slit develops.

In other cases, the chorion splits along a line of weakness such as those which occur at the longitudinal hatching lines of *Calliphora* or at the junction of he body of the shell with the operculum in Heteroptera. *Aedes* has a line of weakness in the serosal cuticle and a split in the chorion follows this passively, perhaps because the serosa and chorion are closely bound (Judson and Hokama, 1965). In a number of insects hatching is aided by cuticular structures, usually on the head, known as egg bursters. These are on the head of the embryonic cuticle of Odonata, Orthoptera, Heteroptera, Neuroptera and Trichoptera, but on the cuticle of the first instar larva in Nematocerca, Carabidae and Siphonaptera. Their form varies, but in the pentatomids they are in the form of a T- or Y-shaped central tooth. Often, as in the fleas, mosquitoes and *Glossina,* the tooth is in a membranous depression which can be erected by blood pressure.

In *Agabus* the egg burster is in the form of a spine on either side of the head, while in Cimicomorpha a row of spines runs along each side of the face from near the eye to the labrium. *Polypax* (Siphunculata) has a pair of spines with lancet-shaped blades arising from depressions above them, while in *Pediculus* (Siphunculata) there are five pairs of these blades and in *Haematopinus* nine or ten pairs. Many Polyphaga have egg bursters on the thoracic or abdominal segments of the first instar larva (van Emden, 1946). For instance, in *Meligethes* there is a tooth on each side of the mesonotum and metanotum, while larval tenebrionids have a small tooth on either side of the tergum of these and the first eight abdominal segments. It is not clear how these various devices function and Jackson (1958) believes that in *Agabus,* the egg of which has a soft chorion, they are no longer functional. In other cases, they appear to be pushed against the inside of the shell until finally they pierce it and then a slit is cut by appropriate movements of the head.

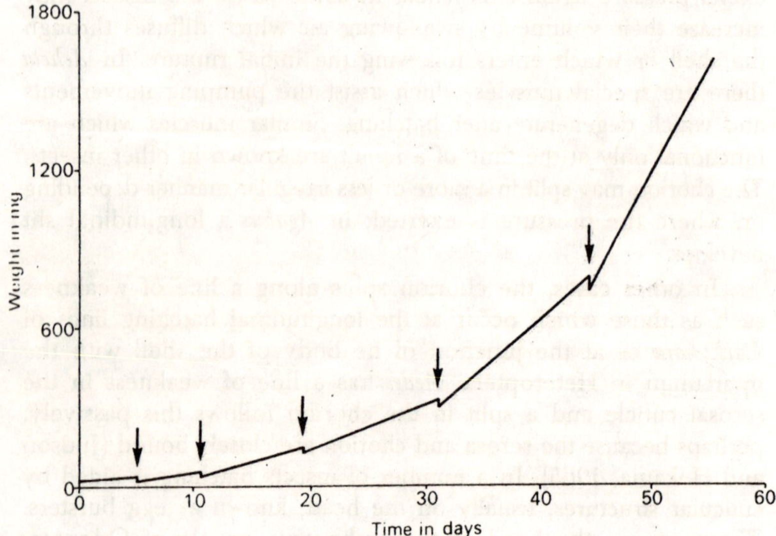

Fig. 4.21. The pattern of increase in weight of female Locusta. The times of the moults are indicated by arrows.

The larva of *Dacus* (Diptera) uses its mouth hooks in a similar way, repeatedly, protruding them until they tear the chorion (D.T. Anderson, 1962). The blades in *Polyplax* and the spines in *Cimex* are used to tear the vitelline membrane, the chorion then being broken by force (Sikes and Wigglesworth, 1931). In Acrididae there is a thin membranous region of the neck which is expanded dorsally by blood pumped in from the abdomen. This cervical ampulla exerts pressure on the serosal cuticle which in these insects represents the main barrier to hatching since the chorion tends to crack away as the egg swells during development. In this group, and probably in the Heteroptera which also develop a thick serosal cuticle, hatching is aided by an enzyme, secreted by the pleuropodia, which digests the serosal endocuticle.

Larval Lepidoptera know their way through the chorion and after hatching they continue to eat the shell until only the base is left. In *Pieris brassicae,* where the eggs are laid in a cluster, a newly hatched larva may also eat the tops off adjacent unhatched eggs (David and Gardiner, 1962). When the egg is enclosed in an ootheca the larva escapes from this after leaving the egg. In the Blattaria

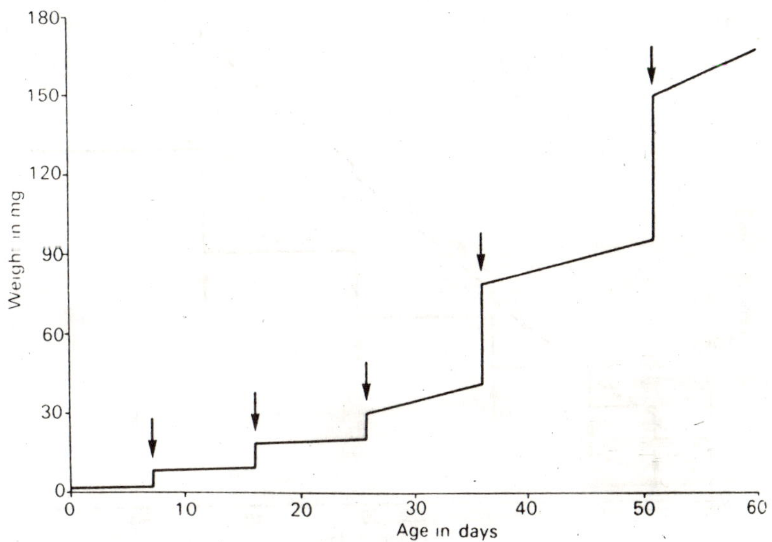

Fig. 4.22. The pattern of increase in weight of Notonecta (Heteroptera). Arrow indicate the times of the moults.

the ootheca is split open before hatching by the swelling of the eggs as they absorb water. Acridids wriggle through the forth of the plug enclosed in the embryonic cuticle and the thoracic dorsal longitudinal muscles may be specially developed to facilitate this since they become non-functional after hatching (Thomas, 1954). Movement is aided by the cervical ampulla which is collapsed as the head pushes into a narrow space and then expanded to give a purchase while the abdomen is drawn up. The larva emerge from the eggs pointing upwards and then move up along the line of least resistance. Mantids probably make their way from the ootheca in a similar manner.

Intermediate Moult

In those insects which possess an embryonic cuticle this separates from the underlying epidermis some time before hatching, but it is not shed so that when the larva hatches it is a pharate first instar. The embryonic cuticle is shed during or immediately after hatching and this process is commonly known as the intermediate moult. As the larva of *Cimex* or a louse emerges from the egg it swallows air, and, by further pumping, splits the

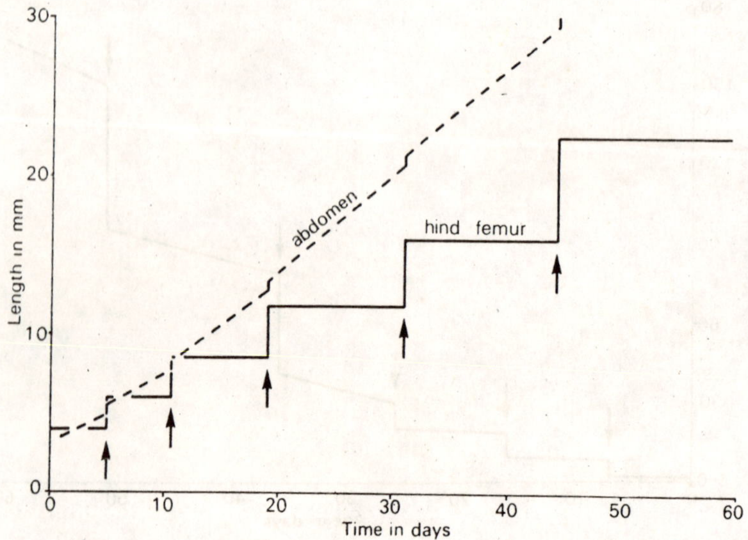

Fig. 4.23. The pattern of increase in length of the hind femur and abdomen of Locusta. Arrow indicate the times of the moults.

embryonic cuticle over the head. The cuticle is shed as the larva continues to hatch and finally it remains attached to the empty egg shell (Sikes and Wigglesworth, 1931).

In Heteroptera the embryonic cuticle is attached inside the chorion at two or three places. The intermediate moult in acridids begins as the larva emerge on the surface of the soil, the cuticle being split by the action of the cervical ampulla so that the first instar larva can free itself. The life history of an insect is divided into a series of stage, each separated from the next by a moult. The form which the insect assumes between moults is known as an instar, that which follows the intermediate moult being the first instar which later moults to the second instar and so on until at a final moult the adult or imago emerges. No further moults occur except in the Apterygota.

Numbers of Instars

Primitive insects usually have more larval instars than advanced species. Thus *Ephemera* and *Stenonema* (Ephemeroptera) moult 30 and 40-45 times respectively, while Heteroptera commonly have five larval instars and Nematocera only four. Even within a group

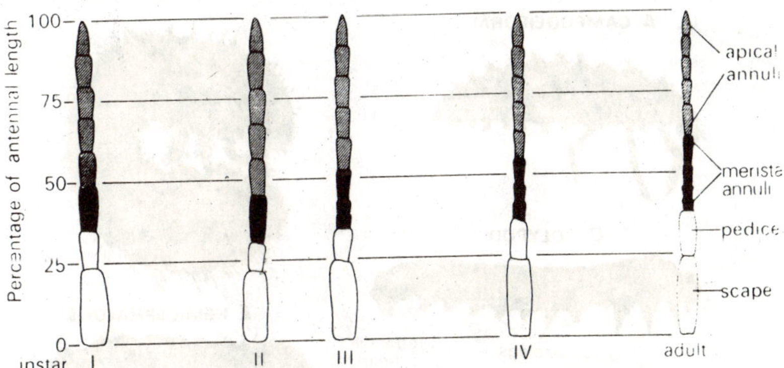

Fig. 4.24. Allometric growth in the antenna of Hemimerus. The diagram illustrates the changes in the relative proportions of the different regions of the antenna in different instars.

of related insects there is variation and amongst the Acridoidea the primitive Pyrgomorphidae have five or more larval instars while the advanced Gomphocerinae, which are usually smaller, have four. The number of larval instars through which a species passes is not absolutely constant. In the Orthoptera, in which the female is bigger than the male, she commonly has an extra larval instar, while larvae emerging from small eggs grow slowly and have an additional instar. *Nomadacris* may have six, seven, or occasionally eight larval instars depending on the treatment of the parents. In *Plusia* and some other Lepidoptera, larvae reared in isolation may pass through five, six, or seven instars, while nearly all of those reared in a crowd have only five (Long, 1953).

Developmental Anomolies

Sometimes eggs are retained by the female after they are fertilised so that they start to develop before they are laid. If this period of internal development is extended the larva may hatch within the parent and in a few species is nourished by her so that finally the female gives birth to a fully developed larva ready to pupate. In other instances eggs which are deficient in yolk are nourished via special placenta-like structures in the ducts of the female or in the haemocoel. Thus, viviparity in insects takes various forms. Amongst parasitic insects eggs sometimes give rise to a number of larvae insected of just one. This is known as polyembryony. Eggs will develop without being fertilised, and sometimes this parthenogenesis is a normal occurrence.

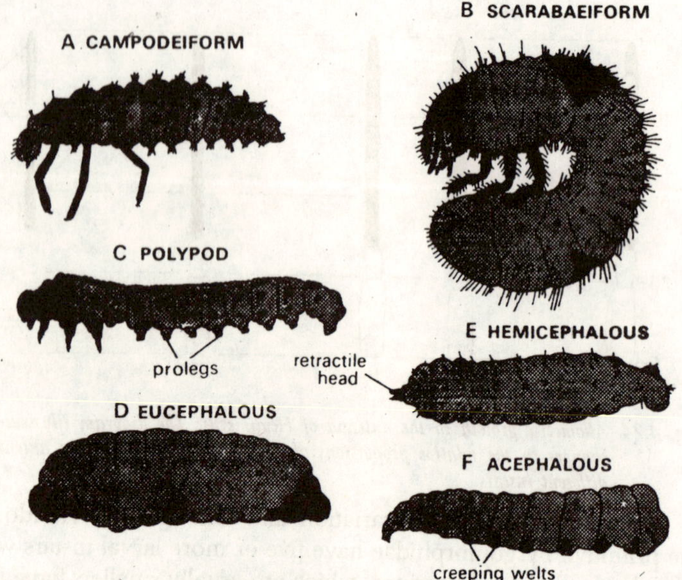

Fig. 4.25. Larval forms. A. Campodeiform–Hippodamia (Coleoptera). B. Scarabaeiform–Popillia. C. Polypod–Neodiprion (Hymenoptera). D–Eucephalous–Vespula (Hymenoptera). E. Hemicephalous–Tanyptera (Diptera). F. Acephalous–Musca.

The sex of the offspring then depends on the behaviour of the chromosomes at meiosis and, in general, haploid eggs are male, diploid eggs female. A disadvantage of parthenogenesis is that it reduces the adaptability of the insect, but in some cases this is overcome by an alternation of parthenogenetic and bisexual generations. A few insects mature precociously and start to produce offspring while they are still larvae or pupae. This is known as paedogenesis. The eggs of some insects are fertilised in the ovary or upper oviduct and in some of these species they are retained within the body of the female for some time before being laid. As a result the eggs start to develop, while they are still within the parent and in *Cimex,* which practices haemocoelic insemination, the embryo has almost reached the stage of blastokinesis by the time the egg is laid. In some other species internal development proceeds until the stage of hatching or even beyond, and such species are said to be viviparous.

Ovoviviparity

Many species retain the eggs in the genital tracts until the larvae are ready to hatch, hatching occurring just before or as the

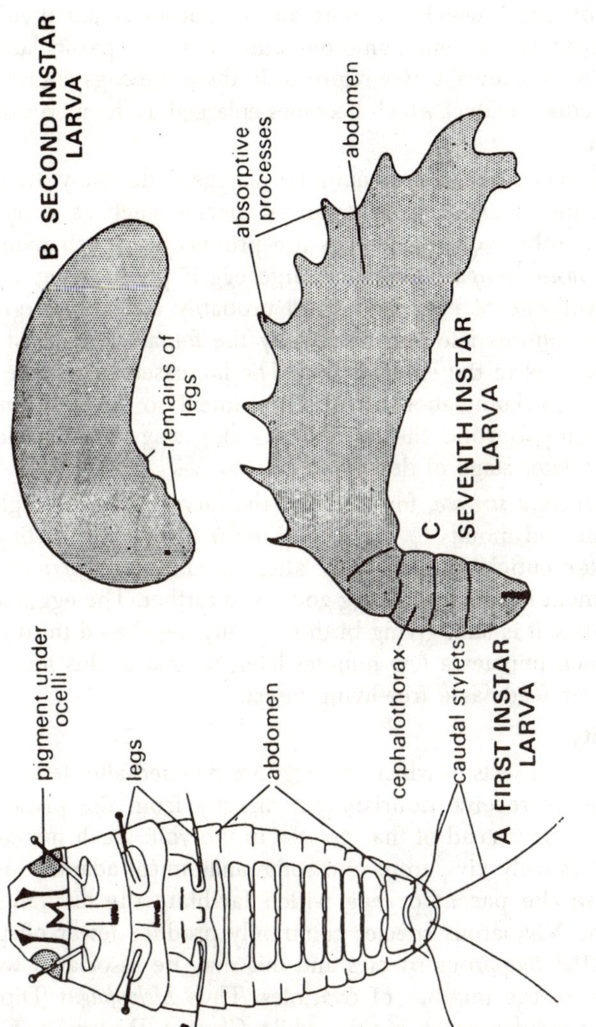

Fig. 4.26. Heteromorphosis. Larval stages of Corioxenos (Strepsiptera). A—Ventral view of the free-living instar larva. B and C—Lateral view of older parasitic larvae.

eggs are laid. All the nourishment for the embryo is present in the egg and no special nutritional structures are developed. Viviparity of this sort is called ovoviviparity and it differs from normal oviparity only in the retention of the egg. Ovoviviparity occurs spasmodically in various orders or insects: Ephemeroptera, Dictryoptera, Psocoptera, Homoptera, Thysanoptera, Lepidoptera, Coleoptera and Diptera, being particularly widespread in the last group from which the following examples are drawn. Sometimes species of *Musca* which are normally oviparous retain their eggs and deposit larvae, but numerous other Diptera, particularly the Tachinidae, are always ovoviviprous. In these the eggs are retained in the median oviduct which becomes enlarged as the uterus during gestation.

Tachinids produce large numbers of eggs so do many oviparous Diptera, but in other ovoviviparous species, such as *Arcophaga,* smaller numbers of bigger eggs are produced at each ovulation and in *Musca larvipara* only one large egg is produced at a time. This lower rate of egg production probably reflects the greater protection afforded to egg carried by the female compared with eggs deposited in the environment. The increased size of the eggs permits the accumulation of more nutriment so that the embryo may develop beyond the normal hatching stage and larvae are born in a later stage of development.

In *Hylemya strigosa*, for instance, the larva passes through the first instar and moults to the second instar in the egg, casting the first instar cuticle immediately after hatching. In *Termitoxenia* development in the uterine egg goes even further. The egg hatches immediately it is laid, giving birth to a fully developed third instar larva which pupates a few minutes later, so that in this insect the larva never feeds as a free-living insect.

Viviparity

In some insects in which the eggs are retained after fertilisation the embryos receive nourishment directly from the present in addition to or instead of that present in the yolk. Such insects are regarded as truly viviparous and some anatomical adaptations are present in the parent or egg which facilitate the exchange of nutriment. Viviparous species commonly produce fewer offspring than related oviparous species and this may be associated with a reduction in the number of ovarioles. Thus *Melophagus* (Diptera) has two ovarioles on each side, while *Glossina* (Diptera) has only one on each side. In the related, oviparous, *Musca* there are about 70 ovarioles in each ovary.

Similarly amongst the viviparous Dermaptera *Hemimerus* has 10-12 ovarioles on each side, but only about half these are functional. *Arixenia* (Dermapera) has only three ovarioles on each side. Sometimes, the eggs are retained and development occurs in the ovariole, as in *Hemimerus,* the aphids and Chrysomedilidae. In other insects, such as the viviparous Diptera, the vagina is enlarged to form a uterus. In Strepsiptera and a few parthenogenetic Cecidomyidae the eggs develop in the haemocoel of the parent. Hagon (1951) recognises three main categories of viviparity and his scheme is followed.

Pseudoplacental Viviparity

Insects exhibiting pseudoplacental viviparity produce eggs, containing little or no yolk, which are retained by the female and are presumed to receive nourishment via embryonic or maternal structures called pseudoplacentae. There is, however, no physiological evidence relating to the importance of these structures. Viviparous development continues up to the time of hatching, but the larvae are free-living. In *Hemimerus* the fully developed oocyte has no chorion or yolk, but is retained in the ovariole during embryonic development. The oocyte is accompanied by a single nurse cell and enclosed by a follicular epithelium one cell thick. At the beginning of development the follicle epithelium becomes two or three cells thick and at the two ends thickens still more to form the anterior and posteriormaternal pseudoplacentae.

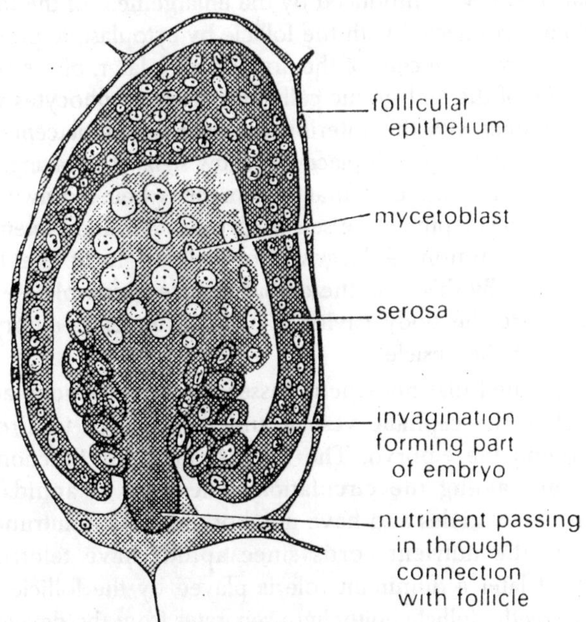

Fig. 4.27. Section through an early embryo of Macrosiphum.

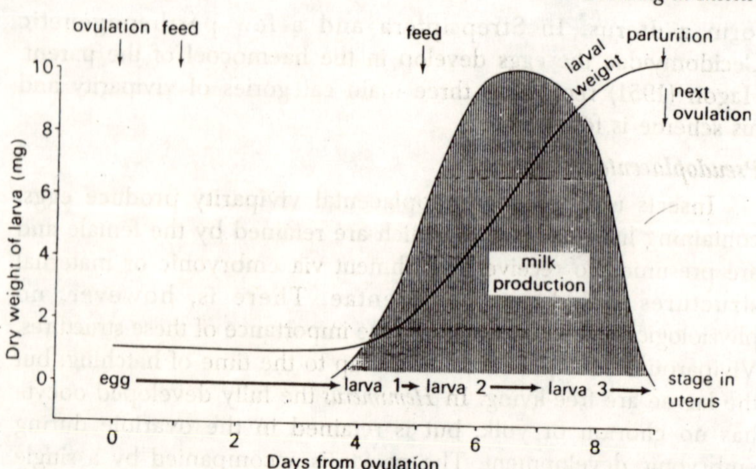

Fig. 4.28. Diagram showing the development of larval Glossina in the uterus of the female in relation to the cycle of milk production and feeding. A feed may occur on any day up to day 6, but feeding after this is unusual.

As the embryo develops it comes to lie in a cavity, the pseudoplacental cavity, produced by the enlargement of the follicle, but it becomes connected with the follicle by cytoplasmic processes extending out from the cells of the amnion and, later, of the serosa. Further, some of the embryonic cells form large trophocytes which come into contact with the anterior maternal pseudoplacenta. The follicle epithelium and pseudoplacentae show signs of breaking down and this is taken to indicate that nutriment is being drawn from them. Later in development the serosa spreads all round the embryo and, with the amnion, enlarges anteriorly to form the foetal pseudoplacenta. By this time the dorsal closure is complete except anteriorly where the body cavity is open to an extra-embryonic cavity, the cephalic vesicle.

It is presumed that nutiement passes from the pseudoplacenta to the fluid in the cephalic vesicle and then is free to circulate into and round the embryo. The heart is probably functional at this time thus aiding the circulation. The eggs of aphids also develop in the ovarioles and have no chorion. At first nutriment is received via the nutrient cords since aphids have telotrophic ovarioles, but later a dominant role is played by the follicle cells. In *Macrosiphum* the follicle epithelium separates from the developing egg, but retain a connection posteriorly. Though this connection reserve materials and symbionts are passed to the embryo, but the growth of the blastoderm resticts and finally serves the link. Possibly there is later some direct transfer of nutriment across the serosa from the follicle cells since the length of the egg increases by about 30 times in the course of development. Pseudoplacental viviparity is also known to occur in a psocopteran, *Archipsocus,* in which the serosa is an important trophic organ, and in the Polyctenidae (Heteroptera) in which first the serosa and then the pleuropodia are important.

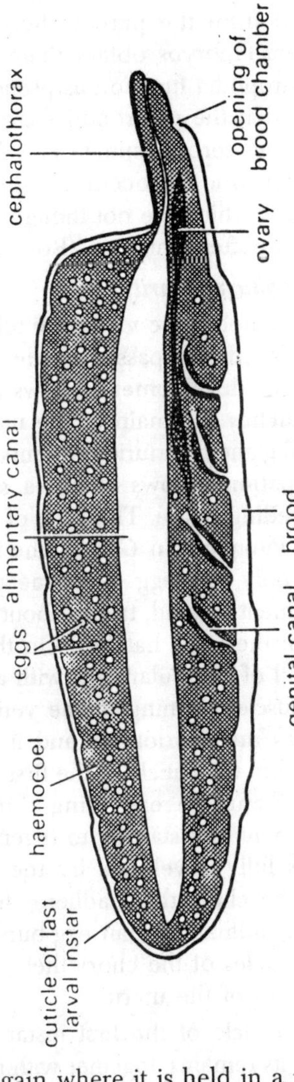

Fig. 4.29. Sagittal section through a female strepsipteran.

Viviparity in Dictyoptera

The position of the cockroaches with regard to viviparity is anomalous. Fundamentally all cockroaches are oviparous, laying their eggs in an ootheca which is extruded from the genital ducts. In some species the ootheca may be carried projecting from the genital opening, but in *Periplaneta* it is finally dropped some time before the eggs hatch. Other species such as *Blattella* continue to carry the ootheca externally until the time of hatching. Some others extrude the ootheca, but then withdraw it into the body again where it is held in a median brood sac which extends beneath the rest of the reproductive system. In this case, the ootheca may be poorly developed and as the eggs increase in size they come to project beyond the ootheca.

In most species the increase in size results only form the absorption of water, but in *Diploptera,* in which the eggs increase in length by five or six times during embryonic development, there is also an increase in dry weight indicating that some nutriment is obtained from the parent after ovulation. It is not certain how *Diploptera* embryos obtain their nutriment, but it is possible that the pleuropodia function as pseudoplacentae. This is suggested by the fact that the pleuropodia are long hollow tubes which extend outside the serosal cuticle and lie beneath the chorion. Similar long pleuropodia occur in some other cockroaches such as *Leucophaea* which are not thought to obtain further nutriment from the parent after ovulation (Roth and Willis, 1958).

Adenotrophic viviparity

In adenotrophic viviparity fully developed eggs with chorions are produced and passed to the uterus where they are retainded. Embryonic development follows as in ovoviviparity, but when the larva hatches it remain in the uterus and is nourished by special maternal glands. Parturition occurs when the larva is fully developed and pupation follows within a short time, there being no free-living feeding phase. This type of viviparity only occurs in *Glossina* and the Pupipara. In *Glossina* the two ovarioles function alternately so that only one egg at a time passes to the uterus. Embryonic development is rapid, taking about three days at 24°C. in *G. palpalis,* and then the larva hatches. On the ventral wall of the uterus is a small pad of glandular cells with a cushion of muscle beneath and other muscles running to the ventral body wall. This structure is known as the choriothete and it is responsible for removing the chorion and the cuticle of the first instar larva. It undergoes cyclical development, degenerating during the later stages of larval development and starting to regenerate just before larviposition so that it is fully developed by the time the next larva is ready to hatch. The choriothete adheres to the chorion and when this is split longitudinally by an egg-burster it is pulled off by the action of the muscles of the choriothete, becoming folded up against the ventral wall of the uterus.

The cuticle of the first instar larva is pulled off in the same way and its remains, together with those of the chorion, are expelled at parturition (Bursell and Jackson, 1957). The first and second instar larvae feed on a secretion produced by 'milk' glands which open by a common duct into the uterus. These glands undergo cyclical development, reaching a maximum during gestation. Their secretion accumulates in the uterus and is sucked up by the larva so that its midgut becomes distended, the contents being used during the latter part of the second instar and throughout the third when the larva grows considerably although it is not feeding. When the second instar moults, its cuticle is not immediately shed, but it is subsequently split by the growth of the pharate third instar and finally it is shed and voided shortly before parturition. The larval respiratory system opens by a pair of posterior spiracles in the first two instars, but in the third instar larva the system is much more specialised.

The terminal segment of the abdomen bears two heavily sclerotised lobes each of which is clossed by three longitudinal bands of perforations leading into the tracheal system. Each of these perforations is guarded by a valve which permits air to be drawn into the system, but not to be forced out. In addition to these openings in the polypneustic lobes the second instar spiracles on the insides of the lobes remain open because the second instar cuticle is not shed. Also as a result of this the trachal system is lined by two layers of cuticle, the second instar cuticle being broken only at the inner ends of the system. Indirectly action dorsoventral muscles produce a piston-like movement in a specialised part of the trachal system in the polypenustic lobes and it is suggested that this movement sucks air in through the valved perforations and forces it forwards between the two linings of the tracheae. An exhalent current flow through the loose second instar linings and out through the second instar spiralces. The respiratory muscles contract 15-25 times per minute. By this mean the larva is able to draw air in through the genital opening of the parent, but this mechanism can only function while the second instar cuticle persists, a period of four or five days at the beginning of the third larval instar. In the earlier instars oxygen may be obtained, as least partly, by diffusion from the female tracheal system which invests the uterus, while in the late third instar the valves in the polypenustic lobes disappear and a two-way airflow through the perforations is possible.

The hindgut of the larva is occulated at its connection with the midgut and again at the anus so that waste materials from the midgut are not voided and the hindgut forms a reservoir for nitrogenous waste. This arrangement prevents the larva from fouling the female ducts. As far as is known the development of the Pupipara does not differ in essentials from that of *Glossina* outlined above, but there is no evidence for a complex air circulation similar to that which occurs in the larva of *Glossina*.

Haemocoelous viviparity

Haemocoelous viviparity differs from the other forms of viviparity in that development occurs in the haemocoel of the parent female. This type of development occurs throughout the Strepsiptera and in some larval Cecidoyide which reproduce paedogenetically. Female Strepsiptera have two or three ovarial strands on either side of the midgut, but there are no oviduct and mature oocytes are released into the haemocoel by the rupture of the ovarian walls. In *Stylops* the eggs contain very little yolk, but some is present in other genera such as *Acroschismus.*

Sperm enter through the genital canals which open in the ventral midline of the female and fertilisation and development continue in the haemocoel with a direct transfer to nutriment from the haemolymph to the embryo. The larvae hatch and find their way to the outside through the genital canals. In *Miastor* (Diptera) the eggs are similarly liberated into haemocoel from simple sacs. The developing egg is nourished via nurse cells, which arise independently of the oocyte, and later via the serosa which becomes thickened and vacuolated. When the larvae hatch they feed on the tissues of the female and any unhatched eggs, finally escaping through a rupture in the wall of the parent.

POLYEMBRYONY

Sometimes an egg instead of giving rise to a single larva may produce two or more, this process being called polyembryony. It occurs occasionally in Acridoidea and probably in other groups, but in some endoparasitic insects it is regular phenomenon. This is true, for instance, in *Halictoxenos* (Strepsiptera) a parasite of *Halictus;* in *Aphelopus theliae* (Hymenoptera) a parasite of Cecidomyidae; and in several genera of Encyritidae and Ichneumonidae which parasitise the eggs and larvae of Lepidoptera. In all these cases, the eggs of the parasite are small and relatively free from yolk, nutriment being derived form the host tissues in which they are situated. When the oocyte of *Platygaster heimalis* matures two polar bodies are produced. The fuse together and the polar nucleus increases in size to form a paranuclear mass.

Some of the cytoplasm in the egg is associated with the paranuclear mass and forms the trophamnion, the remainder is associated with the fusion nucleus and forms the embryonic region. The trophamnion surrounds the embryonic region and the

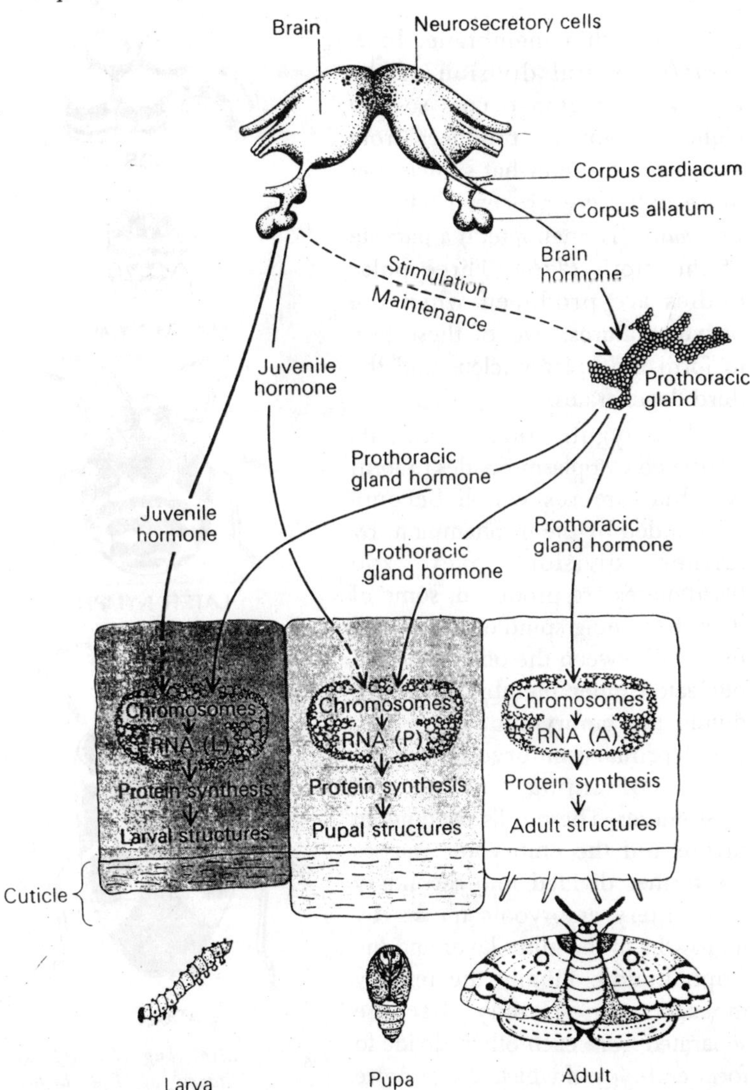

Fig. 4.30. Principal endocrine tissues in the giant silkworm moth, Hyalophora cecropia.

paranuclear mass divides. At the same time cleavage occurs in the embryonic region, but after the second division the whole region divides into two so that embryos are produced. Nutriment is passed to the embryo from the host via the trophamnion, but later the paranuclear masses are absorbed and the trophamnion is represented by a very thin membrane. In *P. vernalis* several division of the embryonic region occurs so that eight embryos are produced from each egg. A somewhat similar, but more extensive process occurs in *Lotomastix* (Hymenoptera) a parasite of the moth *Plusia*. Three polar bodies are produced when the oocyte matures; two of these fuse to form the polar nucleus and the third degenerates.

The zygote nucleus and its associated cytoplasm divides to form two blastomeres which become surrounded by the trophamnion. By further division over zoo blastomeres are produced, some of them becoming spindle-shaped and pushing between the others to form nucleated inner membranes which divide the embryonic region into 15-20 primary embryonic masses each containing up to 50 blastomeres. These cells continue to divide and the embryonic masses are further divided into secondary and teritery embryonic masses by ingrowths of the inner layer and the trophamnion. Finally the tertiary masses, which may become separated from each other, divide to form embryos of which 100 or more may be derived from one egg. The effect of polyembryony is to increase the reproductive potential of the insect, but the net effect is not always much greater than in related monembryonic species because polyembryonic forms tend to lay fewer eggs. Polyembryony may facilitate survival of the species through its relatively long life as a parasite during which time it is subjected to the reactions of the host (Clausen, 1940).

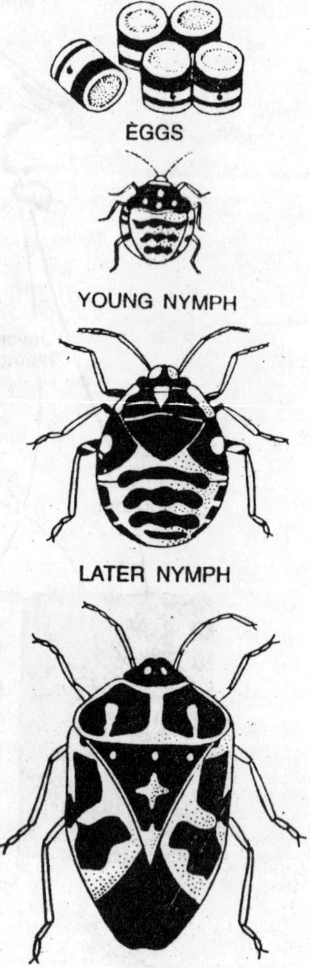

Fig. 4.31. Stink bug (Hemiptera). Stages of life history exhibiting heterometabolic metamorphosis or to say gradual metamorphosis.

Parthenogenesis

Sometimes eggs develop without being fertilised and this phenomenon is known as parthenogenesis. Occasional parthenogenesis, resulting from the failure of a female to find a male, is probably widespread, while in a number of insects parthenogenesis is a normal means of reproduction. It has been recorded from all the insect orders except Odonata, Dermaptera, Neuroptera and Siphonaptera. The sex of the offspring developing from an unfertilised egg is dependent on the sex-determining mechanism of the insect and the behaviour of the chromosomes at the meiotic division of the oocyte nucleus.

In the majority of insects the female is homogenetic (XX) and the male heterogametic (XY or XO), but the Lepidoptera re-exceptional with the females having the heterogametic constitution. Hence, the unfertilised eggs of most insects can contain only X-chromosomes since any Y-chromosome must come from the male. Whether the egg contains one or two X-chromosomes, that is, whether it is haploid or diploid, depends on the behaviour of the chromosomes at meiosis. Sometimes no reduction division occurs or reduction is followed by doubling of the chromosome number so that the diploid and XX composition of the egg is maintained. These eggs will give rise only to females. Eggs which undergo a normal reduction division and in which no chromosome doubling occurs remain haploid and, if they develop at all, become males. Such haploid males are characteristic of some insect groups. Parthenogenesis may be classified according to the behaviour of the chromosomes at the maturation division of the oocyte.

Haplo-diploidy. A normal reduction division occurs in the oocyte, fertilised eggs developing into females, unfertilized eggs into males. This is characteristic of Hymenoptera and some smaller groups.

Apomictic (ameiotic) parthenogenesis. No reduction division occurs so that the offspring have the same genetic constitution as the mother and all are female. That is of common occurrence in blattids, aphids, tenthredinids and curculionids.

Automictic (meiotic) parthenogenesis. A normal reduction occurs but is followed by the fusion of two nuclei so that the diploid number of chromosomes is restored. Often the female pronucleus fuses with the second polar nucleus, or two cleavage nuclei may fuse. In *Solenobia* (Lepidoptera) two pairs of nuclei fuse after the second cleavage division. *Moraba* (Orthoptera) is exceptional in having a pre-meiotic doubling of the chromosomes followed by a normal division so that the diploid number is restored. This type of parthenogenesis, in which only females are produced, occurs in phasmids, coccids and psychids. An alternative classification based on the sex of the offspring produced as a result of parthenogenesis is as follows:

arrhenotoky–Only males are produced.

thelytoky–Only females are produced.

amphitoky–Individuals of either sex may be produced.

Arrhenotoky

Facultiative arrhenotoky, in which the eggs may not be fertilised, is characteristic of a few groups of insects and has probably arise only four or five times. It occurs throughout the Hymenoptera, in some Thysanoptera, the Coccidae Iceryini, some Aleyrodidae and the beetle *Micromalthus*. In all of these the unfertilised, haplōid eggs, produce males. In the Hymenoptera the females determines whether or not an egg is fertilised by controlling the release of sperm form the spermatheca as the eggs pass down the oviduct. The stimuli prompting the female to withhold sperm are largely unknown, but in *Apis* the season and the size of the brood cell in which the female is ovipositing are relevent.

In the parasitic Hymenoptera the size of the host is often important, relatively more unfertilised eggs being laid in small host. The coccid *Icerya purchasi* is of interest in that, apart from a few haploid males, the adult population consists entirely of hermaphrodites which are diploid with diploid ovaries, but which also have haploid testes. No true females occur. When the larva giving rise to an hermaphrodite hatches from the egg all the cells are diploid, but after a time haploid nuclei appear in the gonad. They form a core from which the testis develops surrounded by the ovary. Oocytes undergo a normal reduction division, but the spermatocytes, as in normal haploid males, do not. The hermaphrodites are normally self-fertilising, but they can be fertilised by the occasional males. Cross-fertilisation between the hermaphorodites does not occur. The few eggs which are not fertilised develop into males.

Thelytoky

Thelytokay parthenogenesis probably occurs occasionally in many species of insect and is known to occur in a number of species of Acrididae. Unmated females of *Schistocerca* live for much longer than mated females, but lay about the same number of eggs. Nearly all of these start to develop, but only about 25% hatch and further heavy mortality occurs in the first larval instar. Thus, the viability of unfertilised eggs is much less than that of fertilised eggs, but nevertheless *Schistocerca* has been reared parthenogenetically for six generations. It appears that the only eggs to survive are those in which the chromosomes double after meiosis. Unmated cockroaches also live longer than mated females, but they produce fewer eggs with poor viability (Roth and Willis, 1956). In *Bombyx* the tendency for sporadic parthenogenesis to occur varies in different strains. In other insects thelytoky is a regular occurrence and, for instance, in *Carausius* and some Thysanoptera males are extremely rare, the whole population normally reproducing parthenogenetically.

Sometimes, as in some Psychidae and Coccidae, a parthenogenetic race exists together with a normal bisexual race. Thus *Lecanium* (Homoptera) has one race consisting entirely of females which reproduce automatically and another race which is bisexual and exhibits facultiative thelytoky. In this case fertilised eggs may become males or females while unfertilised eggs develop automatically and so produce females. Commonly such races occur in different areas and in the weevil *Otiorrhynchus dubius,* for instance, a parthenogenetic race occurs in northern Europe and a bisexual race in central Europe. In general parthenogenesis occurs more commonly in the north than further south and in the genus *Otiorrhynchus* 78% of the species occurring in Scandinavia reproduce parthenogenetically, while only 28% of those occurring in the Austrian Alps do so. Constant thelytoky occurs in a few Lepidoptera, such as the psychid *Solenobia*, and this raises the question of sex determination since the females of Lepiidoptera and heterogametic. Thelytoky in these insects may result form the passage of the X-chromosome to the polar body at maturation so that only the Y-chromosomes remain in the egg; or two polar nuclei may fuse to give a female XO or XY constitution, while the egg nucleus degenerates. A completely different explanation supposes that the females are homozygous, YY, and so can give rise only to females. An unusual type of thelytoky, known as gynogenesis, occurs in the form *mobilis* or *Ptinus clavipes* (Coleoptera).

The form *mobilis* exists only as troploid females which reproduce parthenogenetically, but the development of eggs is triggered by healthy sperm of *P. clavipes* or, less successfully of *P. pusillus*.

Thelytoky occurs in many different and unrelated insects and is believed to have arisen on a large number of occasions. It may result from apomixis or automixis and its probable advantages over reproduction involving fertilisation are that the female spends all her time in feeding and reproduction, no time is lost in finding a mate, and, since the whole of the population is female, the reproductive potential is much greater than if half the population are males. These advantages are, however, offset by the absence of genetic recombination which normally occurs at mating. Thus, the long-term effect of thelytoky, in many cases at least, is to prevent a species from adapting to environmental changes so that it is destined to die out or to return to bisexuality.

Alternation of Generations

A number of insects combine the advantages of parthenogenesis with the advantages of bisexual reproduction by an alternation of generations. This occurs, for instance, in the Cynipidae which are commonly bivoltine, a generation of parthenogenetic females alternating with a bisexual generation. *Neuroterus lenticularis* (Hymenoptera) forms galls on the underside of oak leaves in which the species overwinters. Females emerge in the spring and lay eggs. The eggs of some females undergo meiosis but, since, in the absence of males, they cannot be fertilised, they remain haploid and so give rise to males; other females produce eggs in which no reduction division occurs and which give rise only to females. In this way the bisexual generation arises, the insects emerging from catkin galls in early summer. After mating, the females of this generation lay eggs which are fertilised and which produce the females of the following spring generation. Aphids have a more complex alternation of generations with several parthenogenetic generations occuring during the summer. Sometimes, as in *Aphis fabae,* an alternation of host plants also occurs. The first generation emerges on spindle in spring from overwintering eggs.

It consists entirely of females, the fundatrices, which may or may not produce wingless generations of fundatrigeniae before a winged generation, the migrantes, appears. These migrates to bean plant and produce wingless alienicolae of which there may be many successive generations, but ultimately these produce the sexuparae some of which are winged and return to the spindle while others are wingless. The former produce females, the latter winged males which then join the females; they mate and winter eggs are produced.

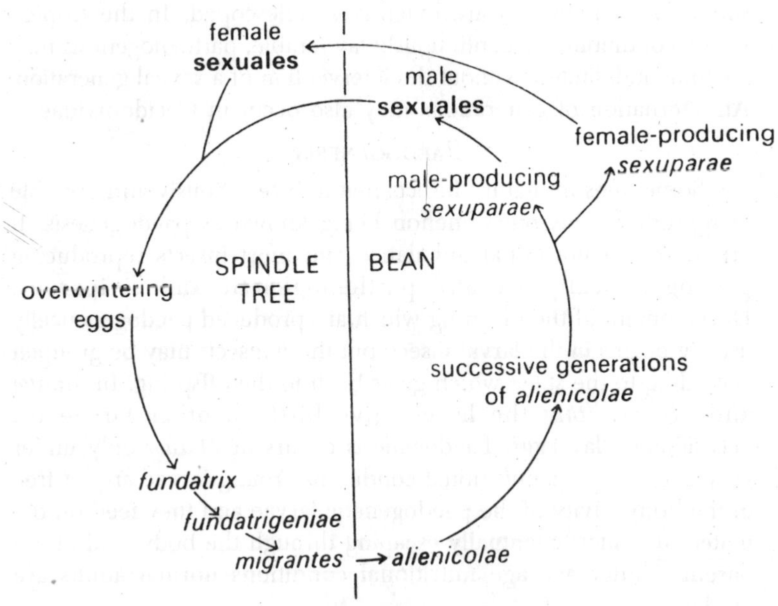

Fig. 4.32. Alternation of sexual and parthenogenetic generation in Aphis. Sexual generation in bold type; wholly female parthogenetic generations in italics.

All these generations except for the last reproduce parthenogenetically and consist entirely of females. In some species such as *Tetraneura* only one class of sexuparae is produced and these individuals give birth male and female sexual forms, an instance of amphitoky. Female aphids are produced apomictially, while males result from the loss of an X-chromosome to a polar body at meiosis, although no reduction of the autosomes occurs. Thus an egg acquires the XO constitution of the male.

The production of males is ultimately under environmental control, but how the environment controls chromosome behaviour is not known (Lees, 1966). Spermatogenesis in male aphids is characterisitc, ensuring that all the eggs produced are female. At the first meiotic division two kinds of spermatocytes are produced: some with an X-chromosome and some without. The latter degenerate so that only the former undergo the second meiotic division and only one type of sperm, containing an X-chromosome, is produced. Hence, the fertilised eggs can only be female.

The aphids reproduce very rapidly, combining the advantages of parthenogenesis with viviparity and paedogenesis so that successive generations are extensively telescoped. In the tropics, where conditions are continuously favourable, parthenogenesis may continue indefinitely without the intervention of a sexual generation. An alternation of generations may also occur in Cecidomyidae.

Paedogenesis

Sometimes immature insects mature precociously and are able to reproduce, this phenomenon being known as paedogenesis. It arises from a hormonal imbalance and most insects reproducing paedogenetically are also parthenogenetic and viviparous. Development of the offspring which are produced paedogenetically usually begins in the larval insect, but these insects may be grouped according to the stage which gives birth to the offspring. In *Miastor* and *Micromalthus* the larvae give birth to other larvae or, occasionally, lay eggs. Paedogenesis occurs in *Miastor* only under very good or poor nutritional conditions. Young larvae are set free in the body cavity of the paedogenetic larvae and they feed on the maternal tissues, eventually escaping through the body wall of the parent. Under average nutritional conditions normal adults are produced.

Micromalthus has five reproductive forms: adult males, adult females, male-producing larvae, female-producing larvae, and larvae producing male and females. The species has a complex heteromorphosis. The form emerging from the egg is a triungulin and this moults to an apodous larva which can develop in one of three ways. It can develop through a pupa to a normal adult female,

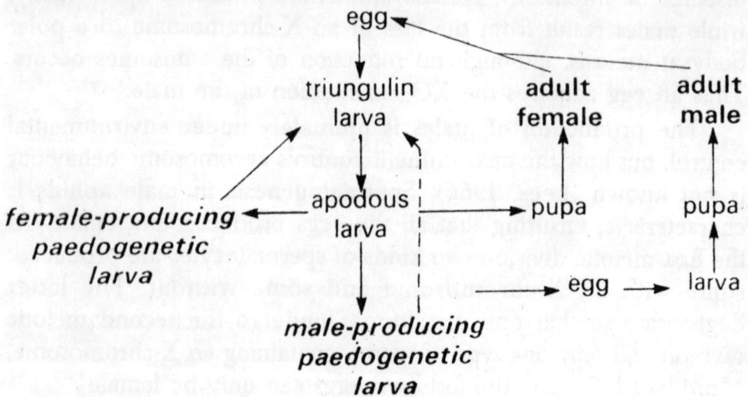

Fig. 4.33. Diagram of the life history of Micromalthus. Reproductive forms in bold type; paedogenetically reproducing larvae italicised.

or it can moult to a larval form which gives rise paedogenetically to a male, or to a paedogenetic larva which produces triungulines. Male-producing larvae lay a single egg containing a young embryo, but the egg adheres to the parent and when the larva hatches it eats the parent larva. If, for some reason the parent larva is not eaten, it subsequenlty produces small brood of female larvae. In the cecidomyids *Tekomyia* and *Henria* papual forms give birth to larvae. The larvae of these insects are of two types; one type produces a pupa and, ultimately, a normal adult, but the other forms a hemipupa, a rounded structure with, in *Henria,* vestiges of wings and legs.

A brood of larvae commonly between 30 and 60 of them, escapes from the hempupa by rupturing the cuticle (I.J. Wyatt, 1961). Wyatt (1963) also suggests that pupal paedogenesis occurs in *Heteropeza* (*Oligarces*). Larvae are released into the haemocoel of the parent larva, but they escape from a form which he interprets as a hemipupa in which they can survive, if the conditions are moist, for up to 18 months. Paedogenesis si the normal method of reproduction in these insects and although normal adults may be produced it is not certain that they are capable of producing viable offspring.

Paedogenesis also occurs in aphids. Although the young are not born until the aphid has reached the adult stage, their development may begin before she is born while she is still in the ducts of the grandparental generation. Development of the offspring continues through the larval life of the parent. The bug *Hesperoctenes* is an example of a paedogenetic form in which fertilisation occurs. Some last instar larvae are found with sperm in the haemocoel as a result of haemocoelic insemination. These sperm fertilise the eggs which develop in the ovaries of the larva.

5

INSECT METAMORPHOSIS

The changes which occur in the transformation of the larva of the adult may be more or less extensive depending on the degree of difference between the larva and the adult. Where the larva and adult are similar metamorophosis is relatively slight, but if the larva differs markedly from the adult a pupal instar may precede the adult. The pupa is probably to be regarded as the equivalent of the last larval instar of hemimetabolous insects and is a prerequisite for the greater divergence of larval and adult forms, permitting the larva to invade entirely new habitats.

During the pupal period reconstruction of the tissues takes place involving particularly the eversion and growth of the wings and the development of the flight muscles. Since the pupa is generally immobile and therefore vulnerable most insects pupate in a concealled cell or cocoon and they employ various means of escaping from this when they emerge as adults. Adult emergence is often synchronised, commonly occurring at night. Insect pupae and their significance are considered by Hinton (1946, 1948b, 1963b), the morphological aspect of metamorphosis by Snodgrass (1954), and the physiological aspects by Agrell (1964) and Wigglesworth (1954b, 1959b, 1964).

THE PUPA

Form of the Pupa

In the pupa of holometabolous insects all the features of the adult become recognisable so that the pupa has a greater resemblance to the adult than to the larva. At the larva pupa moult the wings and other features which have been developing internally in the larva are everted and so become visible although they are not fully expanded to the adult form. In some pupae the appendages are free from the body and this condition is known as exarate, but in many others the appendages are glued down to the body by a secretion produced at the larva pupa moult. This is the obtect condition and obtect pupae are usually more heavily sclerotised than are exarate pupae.

A further differnetiation can be made on the presence or absence of articulated mandibles in the pupa. When articulated mandibles are present, the decticous condition, they have apodemes which fit closely inside the mandibular apodemes of the adult and hence they can be moved by the mandibular muscles of the pharate adult. The alternative condition with immobile mandibles is known as adecticous. Decticous pupae are always exarate. They occur in Megaloptera, Neuroptera, Trichoptera and soem Lepodoptera. Some adecticous pupae are also exarate as in Cyclorrhapha, Siphonaptera and most Coleoptera and Hymenoptera, but others are obtect. Most

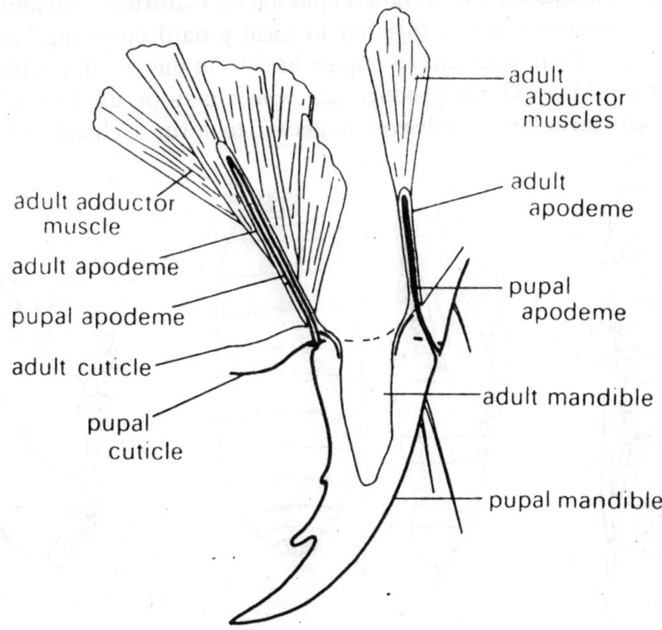

Fig. 5.1. Diagrammatic section through a mandible of a decticous pupa showing the pupal apodemes inside the adult apodemes.

Lepidoptera, Neuroptera, Nematocera, Brachycera, Staphylinidae, some Chrysomelidae and many Chalcidoidea have obtect, adecticous pupae.

Prepupa

The last instar larva is often quiescent for two or three days before the ecdysis to a pupa and in some cases the insect is a pharate pupa for a part of this time. This state is sometimes known as a prepupa, but it does not usually represent a distinct morphological stage. A separate morpholoigcal stage known as the preputa does exist in Thysanoptera and male Coccidae. In these insects the prepupa is a quiescent instar following the last larval instar and it is succeeded by a second, quiescent, pupal instar.

Protection of the Pupa

The pupa of most insects is an immobile and hence vulnerable stage and a majority of insects pupate in a cell or cocoon which affords them some protection. Many larval Lepidoptera construct an underground cell in which to pupate, cementing particles of soil with a fluid secretion. *Cerura* (Lepidoptera) constructs a chamber of wood fragments glued together to form a hard enclosing layer and some coleopterous larvae pupate in cells in the wood in which they bore. Many larvae produce silk which may be used to hold other structures, such as leaves, together to form a chamber for

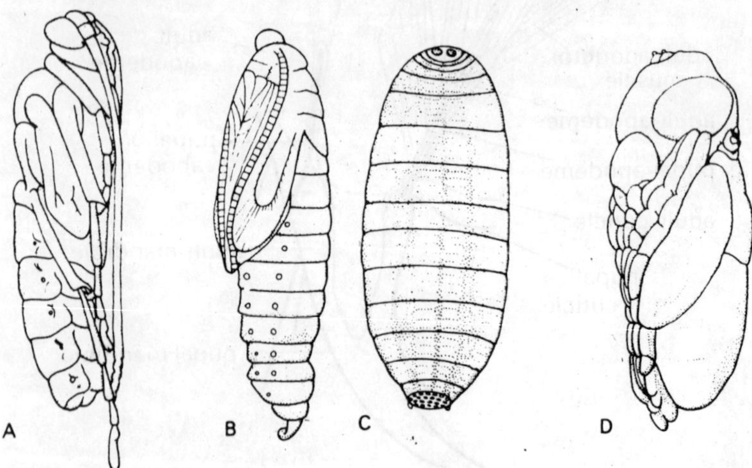

Fig. 5.2. Pupal types. A–Exarate, ichneumon wasp. B–Object, moth, C–Coarctate, house fly. D–Puparial case removed, exposing exarate house fly pupa.

the pupa, while in other species a cocoon is produced wholly from silk. Silken cocoons are produced by Bombycoidea amongst the Lepidoptera and by Siphonaptera, Trichoptera and Hymenoptera. An exceptional protective structure is produced by the larvae of cyclorrhaphous Diptera from the cuticle of the last instar larva. Procuticle is laid down throughout the last larval instar and at the end of this stage the larva rounds off and the outer part of the cuticle is tanned to form a rigid ovoid structure.

The larva moults to the pupa, but its newly tanned and shaped cuticle remains unshed round the outside of the pupa, forming a protective structure known as the puparium. A thin membrane which adheres to the inside of the tanned cuticle probably represents the inner, untanned part of the larval cuticle, but alternatively, it is suggested that the larva undergoes and additional moult within the puparium to produce the pupa, and the thin membrane represnts the cast cuticle from this moult. A few insects form unprotected pupae. There are particularly well known in the Nymphalidae and Pieridae where the pupae are suspended from a silk pad. These exposed pupae exhibit homochromy whereas protected pupae are normally brown or very pale in colour.

Pupae of Aquatic Insects

The behaviour of aquatic insects on pupation varies consisderably. Some larvae, such as the aquatic Arctiidae, Syrphidae and *Hydrophilus,* leave the water and pupate on land, but many others, particularly the aquatic Diptera, pupate in the water. Sometimes, the pupae are fasened to the substratum. For instance, the pupae of Blepharoceridae have ventro-lateral pads on the abdomen with which they hold to stones, which Simuliidae construct open cocoons attached to stones and rocks. The pupa projects from the open end of the cocoon which is constructed more strongly in faster flowing water than it is in weak current.

Chrionomidae pupate in the larval tubes or imbedded in the mud, while *Acentropua* (Lepidoptera) forms a silken cocoon with two chambers separated by a diaphragm. The pupa is in the lower chamber which is air-filled. In all these species oxygen is obtained from that dissolved in the water. Other aquatic pupae obtain oxygen from the air, either directly or indirectly.

The pupae of most Culicidae and Ceratopogonidae are free-living and active. They are buoyant so that, undisturbed, they rise to the surface and respire via prothoracic respiratory horns. If

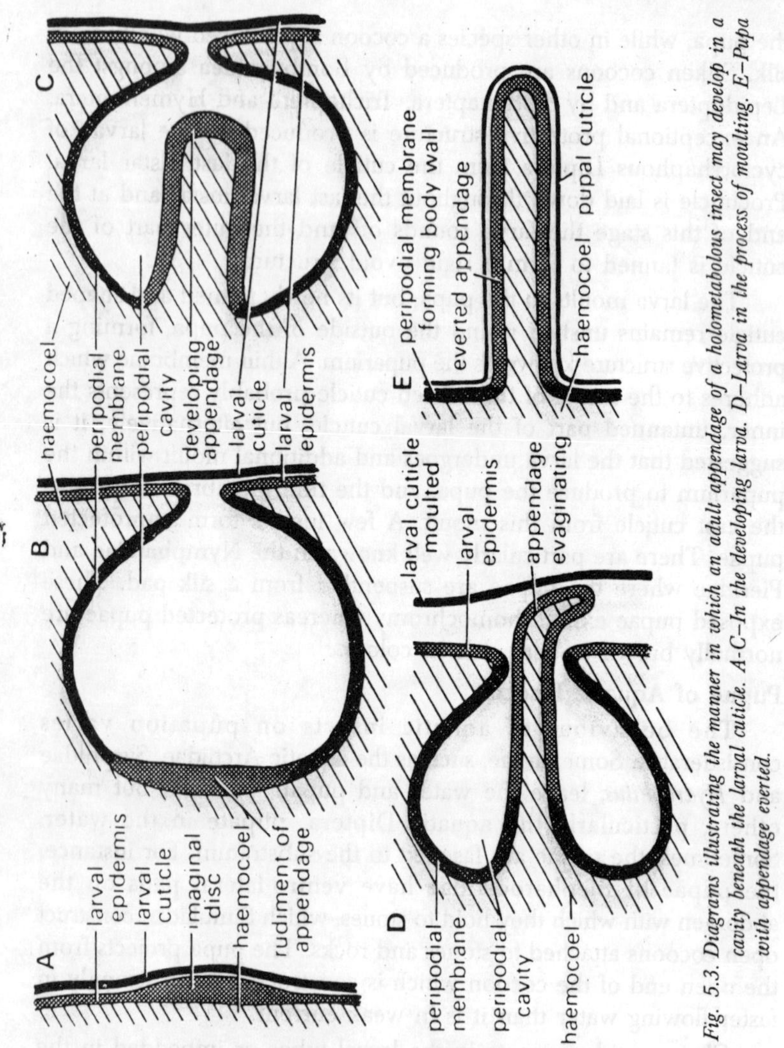

Fig. 5.3. Diagrams illustrating the manner in which an adult appendage of a holometabolous insect may develop in a cavity beneath the larval cuticle. A–C–In the developing larva. D–Larva in the process of moulting. E–Pupa with appendage everted.

disturbed their activity drives them downwards and the anal paddles asist in this movement. The pupae of some Culicidae and Ephydridae have their respiratory horns imbedded in the tissues of aquatic plants, obtaining their oxygen via the aerenchyma.

Significance of the Pupa

The pupa is indicative of the broad differences which occur between larval and adult forms of holometabolous insects. It is a stage during which major internal reconstruction may occur, but possibly its greatest importance is in permitting the full development of the wings. The internal development of the wings within the larva is restricted by lack of space and this problem becomes more acute as the insect approaches the adult condition and the flight muscles also increase in size. Thus, development can only be completed after the wings are everted and for this reason two moults are necessary in the transformation from larva to adult. At the first, from larva to pupa, the wings are verted and grow to some extent. Further growth occurs and the adult cuticle is laid down at the pupa adult moult (Hinton, 1963b). Two moults may also be necessitated by extensive modifications to the muscular system which occur at this time.

Commonly all the muscles of the adult thorax are different from those of the larva and they are not attached to the pupal cuticle. It has been suggested that muscles will only develop in an appropriate form and length if they have a mould in which to do so. The pupa provides a mould for the adult muscles. Hinton (1948b) and Snodgrass (1954), among other authors, believed that the muscles could only become attached to the cuticle when the epidermal cells were active and capable of producing tonofibrillae, that is at a moult. Hence, the second moult, from pupa to adult, was necessary to provide this attachment. If the tonofibrillae are formed from epicuticle which is only produced over a limited period this suggestion remains valid, but in some insects, at least, tonofibrillae are formed long after the moult. These cannot be epicuticular, but may be cuticular since the epidermal cells can go on producing producticle for some time.

Thus it appears that muscle attachment may not be closely dependent on moulting and hence that the pupa/adult moult is not primarily concerned with the attachment with the attachment of the adult muscles (Hinton, 1963b). The importance of the pupa in wing development and associated changes in emphasised by the absence of a pupal instar in the life histories of female Strepsiptera and Coccidae which are wingless and larviform. the males in these groups are winged and have a pupal instar. The pupa is probably best regarded as equivalent to and derived from the last larval instar of hemimetabolous insects,

Development of Adult Features

Adult features may appear at the final moult, but commonly they undergo a progressive development through the larval instars.

This is most obvious in hemimetabolous insects, but is equally true of many features of holometabolous insects.

Hemimetabolous Insects

Epidermal mitosis and expansion only occurs at the time of a moult and in hemimetabolous insects a progressive development of the wing buds occurs at each moult. Apart from their small size the wing buds differ from the adult wings in being continuous sclerotisations with the terga and pleura; the basal region of the wing is not membranous and no accesory sclerites are present. These appear at the final moult. In general the wings arise in such a way that the lateral margins of the wing buds become the costal

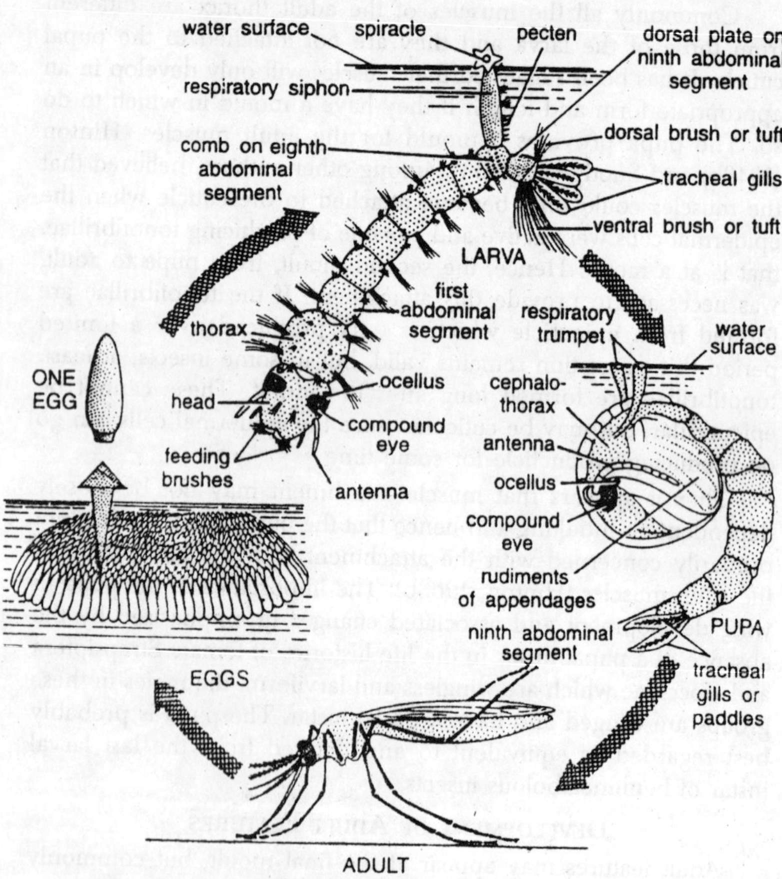

Fig. 5.4. Culex. Life cycle.

margins of the adult wings, but in Odonata the buds arise in an erect position, the margin nearer the midline ultimately becoming the costal margin.

The wing buds of Acrididae originate as simple outgrowth of the terga as in Heteroptera, but at the antepenultimate moult they become twisted into the position found in the Odonata. This twisting results from the lower epidermis growing more rapidly than the upper. At the final moult the wings twist back so that the costal margin of the folded wings in ventral in position. In *Locusta* and dragonflies all the flight muscles are present in the larva although some are histologically distinct, lacking striations and presumably being non-functional. These muscle increase in size in various ways throughout the larval period.

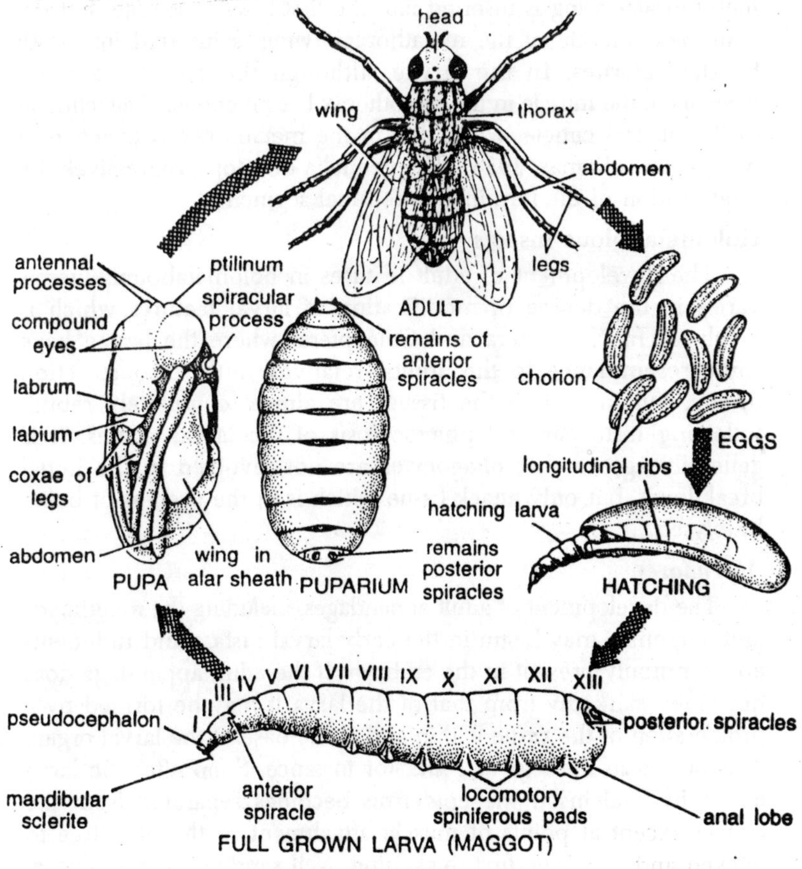

Fig. 5.5. Musca nebulo. Life cycle.

In Orthoptera they grow by the division of the existing elements; in Jassidae and some other Homoptera free myoblasts are incorporated into existing rudiments, while in *Bathylus* (Homoptera) both fibre division and myoblast incorporation occur. The incorporation of free myoblasts apparently take place through a localised gap in the sarcolemma (Hinton, 1959; Tiegs, 1955). Further changes occur in the muscles in the young adult. The phragmata to which the dorsal longitudinal flight muscles are attached become progressively bigger at each moult (Thomas, 1954). Although accessory wing sclerites are not developed in the larva, the muscles which become attached to them in the adult are attached to appropriate positions on the larval cuticle.

For instance, in *Locusta* the promotor-extensor muscle of the mesothoracic wing is inserted into the first basalar sclerite, but the equivalent muscle of the metathoracic wing is inserted into both basalar sclerites. In the larvae, although the sclerites are not developed, the muscle in the mesothorax has no epoint of attachment to the pleural cuticle, while that in the metathorax is attached at two points (Thomas, 1954). The genitalia develop progressively by modification of the terminal abdominal segments.

Holometabolous Insects

The development of adult features in holometabolous insects varies in the degree of modification of larval features which is involved. In Neuroptera and Coleoptera, where the larvae have some resemblance to the adults, relatively little reconstruction occurs, but in diptera the tissues are almost completely rebuilt following histolysis and phagocytosis of the larval tissues. It is generally agreed that phagocytes are not involved in the initial breakdown, but only attack tissue which is in the process of being histolysed.

Appendages

The development of adult appendages, including the mouthparts and antennae, may begin in the early larval instars and rudiments are commonly present in the embryo. If the adult appendage does not differ markedly from that of the larva it may be formed by a proliferation of the tissue within and at the base of the larval organ. This occurs in the legs of *Pieris,* for instance. Soon after the larva enters its final instar the epidermis becomes separated from the cuticle except at points of muscle attachment so that it is free to thicken and fold. The first thickening, well supplied with tracheae,

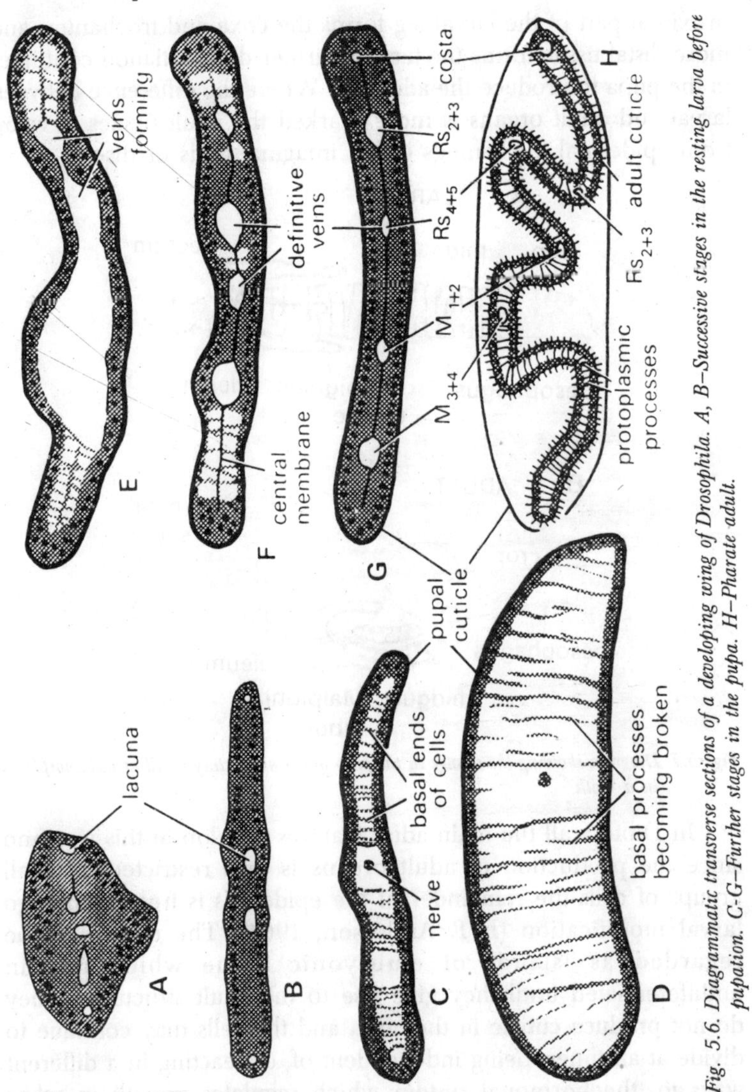

Fig. 5.6. Diagrammatic transverse sections of a developing wing of Drosophila. A, B—Successive stages in the resting larva before pupation. C-G—Further stages in the pupa. H—Pharate adult.

develops at the junction of the second and third leg joints and from this a wave of cell multiplication spread out. As a result of the increase in area of epidermis becomes folded and a particularly large fold develops basally. Later, when the epidermis expands to form the pupal leg, this basal fold is divided by a longitudinal septum to form the femur and tibia. Epidermis from the more proximal part of the larval leg forms the coxa and trochanter, and more distal tissue forms the tarsus. Further differentiation continues in the pupa to produce the adult leg. Where the difference between larval and adult organs is more marked the adult tissues develop from epidermal thickenings called imaginal buds or discs.

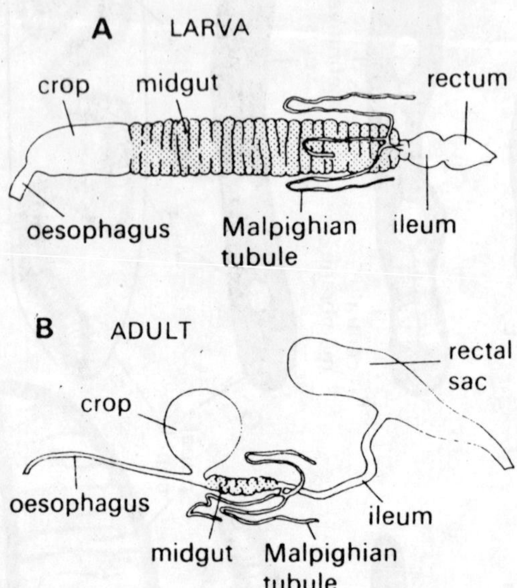

Fig. 5.7. Diagram showing the change in the form of the alimentary canal at metamorphosis in a moth.

In Diptera all the main adult features develop in this way and since the production of adult organs is thus restricted to small groups of cells the remainder of the epidermis is free to undergo larval modification (D.T. Anderson, 1964). The discs may be regarded as islands of embryonic tissue which remain undifferentiated until they give rise to the adult structures. They do not produce cuticle in the larva and the cells may continue to divide at all times, being independent of, or reacting in a different ways to, the hormonal system which regulates growth in other parts of the epiderms. The imaginal disc commonly invaginated beneath the larval epidermis.

In this way a cavity, the peripodial cavity is formed. It is lined with epidermis known as the peripodial membrane and as the imaginal disc enlarges the appendage forms and evaginates into the cavity. As the appendage grows it become folded inside the cavity until finally, at pupation, the rudiment is everted and the peripodial membrane comes to form part of the epidermis of the general body wall. The details of development of the imaginal discs vary from one insect to another and from organ to organ. When an appendage is present in the larva as well as the adult the imaginal disc is closely associated with the larval structure. Thus in *Pieris* the adult antenna is first apparent in the first larval instar as an epidermal thickening at the base of the larval antenna. The cells divide and in the succeeding instars an invagination is produced which pushes upwards deep into the larval head. In the

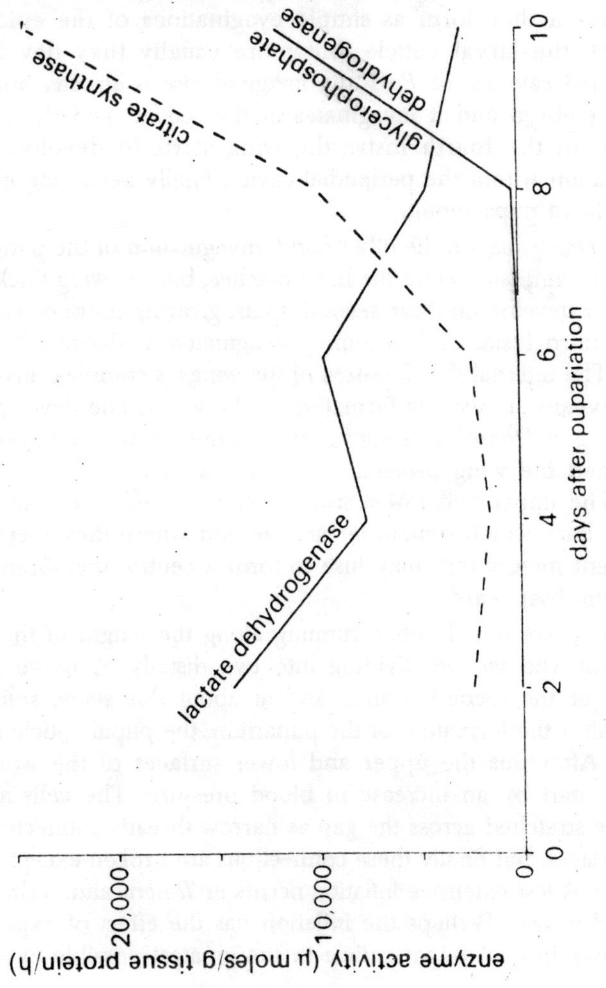

Fig. 5.8. Activity of different enzymes at different times after puparial formation in Calliphora.

fifth larval instar the adult antennal tissue grows more quickly than the peripodial membrane so that it is thrown into folds and towards the end of the instar the larval antenna starts to degenerate and is invaded by imaginal cells. When the peripodial cavity, which opens by a slit on the front of the epidermis of the head, evaginates, the antenna is carried to the outside and peripodial membrane now forms a part of the wall of the head.

The maxilla develops in as essentially similar way, but very little development of the labium takes place until the fifth instar (Eassa, 1953). The wings also develop from imaginal discs. In some Coleoptera they form as simple evaginations of the epidermis beneath the larval cuticle, but more usually they develop in peripodial cavities. In *Pieris* the imaginal disc is already apparent in the embryo and it invaginates in the second and third larval instars. In the fourth instar the wing starts to develop as an evagination within the peripodial cavity, finally becoming everted at the larva pupa moult.

In *Drosophila,* on the other hand, invagination of the peripodial cavity is complete before the larva hatches, but the wing thickening does not develop until the second instar, growing more extensively in the third instar and becoming evaginated at the moult to the pupa. The internal development of the wings is complex, involving great expansion and the formation of the veins. The development of a wing of *Drosophila* is used as an example. When the puparium is formed the wing projects backwards as a hollow cylinder of cells. The upper and lower srufaces come together except along certain lines which remain as lacunae and where they meet their basement membranes may fuse to form a central membrane, but this soon disappears.

There are four lacunae running along the length of the wing rudiment, the second dividing into two distally. A nerve and a traceanter the second lacuna, and at about this stage, some six hours after the formation of the puparium, the pupal cuticle is laid down. After this the upper and lower surfaces of the wing are forced apart by an increase in blood pressure. The cells at first become stretched across the gap as narrow threads connecting the two surfaces, but finally these connections are broken except at the margins. A less extensive inflation occurs in *Tenebrio* and *Habrobracon* (Hymenoptera). Perhaps the inflation has the effect of expanding the newly formed pupal cuticle to the greatest possible extent so

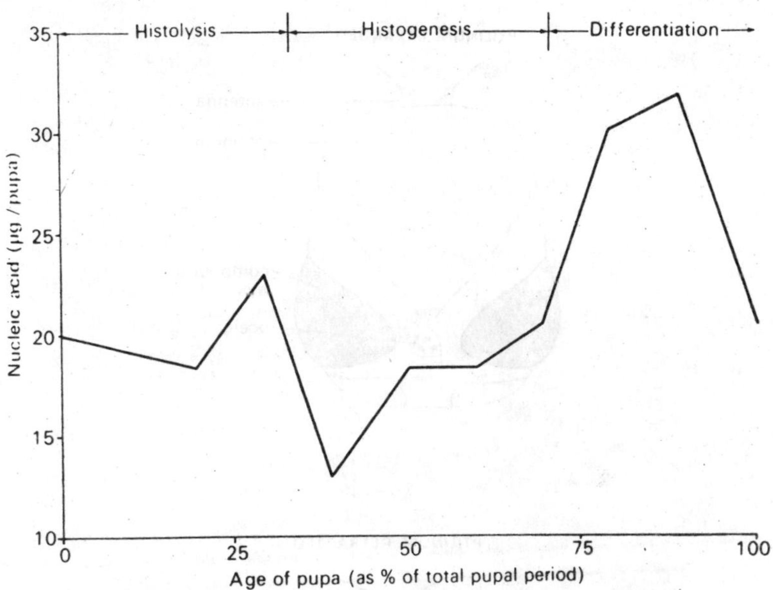

Fig. 5.9. Changes in the amount of ribosomal RNA present in different stages of the pupal instar of Calliphora.

that the development of the adult wing an proceed. Following the inflation, the wing contracts again. The epidermal layers on the two sides first become apposed round the edges and then the contraction spreads inwards so that a flat double membrane is produced. During this process the definitive wing veins are formed along lines where the two epidermal layers remain separated.

The veins are at first wide channels, but ultimately they become narrower as the membrane continues to expand. Cell division proceeds actively, especially above the veins so that there the cells become crowded and columnar, while elasewhere they are flattened. The fully developed wing finally secretes the adult cuticle.

Epidermis

When the imaginal appendages are everted from the peripodial cavities at the time of pupation the peripodial membrane contributes to the general epidermis of the adult body wall. The extent to which the larval epidermis is replaced varies. In Coleoptera there is no extensive replacement, but in Hymenoptera and Diptera the epidermis is completely renewed from imaginal discs. The epidermis of the head and thorax are formed by growth from the imaginal

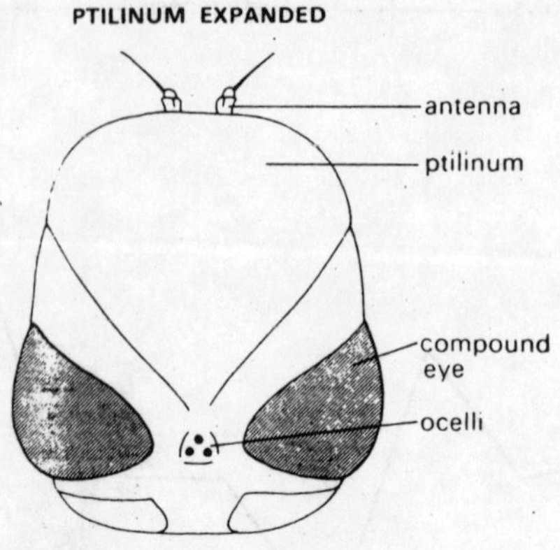

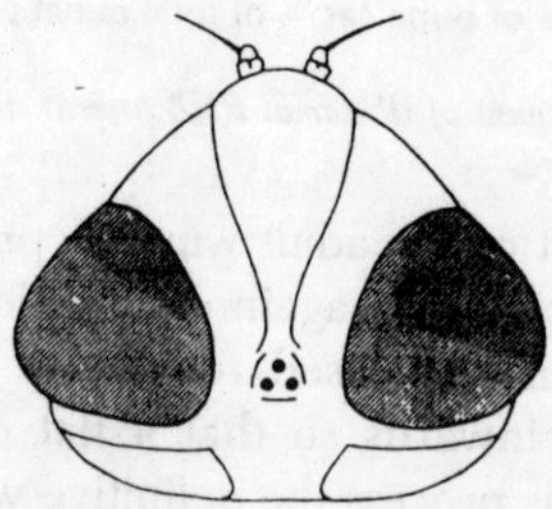

Fig. 5.10. Dorsal view of the head of a cyclorrhaphous fly showing the ptilinum expanded and retracted.

appendage discs, while the abdominal epithelium is formed from special imaginal discs. In *Drosophila* most abdominal segments have pair of dorsal, ventral and spiracular discs which expand to form the adult epidermis. As they do so the larval cells are sloughed off into the body cavity and are phagocytosed.

Muscles

The muscular system usually undergoes extensive modification at metamorphosis and the muscles fall into five categories according to their fate at this time:

1. Larval muscles may pass unchanged into the adult. This applies to some abdominal muscles.
2. Existing larval muscles are reconstructed.
3. Larval muscles may be destroyed and not replaced.
4. Larval muscles are destroyed, but are replaced in the adult by new muscles.
5. New muscles, not represented in the larva, may be formed.

In general larval muscle are histolysed and adult muscles rebuilt in the pupa, but the precise timing varies because some larval muscles have specific functions in pupal development and are destroyed much later than some other muscles. For instance, in *Drosophila* most muscles of the head and thorax start to break down before puparium formation and are fragmented before the larva pupates, but the dialator muscles of the pharynx remain unchanged until after pupation. They apparently are important in the evagination of the head region of the insect at pupation and after this they degenerate.

In addition, one pair of muscles persists in each abdominal segment for about half the pupal period. It is suggested that they help to establish the segmentation of the pupal abdomen by telescoping each segment into the preceding one. The first sign of msucle degeneration is liquefaction of the peripheral parts of the fibre. This is followed by a separation of the fibrils and in *Ephesiia* phagocytes penetrate the sarcolemma and assist the destruction. The sarcolemma braks down and hte msucles separate from their attachment and fragment, the remains being consumed by phagocytes. New muscles are always formed by free myoblasts, but muscles may be reconstructed in two ways.

In the Neuroptera and Coleoptera the larval muscles contain two sets of nuclei, the functional larval nuclei and other small nuclei which are scattered through the cytoplasm. At metamorphosis the small nuclei multiply and, with associated cytoplasm, form myocytes. These migrate into the body of the muscle and associate in strands to form new fibres. In Diptera and some Hymenoptera, on the other hand, myoblasts which originate outside the larval msucle are concerned in the production of adult muscle, adhering to the outside or penetrating the sarcolemma in order to form new fibres. Sometimes, as in *Simulium* and chironomids, the adult muscles are already present in the larva as rudimentary non-functional fibres.

The dorsal longitudinal muscles in *Simulium*, for instance, are only about four microns in diameter in the first larval instar. They grow throughout the larval period and their nuclei increase in number. During the pharate pupal period they become divided up to produce the definitive number of fibres and at this time also myofibrils appear for the first time. They continue to grow until some time after the final molt (Hinton, 1959).

Alimentary canal

The alimentary canal is extensively remodelled at metamorphosis in species which have different larval and adult diets. In Coleoptera the reconstruction of the stomodaeum and proctodaeum is carried out by the renewed activity of the larval cells without any accompanying cell destruction, but in Lepidoptera and Diptera new structures develop from imaginal rings which are proliferating centres at the tips of the foregut and hindgut. The larval cells are sloughed into the body cavity. The midgut is probably completely renewed in all holometabolous insects, usually being reformed from the regenerate cells at the base of the epithelium. These cells proliferate and form a layer round the outside of the larval cells which thus come to lie in the lumen of the new alimentary canal. Sometimes this process occurs twice, once on the formation of the pupa and again when the adult tisues are forming and it is suggested that the special pupal midgut so that these can be assimilated and used in the reconstruction.

Malpighian tubules

Sometimes the larval Malpighian tubules pass unchanged to the adult or slight modifications may occurs as in the Lepidoptera. Here the larval tubes have a cryptonephridial arrangement, but at metamorphosis the parts associated with the rectum are hostolysed while the more proximal parts form the adult tubules (Srivastava and Khare, 1966). In Coleoptera the tubuels are rubuilt from special cells in the larval tubules, while in Hymenoptera the larval tubules break down completely and are replaced by new ones developing from the tip of the proctodaeum.

Fat Body

The fate of the fat body at metamorphosis depends on the degree of reconstruction of the other tissues. Where, as in Coleoptera, many larval tissues remain unchaged the fat body shows little depletion, but where reconstruction is extensive the fat body may be almost or completely destroyed. It is then reformed in the adult from the few remaining larval fat cells or, as in *Musca,* from mesenchyme cells on the inside of the imaginal discs.

Other systems

In general the tracheal system shows little change other than the development of new branches to accommodate the particualr needs of the adult, such as the supply to the flight muscles, and the elimination of some specifically laral elements. Sometimes, however, extensive renewal of the system occurs and in *Calliphora* this takes place from small groups of cells scattered through the wall of the larval tracheae. The circulatory system undergoes little changes from larva to adult. In most holometabolous insects, particularly those that are more specialised, the central system becomes more concentrated at metamorphosis. This concentration is accompanied by a forward movement of the more posterior ganglia resulting from the shortening of the interganglionic connectives.

For instance, the larva of *Pieris* has, in addition to the head ganglia, three thoracic and eight separate abdominal ganglia. In the adult the meso- and meta- thoracic ganglia are fused with the first two abdominal ganglia to form a compound ganglion close behind the prothoracic ganglion. The next three abdominal ganglia remain separate, but the last three fuse together to form another ganglion. In the course of these changes the perineurium is histolysed and the neural lamella digested, the former being redeveloped from remaining glial cells. The nerve cells increase in number and this involves an accompanying increase in the numbers of glial cells (Heywood, 1965). The higher Dipetera are exceptional in having a more concentrated system in the larva than in the adult.

Biochemical changes

During the pupal period energy metabolism, as measured by the oxygen consumption, at first falls and then rises again, following a characteristic U-shaped curve. The fall in metabolic rate corresponds with the period of histolysis, while the subsequent rise occurs during the period of histogenesis and differentiation. These changes are produced by corresponding changes in the activity of the oxidative enzymes. The main substrates utilised during the pupal period are fats supplemented by small amounts of carbohydrate, but in *Apis* carbohydrate utiliation is high throughout. There is relatively litte information on nitrogen metabolism during metamorphosis despite its importance in histolysis and histogenesis.

The total amount of nitrogen is constant throughout the pupal period and there are only relatively small changes in the ratio of protein nitrogen: nitrogen in compounds of low molecular weight such as peptides and amino acids. There is a small increase in the haemolymph concentration of some of the free amino acids during histolysis and a decrease during histogenesis, but these changes are smaller than would be expected considering the extent of tissue reconstruction. Other free amino acids in the haemolymph do not increase in concentration and it is possible that the larval proteins are only broken down to relatively complex peptides which are then rebuilt into the adult proteins (Chen and Levenbook, 1966). Ribosomal RNA, which is concerned in protein syntehsis, increase

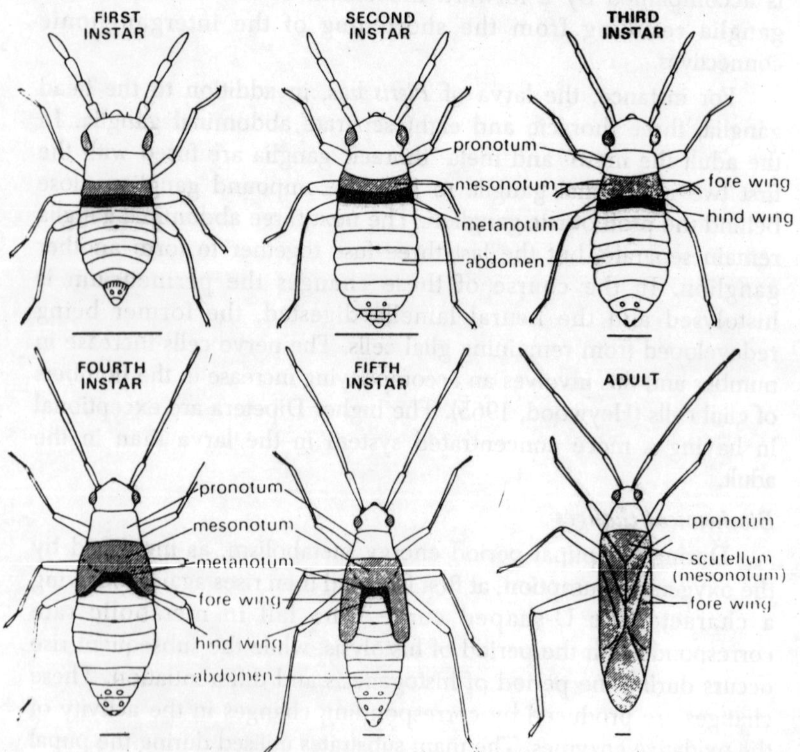

Fig. 5.11. Larval development of a hemimetabolous insect. The larva and adult stages of Cyllecoris (Heteroptera). The horizontal line under each stage represents 0.5 mm.

during histogenesis. There is a marked drop in the concentration of haemolymph protein at pupation due to the passage of the proteins into the tissues, some of which absorb particular proteins preferentially. It is suggested that these proteins may be involved in the transport of lipids and carbohydrates as conjugated groups (Loughton and West, 1965).

The waste products of pupal metabolism are discharged as the meconium when the adult emerges. Uric acid accumulates throughout the pupal period, but especially during histolysis, while in Lepidoptera and Hymenotpera allantoic acid comprises an appreciable part of the nitrogenous waste of the pupa. In *Pharmia* (Diptera), urea accumulates during the development of the adult suggesting that it is the end product of nitrogen metabolism in this insect.

Control of Metamorphosis

The development of adult characters is controlled by hormones, but while the growth and moulting hormone causes the epidermis to adopt adult features it does not order the development of particular parts of the body. Some degree of determination of the adult tissues may be apparent in the embryo, the presumptive adult area coinciding with the presumptive larval areas, but in *Drosophila* some lability of the adult tissues persists up to the time of pupation and even later (Waddington, 1956). More detailed features are not determined until after the more basic characters. For instance, the outline of the wing pattern of *Philosamia* (Lepidoptera) is determined before the details of the pattern. There is some evidence that the various organs each have a differentiation centre which orders their development. For instance in the leg of *Pieris* a basal thickening appears as an area of rapid cells division and mitosis spreads outwards from this point (Kim, 1959).

Similarly the eye of *Aedes* begins as a thickening at the back of the future eye region from which a wave of mitosis passes forwards and a thickening, the optic placode, develops in an area which is already physiologically delimited. It is suggested that development of the optic placode is initiated by some factor which spreads forwards from the posterior region, while in the final larval instar a wave of ommatidial differentiation also starts at the back of the eye and spreads forwards (R.H. White, 1961). Finally, the pattern on the wings of *Ephestia* is determined in a regular sequence which again might be due to the spread of some factor from a differentiation centre in the middle of the wing.

Adult Emergence

The escape of the adult insect from the cuticle of the pupa or, in hemimetabolous insects, of the last larval instar is known as eclosion. The thorax of the enclosing cuticle splits along a line of weakness which in the pupa is T-shaped. To produce the split the adult swallows air to increase its volume and then further increases its thoracic volume by pumping blood forwards from the abdomen. In Lepidopatera and Diptera with an object pupa the mouth is sealed by strongly sclerotised plate so that the adult insect cannot suck air directly into its gut. However, although some of the spiracles of the pharate adult connect with the pupal spiracles other do not, but open beneath the pupal cuticle. It is thus possible for the insect to pump air out of the tracheal system into the space between the adult and pupal cuticles and this air can be swallowed so as to increase the volume of the body (Hinton, 1946). Having split the cuticle the adult pulls itself out, expanding the wings by pumping blood through them. In many insects the newly emerged adult hagus upside down so that the force of gravity assists the unfolding of the wings.

Escape from the Cocoon

Where the pupa is enclosed in a cell or cocoon the adult also has to escape from this. Sometimes the pharate adult is sufficiently mobile to make its escape while still within the pupal cuticle. This is the case in species with decticous pupae which use the pupal mandibles, actuated by the adult muscles, to bite through the cocoon. Sometimes, as in Trichoptera, the adult mouthparts are non-functional and the sole function of the adult mandibular muscles is to work the pupal mandibles at emergence; subsequently they degenerate. The pupa moves away from the cocoon before the adult emerges and this is facilitated by the freedom of the appendages together with backwardly directed spines on the pupal cuticle which assist forward movement. In species with adecticous pupae other methods are employed in escaping from the cocoon.

In Monotrysia and primitive Ditrysia the pupa works its way forwards with the aid of backwardly directed spines on the abdomen, forcing its way through the wall of the cocoon with a ridge or tubercle known as a cocoon cutter on the head. The pupa does not escape completely from the cocoon, but is held with the anterior part sticking out by forwardly directed spines on the ninth and tenth abdominal segments. With the pupal cuticle fixed in this way the adult is able to pull against the substratum and so drag itself free to the pupal cuticle more readily. Cocoon cutters are also present in Nematocera although in this group they are usually multiple structures. In many insects with adecticous pupae the adult emerges from the pupa while it is still in the cocoon, making its final escape later, often while its cuticle is still soft unexpanded. This is true to higher Ditrysia whose escape is facilitated by the flimsines of the cocoon or the presence of a valve at one end of the cocoon through which the insect can force its way out, while the ingress of other insects is prevented.

The cocoon of *Saturnia* is of this type, while in megalopygidae a trap door is present at one end. Some Lepidotpera produce secretions which soften the material of the cocoon. *Cerura,* for instance, produces from its mouth a secretion containing potassium hydroxide which softens one end of its cell of agglutinated wood chips. This enables the insect to push its way out protected by the remnants of the pupal cuticle. The silk moth, *Bombyx,* produces a protease which attacks the sericin of silk and a few Noctuidae also produce softening secretions. The Cyclorrhapha have a special structure, the ptilinum, which facilitates escape from the puparium and helps the insect to burrow to the surface of the debirs in which the puparium is often buried.

The ptilinum is a membranous sac which can be everted by blood pumped in by compression of the abdomen so that is presses against the puparium and this splits along a line of weakness. The ptilinum is well developed in Schizophora, but is only rudimentary in Syrphidae. The degree of hardening which these insects undergo before escaping from the cocoon varies. In some, most of the cuticle remains soft until after eclosion, but some parts, particularly those involved in locomotion, harden beforehand. Thus, in *Calliphora,* the legs and apodemes harden, so do the bristles which protect the soft cuticle, and such specialised parts as the halters, antennae and genitalia. The remainder of the cuticle does not harden until after it is expanded when the insect is free.

In Lepidoptera, however, the body does not expand greatly after emergence and here hardening of the cuticle is extensive before the insect emerges from the cocoon, although the wings remain lump. Other insects emerge from the pupa and harden fully before making their escape from the cocoon and they may have

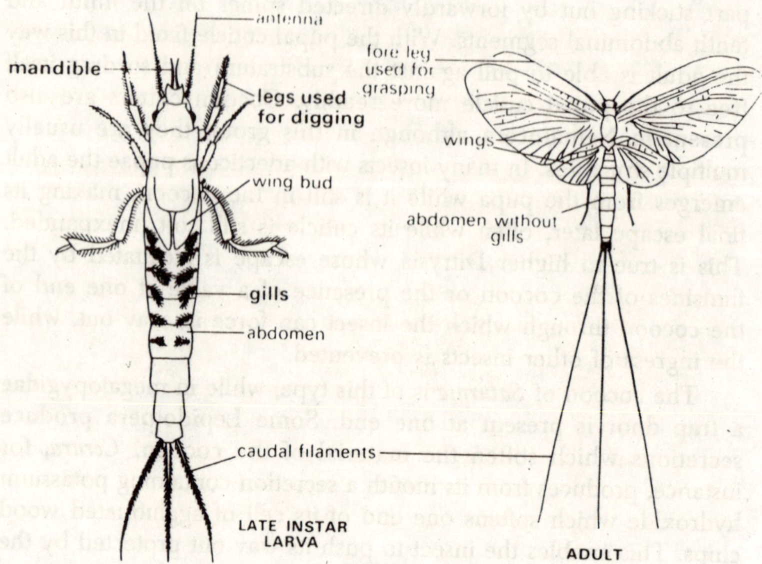

Fig. 5.12. Late larval instar and adult of Ephemera, a hemimetabolous insect showing conspicuous adaptive features in larva.

specialised features to assist this. Coleoptera and Hymenotpera use their mandibles to bite their way out. Some weevils of the subfamily Otiorrhynchinae have no the outside of the mandible and appendage known as the false mandible which is used in escaping from the cocoon and then, in most species, falls off. Amongst the Cynipidae which do not feed as adults, escape from the host in which the larva pupated in the sole function of the adult mandibles.

The cuticle of fleas also hardens before they escape form their cocoons and they may remain in the cocoon for some time after emergence. Their escape is stimulated by mechanical disturbances and in many species facilitated by a cocoon cutter on the frons. In *Trichopsylla* the cocoon cutter is deciduous. Finally in the males of Strepsiptera the mandibles the mandibles are used to cut through the cephalothorax of the last larval instar in which they pupate. The larval cephalothorax is earlier extruded through the cuticle of the host so that the adult insect can easily escape (Hinton, 1946).

Emergence of Aquatic Insects

Of the insects which pupate under water some emerge under water and swim to the surface, while in others the pupa rises to the surface before the adult emerges. In the Blepharoceridae the adult undergoes some degree of hardening within the pupal cuticle so that as soon as it merges it rises to the surface and is able to fly. *Simulium* and *Acentropus* also emerge beneath the surface, but come up in a bubble of air. In *Acentropus* this is derived from the air in the cocoon, while *Simulium* pumps air out into the gap between the pupal and adult cuticles. Thus, the adult emerges into a bubble of air and is able to expand its wings before rising to the surface. The pupae of Culicidae are buoyant, while some other insects, such as *Chironomus*, whose pupae are normally submerged, increase their buoyancy just before emergence by forcing air out beneath the pupal cuticle or increasing the volume of the tracheal system. Aided by backwardly directed spines these pupae then escape from their cocoons or larval tubes and rise to the surface.

Many Trichoptera swim to the surface as pharate adults, the middle legs of the pupae of these species being fringed to facilitae swimming and the insect may continue to swim at the surface until it finds a suitable object to crawl out on. In other Trichoptera the pharate adults crawl up to the surface, while the last instar larvae of Odonate and Plecoptera crawl out on. In other Trichoptera the pharate adults crawl up to the surface, while the last instar larvae of Odonate and Plecoptera crawl out on to emergent vegetation so that the adult emerges above the water. Larval Ephemoroptera also come to the surface, but the form which emerges is a subimago, not the imago. The subimago resembles the imago, but its legs and caudal filaments are shorter and the wings are translucent instead of transparent and are fringed with hairs. The subimago flies off as soon as it merges, but settles a short distance away and soon moults to the imago.

Timing of Emergence

It is important that the emrgence of species is timed so that the life history is synchronised with suitable environmental conditions and so that the meeting of the two sexes is facilitated. Synchrony with the environment results from the common reaction to the environment of the members of a species. Termperature is particularly important in this since to a large extent it governs the rate of development and the activity of the insects. Often a diapause is involved and in long-lived species a dipause may also be important in synchronising emergence so that the sexes meet. It is common for male insects to emerge as adult a little before the females although the difference is not great. This is often the case with locusts and mosquitoes (Clements, 1963), for instance, and it probably reflects, in part, the smaller size of many male insects.

Apart from these general seasonal effects may insects emerge mainly at particular times of day, often at night or in the early morning. This may have some adaptive significance in given the insect some degree of protection against the predators while it is vulnerable in the period before it is able to fly. Thus in Britain, the last instar larvae of *Anax* (Odonata) which are ready to molt leave the water between 20.00 and 21.00 hours and 23.00 hours molt adults have emerged and are expanding their wings. The timing of emergence of *Sympterum* (Odonata) and most trophical dragonflies is similar, although some temperate species emerge during the day, perhaps because activity is limited by low temperature at night. Adult *Drosophila* emerge mainly in the first three or four hours of light in 12 hours light : 12 hours dark regime and it has been suggested that this is due to a failure to emerge in the dark so that numbers of insects ready to emerge are inhibited from doing so until the light phase. However, Harker (1965), suggests that the lengths of various stages of development in the pupa are affected by the point in the light cycle at which they enter each stage and that a summation of these effects causes a mass emergence in the first few hours of light.

6

MORPHOLOGICAL CHANGES DURING METAMORPHIC

The aim of this chapter is to set the scene for the detailed accounts that will be given in the following three chapters. Metamorphosis has long been recognized as a characteristic of insects. Even the earliest systems of classification distinguished between those insects that metamorphosed and those that did not. From paleontological records the oldest insects determined with certainly are winged forms. However, it is fairly firmly established that the origins of the insects lie with the symphylan group of the Myriapoda, and that ancestral primitive insects were wingless and therefore would have lacked a metamorphosis. What exactly do we understand by "metamorphosis"?

A dictionary definition may be as follows: "A change in form structure ,or function, as a result of development, specifically the physical transformation during postembryonic, as of the larva of an insect to the pupa or the pupa to an adult or the tadpole to the frog." We speak, within the insects, of the absence of metamorphosis in the groups Collembola, Diplura, and Thysanura, which are described as ametabola since no dramatic changes in appearance occur during development. These insects are wingless and therefore are apterygotes. The rest of the insect orders are primitively winged or pterygote. Of these, the exopterygote insects–the bugs (Hemiptera), the may flies (Ephemeroptera), the stone flies (Plecopterà), the termites (Isoptera), the dragonflies (Odonata), cockroaches, and locusts (Orthoptera), to name a few, have their wings developing externally, hence their name Expoterygota. These are characterized by a series of larval stages or nymphal stages hat lack wings and have immature reproductive systems. They show "incomplete" metamorphosis, i.e., a gradual transition to the winged, reproducing adult. Each nymphal stage molts into the next stage, which may have slightly more developed wing buds and reproductive system. However, even here the greatest "jump" or change is generally seen at the last nymphal-adult molt. At this time the most dramatic growth in wings and reproductive systems occurs. The second group of pterygote insects are endopterygotes with wings developing internally and include the scorpion flies (Mecoptera), the alder flies and lace wings (Neuroptera), the caddis flies (Trichoptera), beetles (Coleoptera), moths and butterflies (Lepidoptera), bees and wasp (Hymenoptera), and flies (Diptera). In these insect orders metamorphosis is described as "complete" .i.e., the larval stages completely lack any external evidence of wings and reproductive system, and the transformation into the adult is a two-step process, involving first a most from the larva into a pupa, and then a molt from a pupa into an adult.

Often, as in the more specialized forms of the Lepidoptera, Hymenoptera or Diptera, the larva is completely unlike the adult in appearance. Thus, in the Lepidoptera we have the larva or caterpillar metamorphosing into the chrysalis or pupa from which the very different adult moth or butterfly emerges. In this last category there are degrees of change. At the least dramatic, we have cases like Sialis the alder fly, or the lacewing or ground beetles, where the larva, pupa, and adult still show strong resemblances, particularly in the abdominal segments and the pupa is capable of movement. Larva and adult differ mainly in the development of compound eyes, wings, and external genitalia.

Significance of the Pupal Stage

Much has been written of the "pupal stage" and its significance in, the evolution of the insects. One of the more recent discussions is that by Hinton (1963), who compares some of the older theories of Berlese and Poyarkoff. Hinton favours the argument that a pupal stage has been necessitated by the development that a pupal stage has been necessitated by the development between larva and adult of flight muscles that require a "mold". He homologizes the various nymphal stages of exopterygotes with the larval or nymphal stages of the endopterygotes, the last nymphal stage of exo-with the pupa of endo-, and the adults of both. Berlese was of the opinion that the endopterygote larval stages corresponded to embryonic forms and that all of the exopterygote pupa. Poyarkoff suggested that the adult exopterygote corresponded to the pupal endopterygote and the adult endopterygote was an *extra* adult stage.

Hinton draws attention to examples of "pupal" forms among the exopterygote orders–Thysanoptera (thrips). Coccidae (mealy bugs), and Aleurodidae (greenhouse white flies). In these, the general structure of the feeding larva or nymph has departed very widely from that of the adult; the structural differences are bridged in he last two nymphal stages. As the differences between the stages have become greater, this has involved a greater degree of structural reorganization to bridge the gap of the internal tissues. As a result, the last two nymphal stages. Instars have become more and more quiescent and have eventually ceased to feed, becoming closely analogous to endopterygote pupae. The structural reorganization required to bridge the gap, in some cases, may be greater than is found in some primitive endopterygotes.

General Changes Involved During Metamorphosis

Besides the development of wings, compound eyes, and of external and internal genitalia, one of the other mains changes that occurs in metamorphosing insects may be in their feeding habits. Thus, a caterpillar and a moth or butterfly, or a fly maggot and the adult fly may have completely different habits, and these are reflected in structural changes in the mouth parts, gut, salivary glands, and gut muscles. In some frogs a change from the larval herbivorous diet to the adult carnivorous one is accompanied by a shortening of the gut at metamorphosis. In a fly, the gut shortens also, and the larval crop, larval caecae, and larval salivary glands all disappear. The very much shorter adult gut develops a crop and an elaborate rectum with characteristic rectal papillae. Clearly, changes in mouth parts accompany changes in feeding habits, as can be seen in a chewing caterpillar and a sucking moth or butterfly. Also, changes in feeding habits and mouth parts are accompanied by changes in gut musculature. General body musculature may change considerably, as the case with the higher flies where all larval muscles are replaced, or these changes may be far less extreme, involving only the musculature to the newly developing wings. For example, insects such as the alder fly (Neuroptera), scorpion fly (Mecoptera), or even the generalized members of some of the more specialized families such as Lepidoptera and Diptera, retain essentially larval musculature in the abdomen.

The motile pupa of a mosquito or of a chironomid nonbiting midge is very different from the immobile pupa of a higher fly in which all larval muscles are broken down. In the case of an amphibian such as the frog, metamorphosis involves the migration from water to land, and correlated changes in the respiratory system occur. In the pterygote or winged insect there are changes in the respiratory system of the terrestrial larval form during metamorphosis to the flying adult form. The elaborate adult thoracic musculature is heavily tracheated, often with an equally elaborate system of air sacs. A simpler system of air sacs is also found in aquatic forms, where one function would be to provide buoyancy. Insects appear to have evolved from terrestrial ancestors with well developed tracheal systems, paralleled today in the Myriapoda. Aquatic forms therefore are secondary and are found with normal tracheal systems that may have closed spiracles; these become open and functional in the pupal and adult stages. Aquatic larvae may have tracheal gills or a superficial dense tracheation beneath the thin body wall, but these are lost during metamorphosis to the terrestrial flying adult form.

The tracheal system is intimately involved in the metamorphic changes undergone by the other organ systems. For instance, it has been found within the Diptera that the degree of breakdown of the respiratory system follows closely the degree of histolysis of the abdominal musculature (Whitten, 1960). The most specialized examples among the Endopterygota will be considered in most detail in the present survey, for it is here among the specialized lepidopteran, hymenopteran, or dipteran that one finds the extremes in metamorphic change as one proceeds from larva to adult. Not only are reproductive systems, wings, and compound eyes present in adult and absent in younger larval stages, but with changes in the feeding habits extensive changes in mouth parts and gut may occur, and muscle requirements may be drastically different in lava and adult.

With the acquisition of a flying habit the sensory environment is completely different (this occurs to a lesser or greater extent in various exopterygotes also) so that larval sensillae are broken down and new adult ones appear. Thus, drastic internal changes are reflected in changes in the respiratory and nervous system, and recent evidence shows equally elaborate changes occurring in the hemocytes (blood cells),associated with changes in other tissues. In fact, there would appear, in the specialized endopterygote, to be no system that remains unaffected by the metamorphic changes occurring during the pupal stages. The specialized endopterygote is however best considered first in relation to its generalized relatives, and then to other endopterygotes. Since all of the endopterygote orders cannot be given equal emphasis, the Diptera will be treated in some detail in the present chapter. Within the Diptera, the specialized problems found in the Diptera Cyclorrhapha (higher flies) are placed in better perspective when compared with those of the more generalized forms within the order.

The mosquito is a member of the more generalized Diptera Nematocera and it is significant that the extremes of larval tissue histolysis and reorganization are absent here. Clements (1963) has stated for the mosquito that "metamorphosis starts early in larval life with slow development of imaginal buds, compound eyes and gonads, processes accelerated in the fourth instar, and largely completed in the pupa. It is clear that important changes continue to take place in the first days of adult life that only in the older adult is metamorphosis completed." Pigmentation of the adult eye ommatidia apparently occurs in *Culex pipiens*, by the second larval instar!

The evolution of extremes in metamorphosis is therefore seen in many cases to have occurred within the higher endopterygote orders, just as there has been a general progressive evolution in the insects as a whole form the a-metamorphic apterygotes, through the exopterygotes to endopterygote forms. It should be noted that apterygote (Watson 1963). It would appear that the juvenile hormone has the same relative role–that of blocking the factors responsible for differentiation of tissues into adult forms, and a second role, that of stimulating yoke deposition in adult females. In all insects ecdysone as the same function also, that of inducing molting. Changes in hormone balance influence the *progress* of metamorphosis as will be seen in this and succeeding chapters but the endocrine organs themselves do not *cause* the differences between a non-metamorphosing and a metamorphosing insect.

The question of why the cockroach begins life as a wingless individual or why a fly has evolved into a form that begins life as wingless, legless, eyeless maggot remain unresolved. It is possible that, as Berlese conjectured, the latter represents an embryonic form ,hatching prematurely.

In the vertebrates the old concept of "ontogeny recapitulating phylogeny," where embryonic features tend to be retained while the adult form is more susceptible to change, is seen for example in the transient development of gill arches and circulation in the embryos and early stages of development of higher vertebrates. In an amphibian such as the frog we find the aquatic tadpole, with its fishlike respiratory and corresponding circulatory systems, "metamorphosing" into the adult, terrestrial amphibian with lungs, no gills, and modified gill-arch circulation. If the ancestral insect is a symphylan-diplura type, and the pterygote insect larva is prematurely hatched in the ancestral wingless form, then the acquisition of wings at metamorphosis closely parallels the amphibian condition, with its acquisition of legs at metamorphosis. These are academic questions , the answers highly speculative, but the facts of metamorphosis remain significant to the biologist, with the insects and amphibian remaining striking examples with which to study the process.

Problems Posed by Insect Metamorphosing Tissues

In a metamorphosing pupa one finds various larval tissues degenerating at specific times. What initiates cells death? Can this predetermined event be experimentally altered, delayed, or permanently prevented? Also in the developing pupa we have tissues seemingly regress dedifferentiate, and new adult tissues redifferentiate from them. Do we know what happens to individuals cells? Do cells that are differentiated for a particular function in the larval stage dedifferentiate and then are differentiate into cells performing like or dissimilar functions in the adult stage? Are some adult cells developed from undifferentiated "reserve" larval cells? In the metamorphosing pupa we may have tissues that appear as new structures not represented previously.

Some of the insect flight and leg muscles come into this category. Since muscles are characteristically mesodermal, can the leg muscles develop from epidermal, ectodermal elements of the imaginal discs or mesodermal myoblasts present in the discs? It has been shown that radioactive proteins in the cuticle of larval insects reappear in the developing adult cuticle. In this case, how are these materials removed from one and transferred to the other in space and time? The mass of larval muscle in a fly is relatively enormous. At the time of muscle breakdown are the materials lost to the body or is there conservation and use in the build-up of adult tissues? Are the blood cells involved o any great extent in tissue breakdown? Are the cells destined to cell death autocytolyzed or are living cells attacked by phagocytic blood cells? Are there ultracellular components comparable to focal degradation points in other animals, and how general are lysosomes? How does a tracheal system laid down at a previous instar solve the problems of providing respiratory gases, firstly to larval tissues destined to die, secondly to larval tissues to be retained, and thirdly to adult tissues not yet differentiated...all requirements within the same pupal instar? A vital and fundamental problems involves the role of the central nervous system. How much metamorphosis occurs in this system? Are we further forward now than we were in 1952 when power, in one of remarkable series of papers on the *Drosophila* nervous system stated, "The morphogenesis of the insect's nervous system is still largely unexplored in spite of a large bibliography."

Among the problems that remained unanswered he named the basic sequence of events in the origin and differentiation of the component parts of the central nervous system, the fate of the peripheral nervous system at the time of metamorphosis, the development of the innervation of the imaginal organs, and the changes that occur in the nervous system correlated with the changes that occur in the various end organs. An understanding of the events occurring in the nervous system is basic to any understanding of metamorphosis of individual organs, yet work on the central nervous system during metamorphosis is far behind that on the endocrine system. Even here it has been shown possible to change, enhance, or retard metamorphic events by altering the hormonal environment, yet we are still left with the basic problem of control of the endocrine organs themselves and, ultimately, of the brain neurosecretory cells. What, e.g., are the controlling mechanisms behind the circadian rhythms of development in the metamorphosing *Drosophila* pupa, shown so elegantly by Harker (1965a, 1965b)?

The Significance of Metamorphosis to General Biological Problems

Embryology

In the more extreme cases of metamorphic change there is a virtual return of most tissues to an embryonic state with the constituent cells in a undifferentiated form. As metamorphosis proceeds, differentiation of individual cells of the various tissues occurs. Such is the case in a higher fly or bee where the muscular system is entirely broken down and reorganized, and there are correspondingly elaborate changes in the gut, general epidermis and tracheal system, with correlated reorganization of the nervous system. One of the most valuable aspects of insect metamorphosis is as subject material for study of embryological and general developmental problems. Problems of cell and tissue differentiation, of pattern formation, of cell determination and of cell death can all be studied in the metamorphosing insect. Of primary significance also is the fact that insects are amenable to experimental techniques such as those of implantation (e.g., Bodenstein 1950 and Hadron 1951).

Genetics

Studies using the fly *Drosophila*, and to a lesser extent, the hymenopteran *Habrobracon*, have played a large part in our understanding of genetical problems; these are essentially studies of metamorphic events in these insects. The various eye mutants or wing mutants or mutants showing abnormal bristle pattern result from the action of various mutant showing abnormal bristle pattern result from the action of various mutant genes. However, an understanding of the steps which intervene between the formation of the initial protein and the manifestation of the visible effect, is in essence an understanding of the events of differentiation within the metamorphosing pupa, at least where the effects are on the adult phenotype. There may be a large number of developmental sequences between the activity of the mutual gene and the expression of for example, "vestigial wing" in *Drosophila*. Studies on the formation of biological "patterns" are often also metamorphic problems, as in the case of bristle patterns in *Drosophila* (e.g., Sondhi, 1963).

Chromosomes Structure and Gene Action

Insects have another tremendous advantage over other organisms in possessing giant chromosomes. It is unfortunate that they are known only from the order Diptera among the insects, but at the same time it makes this order of greater interest. These giant polytene chromosomes arise by chromosomal DNA replication in the absence of corresponding cell division. They are best known from larval salivary glands; in *Drosophila* has been possible o "map" the different bands of the salivary cell chromosomes, so that specific regions represent the positions of particular genes whose action has been determined by genetical means. Giant chromosomes are characterized by regions of intense "puffing" activity. These regions represent areas of active RNA synthesis.

The RNA us thought to be messenger RNA which passes into the cytoplasm, where in association with the ribosomes, it organizes the amino-acids into particular sequences to form specific proteins. In this way the origin of a specific protein could be traced back to the activity of a specific puff at a definite position on the giant chromosomes. Cells other than salivary gland cells may contain polytene chromosomes, although they are relatively smaller. These include cells of Malpighian tubules, male seminal vesicles, nurse cells of the ovary, heart cells, pericardial cells, and even fat body cells. Recently, cells having chromosomes as large as salivary gland chromosomes have been recorded for epidermal cells of the fly foot (Whitten, 1964a). They are responsible for secretion of the various layers of cuticle forming the dorsal surface of the adult foot pad of the higher flies. These cells are thus intimately involved in general adult "development", whereas salivary gland cells have the disadvantage of undergoing cell death at the beginning of the pupal stage.

Changes in salivary chromosome puffing, brought about by changing the hormonal milieu, have been correlated with effects on "general development". Titles such as "Regulation of gene action in insects development" (Kroeger and Lezzi, 1966), and "Chromosome changes associated with differentiation" (Clever, 1965), indicate the extent to which studies of salivary gland chromosome puffing activity have attempted to relate events at the chromosomal level to events at the level of the whole organism, in this case the metamorphosing pupa. Giant epidermal cells, such as those of the foot pads in the higher flies, should prove of particular interest in pursuing these problems.

Questions of growth by cell division, as opposed to growth by cell size, have also been raised has a result of studies involving metamorphosing insects, particularly the mosquito (e.g., Trager, 1937 and discussed in clements, 1963). It has been found that as a role, cells which increase in size (for example of salivary glands, Malpighan tubules, rectum, abdominal muscles, trichogen cells, larval oenocytes) are incapable of cell division, and expect for Malpignian tubules are incapable of transforming into adult organs, so that these then are the cells which histolyse in the metamorphosing pupa. During growth, replication of chromosomes proceeds without separation and without cell division, resulting in the characteristic polytene chromosome. On the other hand cells that multiply by cell division remain small, and are retained and pass through into the adult; presumably they are undifferentiated and therefore capable of subsequent differentiation.

A seemingly unique condition is found in the various mosquitoes–that of *endopolyploidy* with somatic reduction. Here, in the small intestine, and to lesser extent in other parts of the gut, epidermis, tracheae and neurilemma, cells become polytene but apparently not differentiated. In *Aedes* the cells may be 16 N with 48 instead of the normal 6 chromosomes, with the homologues closely associated in a polytene chromosomes. However, at the onset of population the polytene chromosome splits up and cell division occurs, without chromosome replication. In this way there are produced many small diploid cells which differentiate to reconstitute the small intestine and rectum. An answer to the question of why this should happen is not at present available.

Gross Morphological Changes at Metamorphosis

One of the most striking aspects of insect metamorphosis, particularly to the casual observe, is the dramatic change from larva to pupa at the larval-pupal ecdysis. This is all the more remarkable when one realizes that the pupal "mold" produced at the larval-pupal molt follows the future adult shape, particularly with respect to the large muscle-filled thorax. Yet at the time of its formation none of these adult structures may be developed in some instances, as is the case with the higher flies. The mechanical aspects have been discussed at length recently in a review article by Cottrell (1964): The external shape of the insect is largely limited by the inherent form of the different parts and more particularly by the inherent form of the different parts and more particularly by the extent to which the outermost layer of cuticle–the epicuticle–can be unfolded. Then again, the unfolding of the epicuticle and the stretching of the underlying endocuticle is brought about by hydrostatic pressures produced by two activities: swallowing of air or water and the contraction of muscles.

The relative importance of these two may differ in different insects. Final form is also attained by the expansion of soft, not yet expanded parts, but not of any prehardened areas. There is a very exact relationship between the two. Kohler has shown that the eversion of the wing buds in *Ephestia,* the clothes moth, is brought about by hydrostatic pressures and by muscles opening the mouths of the peripodial cavities. Fraenkel showed that in *Calliphora* eversion of the pupal head is brought about by peristaltic abdominal contractions. Cottrell discusses other examples also. Personal observations with *Drosopila* and *Sarcophaga* have shown the presence of a series of larval suspensory muscles that seem to play a role in attaining pupal form.

Typically, the cyclorrhaphan nephrocytes are shown as slung from the two lobes of the salivary glands. In fact, their attachment in both *Drosopila* and *Sarcophaga* (and probably other Cyclorrhapha also) is by way of a muscle that runs along the proximal third of each lobe of the salivary glands, with firm attachment between gland, muscles,andnephrocytes being maintained by the continuous connective tissue basement membranes (micrographs in Whitten, 1964). These muscles are attached at one end near the spiracles and at the other on the gut. Two other muscles, with origins in the prothoracic region and insertions asymmetricaliy on the midgut ,pass through the body cavity to their points of attachment on the midgut. These muscles, show rhythmical contractions and resemble the alary muscles, being clearly striated under phase optics. In the fly, a "cryptocephalic pupa" has its head still invaginated, but leg and wing discs are everted though flaccid and unexpanded ; the abdomen is larval in shape being disproportionately long, and larval tracheae are still within the body. Simultaneously with eversion of the head, there is extension of the wings and legs and contraction of the abdomen. Contraction of the abdomen must reduce the internal volume and help force blood into the legs and wings and head. A large air bubble present before, and absent after eversion of the head, also probably plays some part. Observations of internal changes reveal a dramatic overall contraction of the entire system of longitudinal visceral muscles and unpaired suspensory muscles.

This simultaneous contraction of the ;muscles effects the shortening of the larval gut to the dimensions of the midgut caecae reduces these to four small knobs on the gut. Their cells later histolyze, and these structures are not represented in the adult gut. At the same time, the contraction of the "slinging" or suspensory muscles plays a role in the form of the pupa and future adult. By anchoring the gut and salivary glands they seem to help prevent their being everted into the head along with the brain. This is probably also helped by muscles such as the alary muscles, which also attach to tracheae. At the same time the muscles, whose origins are on the body wall behind the prothoracic spiracles, may help during their contraction to draw in the "neck" region between head and thorax. Nothing is fortuitous, and the establishment of shape in the pupa is a very precise process. Attainment of adult shape is equally intriguing.

Clearly, it is genetically determined that the slightly bilobed saclike pupal leg appendage of the higher fly should develop by epidermal cell retraction, cell division, and differential growth into an adult leg, yet little is known of the factors that determine this shape. There is a suggestion in the development of the fly foot that the ecdysial membrane (c.f. Taylor and Richards, 1965), the *first* secretion of the adult epidermal cells that form hairs of various sizes. Once the adult structures are laid down, it can be seen from Cottrell's account that the adult shape is attained by a dual process–by the prehardening of structures that will not expand after ecdysis (the various leg bristles and the terminal pretarsal structures shown in the fly foot) and the postecdysial expansion of soft structures that subsequently harden, e.g., the wings and general abdominal cuticle. Evidently, timing is delicate and there are elaborate control mechanisms involved.

Again it is with the flies that much work has been done. Fraenkel and Hsiao (1962) and Cottrell have separately found a "darkening factor", named by former "bursicon". More detailed discussion is given in Cottrell's review. It is sufficient here to note that the factor is released after the flies have ceased "digging" their way through the soil and that darkening can be produced in these "digging" flies by injection of blood from air swallowing, darkening flies. The cessation of digging and the onset of air swallowing and darkening are closely interrelated; "shut off" of digging movements is permanent. It cannot be related to the hardening process and, according to Cottrell, the mechanism of "shut off" is probably nervous. Push through the ground is accompanied by and partially brought about by alternate eversion and retraction of the blood filled sac at the anterior end of the head–the ptilinum. The muscles producing the movement breakdown within three days of emergence, demonstrating the remarkable co-ordination of events.

The nervous stimulation for cessation of digging may well result from signals received (or no longer received) by the tiny tactile hairs dispersed over the surface of the ptilinum (Whitten, 1963c;). Each hair possesses a sensory nerve fiber, and avoiding reactions with soil particles are seen to result from stimuli received by these hairs. When the fly appears above ground, no resistance is felt by the ptilinum, and signals cease to be received by the sensory nerve fibers. This in itself could *initiate* the chain reaction which involves retraction of the ptilinum, air swallowing, darkening, and finally breakdown of the ptililnal muscles. Particularly puzzling is that if the darkening factor is a hormone whose release is necessary for the final synchronous darkening of all areas to be hardened and darkened following expansion, what is responsible for the progressive predarkening of areas such as the legs before ecdysis? This occurs from thorax forward and backward and proximodistally along the legs. The predarkening process has been used as a very reliable aging device, e.g., for general development aging in *Drosophila (*Harker, 1968a, 1965b) and in the development of the fly leg (Whitten,1968). Are the preecdysial and the postecdysial hardening and darkening processes really controlled and effected by different Agnets?

Developmental Rate in Metamorphosing Pupae: Internal Clocks

A very different but equally intriguing aspect of the metamorphic process occurring between larval and adult stages can be followed in two papers by Harker in a study of developmental rates and their controlling mechanisms. Working with the fruit fly *Drosophila melanogaster,* Harker (1965a) first studies the course of development in two strains in an attempt to establish the factors affecting the time taken to complete three stages of: head eversion to yellow eye, yellow eye to wing pigmentation; wing pigmentation to eclosion. Time measurements were taken for pupae entering each stage at each particular hour of the day when insects were kept at 12-hour light; 12-hour darkness; 12-hour bright light: 12-hour dim light; or in continuous darkness. She found that the duration of each stage in both strains is affected by the time of day relative to the *light cycle at which the stage is entered.* The duration of each stage for pupae kept in continuous darkness is affected by the time of day at which the stage is entered, relative to the light cycle to which they had been exposed as larvae. She further found that the time interval curves for all three stages of any one strain take the same form.

Consequently, because of the very wide range of development rates, dependent on the time within the light cycle at which each stage begins, a population in the larvae all pupate within a 24-hours period will continue to produce adult flies over several days. The eclosion rhythm is a population effect and does not reflect the phasing of individuals at a dawn eclosion; the majority of individuals emerge at dawn because of the summation effect of circadian rhythms of development at earliest stages. In a second series of experiments, Harker (1965b) studied the times taken for *D. melanogaster* to complete the same three stages of development for pupae entering each stage at each particular hour of the day in cycles of 12-hour light; 12-hour darkness; 4-hour light: 20-hour darkness; 18-hour light; 6-hour darkness. Similar experiments were carried out on insects in which the larvae were subjected to the light cycles but the pupae remained in constant darkness. It was found that the duration of each stage is affected by the time of day, relative to the light cycle, at which the stage is entered. When pupae are in constant darkness, the development rate of each stage is entered relative to the *particular light cycle* to which the larva had been exposed.

It was further shown that the eclosion rhythm, arising as a "summation effect" of rhythms of development at earlier stages, may become bimodal in light cycles with suitable photofractions. It was further found that the rate of development of a pupa entering a stage during the light "on" signal; the preceding dark period has no effect. Finally, the development rate of a pupa entering a stage during the dark period is affected by the time interval elapsing since the light "off" signal, but may also be affected by the previous light "on" signal, although there is no simple relationship between them.

Harker concluded that as the development rates are maintained in constant darkness, the rate is affected by factors following a *diurnal rhythm.* The *form* of the rhythm is determined by both light "on" and light "off" signals, but the *timing* of the rhythm is determined by the two signals acting independently of each other.

Metamorphic Changes in Individual Systems

In order to follow the types of problems presented by individual tissues one needs to look at each system in turn, for each presents its own characteristic problems. Cell differentiation and pattern formation may be best studied in various epidermal cell types, cell death perhaps in muscle systems, along with considerations of blood cell relationships, cell determination in the various imaginal rudiments, and controlling mechanisms in endocrine and nerve-tissue relationships. As general a picture as possible will be given in the space available, but slightly more emphasis will be given, in tracing sequences in individual systems, to members of the Diptera Cyclorrhapha, relating them as far as possible to events occurring in the lower Diptera Nematocera and other insects. For each system the same question will be asked. "In what way does the mosquito (Diptera Nematocera) differ from the fly (Diptera: Cyclorrhapha), and how then do other insects compare with both?".

Metamorphic changes in the Alimentary Canal

Changes occurring in the alimentary canal vary from one insect type to the next and cannot be covered comprehensively here. The cyclorrhaphan system again will be taken as an extreme example of the type of changes that occur. The feeding apparatus in larva and adult are quit different: It has been mentioned earlier that the larval gut is longer, possesses a larval "crop" and four midgut caecae and is otherwise more or less a long, straight tube. The adult gut lacks the crop and midgut caecae, and is characterized instead by the development of an adult crop and of a rectum containing four "rectal papillae" of characteristic appearance but until recently of little known function.

The mouth parts of the larva consists of a pair of strong mouth hooks; the adult mouth appendages–labial, maxillary, and mandibular–are highly specialized and arise from the median portion of the frontal sac of the larva, an imaginal rudiment lying between the brain lobes. Changes in the gut during metamorphosis involve degeneration of the midgut epithelium, disappearance of the midgut caecae, disappearance of the labial "crop" and the development of adult crop and rectum from two "imaginal disc" areas located at the junctions of foregut and midgut and midgut and hindgut, respectively. Also, the larval salivary glands breakdown, and adult salivary glands develop from an "imaginal disc" area located between the base of the larval gland and the duct. From the undifferentiated cells of this area the adult glands arise by cell mitosis, growth and differentiation. Other imaginal areas may include labial and a ring of peri-anal cells.

Co-ordinated longitudinal contraction of the gut muscles and those of the midgut caecae bring about shortening of the gut and reduction of the caecae to stumps. This contraction throws the basement membrane into elaborate folds. Slightly later, the basement membrane breaks down as does that the other tissues in the pups (Whitten, 1962a), and the gut lining cells degenerate, adult epithelium being formed from small interstitial cells in the gut wall. The age of a midpupa can be fairly accurately followed by determining the length and degree of development of the adult crop. In the cyclorrhaphan pupa the Malpighian tubules are retained throughout pupation and evidently function at least in the late pupa, where they are filled with a milky material. This is passed into, the accumulates in, the rectum and later partially contributes to the gut content or "meconium", which is eliminated from the gut at the time of emergence.

Virtually nothing is known of muscle reorganization. Do the adult visceral muscles arise from dedifferentiated gut muscle cells or do some or all arise from myoblasts, as has been suggested in older literature and again recently by Crossley (1965) for certain of the abdominal skeletal muscle cells in *Calliphora,* and by Daly (1964) for a large proportion of the thoracic muscles in *Apis,* the bee? In the case of the fly gut, the muscles apparently disappear, with adult muscles appearing in the late pupa. However, it is very probable that the muscles dedifferentiate, to redifferentiate later in a fashion similar to the cyclical muscle changes show in *Rhodmus* by Wigglesworth (1956). Here the muscles fibers disappear between molts, with only the sheath and nuclei remaining; then at the onset of each molt the muscles are completely differentiated again.

Changes in the mosquito gut are essentially similar (Clements, 1963). The phenomenon of endopolyploidy, discussed earlier, is of particular interest.

Metamorphic Changes Undergone by the Heart

The heart during metamorphosis is an organ that has received relatively little attention. It reportedly ceases to function in many insect pupae, and in some it has been described as reversing the direction of heartbeat. In *Sialis,* the alder fly, a very fine study has recently been made, by Selman (1965), on the circulatory system. In this species he notes that, as is the case in other larvae including those of dipteran flies, the larval heart is not innervated, whereas for some at present inexplicable reason the adult heart is innervated.

In general, the heart changes relatively little during metamorphosis. For instance, in the mosquito (Clements, 1963) the heart and associated pericardial cells are said to pass unchanged from larva to adult.

Personal observations on *Sarcophaga* and preliminary work on *Drosophila* suggest a closely similar pattern of events for these cyclorrhaphans. The heart is seen to pass without dissolution through to the adult stage, but in the midpupa metamorphic changes do occur. Also the larval posterior tracheal plexus is absent. The heart wall itself becomes somewhat thickened, and this thickening does not extend to the aorta. Correlated with the increase in wall material, the intrinsic heart cells become active and display well-developed polytene chromosomes (Whitten, 1963a). The smaller anterior and the large posterior pericardial cells undergo changes in the midpupal stage, and the larger cells also develop well defined polytene chromosomes. The thickening of the heart but not of the aortal region produces a humped effect at the junction of the two regions, and this comes to lie in the raised abdominal region immediately posterior to the constriction between adult thorax and abdomen. In a midpupa the heart muscles have degenerated and degeneration is also seen in the alary muscles. In these latter, there would appear to be events occurring that are comparable with the cyclical muscle dedifferentiation and redifferentiation in the abdomen of *Rhodnius*. In a fly midpupa, heart pumping ceases and there is virtually no circulation of blood. Developing adult nerves can be followed migrating over the heart surface of a mid to late pupa. An attempt to review events recorded for other insects will not be made; the scarcity of information concerning metamorphic changes can be seen from the recent review of Jones (1964). The absence of circulation during the midpupal period would seem to be pertinent to questions of hemocyte relationship in the pupal stages, and these will be discussed later in more detail.

There are many other peripheral issues in connection with the developing adult circulation during metamorphosis, including the development of pulsatile organs, dealt with in current textbooks on entomology. As with all the other systems, the more one looks into the problem the more one realizes how little, relatively, is known of the subject. Few insects have their circulatory systems as thoroughly investigated as has the alder fly *Sialis* in Selman's recently study.

Metamorphic Changes in the Fat Body, Oenocytes and Nephrocytes

The fate of the fat body varies in different insects. For instance within the Dipters, in a generalized dipteran such as the mosquito, it is said to pass from larva to newly emerged adult unchanged and little if at all diminished in size (Clements, 1963). Changes in the higher flies are however somewhat more complex. It was formerly considered that the larval fat body of the fly broke down in the early pupa and thereafter ceased to function (e.g., Bodenstein, 1950). However, what happens at pupation is that the connective tissue sheath (basement membrane) that binds the cells into a composite organ breaks down and release the component cells (Whitten 1962a;). The breakdown, as with all the events in the metamorphosing pupa, is perfectly timed and co-ordinated. Breakdown of thoracic segments of the fat body occurs in the cryptocephalic pupa while the abdominal portion breaks down later, in the early pupa. Although appearing diffuse, the fat body arises segmentally and functions as such in later life. A branch of a middorsal nerve has been traced to the fat body in *Sarcophaga,* and the basement membranes of the nerve and fat body are confluent. The fat body in this fly would therefore seem to be innervated.

The cyclorrhaphan larval fat body cells function right into the adult stage. They have been the subject of much recent research, particularly in *Drosophila,* where they seem to play an important part in intermediary metabolism in the developing pupa. Changes in the fat body are characteristic of certain *Drosophila* mutants.

In many insects including flies and moths a separate adult fat body originates within the pupal stage. Older workers have suggested that adult fat body cells may have originated from hemocyctes. The whole subject is by no means clear. Evidence in the Diptera Cyclorrhapha recently has suggested that this is not as unlikely a possibility as might at first be supposed (Whitten, 1964b). Phagocytic hemocytes, which at time of pupation ingest larval fragments, later undergo a second cycle of activity. They persist right into the adult stage, and at this time appear very like the adult fat body cells. The mosquito is seen not to have a distinct adult fat body and the larval fat body is said not to histolyse in the early adult [Clements, 1963].Certainly phagocytic hemocytes would be relatively much less active in the mosquito where tissues histolysis is less extensive.

Oenocytes are cells which are though to be involved in intermediary metabolism, in growth and in reproduction (Wigglesworth, 1965). In many insects including the mosquito and fly there are two generations–larval and adult. The larval oenocytes in the fly are large cells with polytene chromosomes located segmentally in groups along the side of the body. Adult oenocytes are smaller, appear in the pupal stage and are often associated with the adult fat body cells. In the mosquito the larval cells grow in size, develop polytene chromosomes, and break down in the pups. The smaller adults oenocytes appear in the abdomen during the fourth larval instar (Clements, 1963).

In the fly, *ventral nephrocytes* thought to be excretory in function and related to pericardial cells , are slung as a garland between the lobes of salivary glands (attached by muscles and basement membrane material). They break down in the early pupa, at the disappearance of the salivary glands. Comparable cells are present in the mosquito where in some species the also form a chain between the salivary lobes.

Metamorphosis of the Muscular System

Among the Endopterrygota there is a considerable range in the degree to which muscles are retained from larva to adult. In some of the more primitive orders, but also in some of the more primitive members of the more specialized orders including the Diptera, the majority of the larva muscles are carried over into the adult stage. The muscles, particularly of the abdomen, are carried over and changes occur mainly in the muscles of the thorax that are responsible for flight. In the mosquito for example, larval abdominal muscles are used in the pupa for swimming and these degenerate early in adult life. The adult abdominal muscles arise from myoblasts in the pupal stages. However, among the specialized from larva to adult. In the thorax of the honeybee, a very comprehensive survey made recently by Daly (1964) should that no larval muscles pass unchanged into the adult, that 85 percent of larval muscles are associated with the development of adult muscles, and that 15 percent of the larval muscles degenerate and have no relationship to adult muscles. Some 61 percent of adult muscles arise from aggregations of myoblasts, including those of appendages and direct flight muscles. Crossley (1965), in a recent study on the blowfly *Calliphora erythrocephala*, similarly showed the origin of certain of the abdominal muscle from myoblasts, an observation substantiating the opinions of several of the older workers at the turn of the century (e.g., Perez, 1910). The description by nuclei that subsequently differentiate into adult muscles in the same location as the larval is difficult to conceive but is all more fascinating if actually the case.

In recent years interest has centered largely on the breakdown of muscles and the cause of the breakdown. This remains one of the most intriguing problems of insect metamorphosis. Of particular interest is the fact that not all larval muscles degenerate at the same time. Some are retained while their neighbours degenerate. Initial work on this subject was in the Leopidopetra (see e.g.,Finlayson, 1956). More recently, an extremely interesting series of papers have appeared by Lockshin and Willams (1965), also on the Lepidoptera. Whereas the earlier workers concentrated on muscles that were carried over from the larva into the early pupa and then degenerated three days later under normal conditions. Similar differential breakdown occurs in other insects: e.g., in the cyclorrhaphous. Diptera certain larva muscles do not degenerate with the remainder after puparium formation (Cottrell, 1964), but they do so some 48 hours later. Likewise, there are muscles that degenerate in the early adult, including those of the ptilinum and other muscles expressly involved in the process of eclosion. The ptilinum muscles degenerate following the sequence of events that starts with cessation of digging. Similar controlling mechanisms to those shown by Lockshin and Williams for the lepidoptean may apply in these and other cases. They have concluded that the initiation of cell death in the muscle cells is brought about by the coordination of several factors-humoral, nervous, and an ability of the cell to respond to the factors. Thus, the breakdown of certain intersegmental muscles in the adult of first potentiated by exposure to the molting hormone ecdysone during the first few days of adult development. There weeks later, when development is completed, actual breakdown is triggered by a neural mechanism. The muscle cells attain the capacity to respond to this during the three days before emergence.

Johnson (1959) has shown that where degeneration of flight muscles normally occurred in adult aphids this was prevented by decapitation. Further, the thoracic muscles of decapitated aphids broke down when they were joined in parabiosis with intact aphids. The complete story may well be similar to the postadult muscle remains to be done on this subject. It is an important biological problem because the whole question of "programmed cell death" is involved.

The problem of muscle dedifferentiation and subsequent redifferentiation are equally interesting. Why should the muscles of *Rhodius* (Wigglesworth; 1956) dedifferentiate between molts, and shortly before molting redifferentiate contractile muscle fibers? The same problem is posed in the metamorphosing pupa by the larval muscles that the transformed into adult muscles, constituting the 39 percent of the total thoracic musculature in the adult honeybee (Daly, 1964).

Reorganization of the alary muscles in the fly pupa presents a similar problem. Another problem is posed by muscles that develop in the pupal stage and degenerate before emergence. The functions of muscles of this type are discussed at length by Daly for the honeybee, and their possible involvement in determining shape is also considered.

Hinton (1961) has described how the disposition of the indirect flight muscles of insects can be brought about at the time of the larval-pupa molt by activity of the epidermal cells. He demonstrates this with *Simulium,* which as a generalized mematoceran member of the Diptera has (unlike the more specialized Cyclorrhapha) its epidermal cells carried over from larva to adult. The indirect flight muscles at the time of the larval -pupal molt are undifferentiated strands, attached to the epidermis but in relative position quite unlike the relative positions of the corresponding future adult muscle. He concludes that the definitive shape of the adult mesothorax and the final position and orientation of the skeletal muscles are determined by the way the epidermis grows between the time of secretion of pupal cuticle (larval-pupal molt) and the time that the larval cuticle is shed (larval-pupal ecdysis).

The Controlling Centers of Metamorphic Change: Endocrine Glands and the Nervous System

A brief look will be taken of the hormone system that controls the metamorphic changes in the various organs. As, was seen earlier, the endocrine system is essentially similar in all of the groups of insects. The corpus allatum secretes juvenile hormone, the prothoracic (ecdysial) gland cells (and the ventral gland cells, their homologues in the apterygote thysanurans) secrete ecdysone, the molting hormone, while the corpus cardiacum, which has been assigned various functions, contains distinct glandular and storage region. The cephalic endocrine system in the Thysanura Apterygota is seen to resemble closely that of the generalized pterygote (Watson, 1963). It is currently thought that the brain hormone, secreted by neurosecretory cells of the brain, stimulates the prothoracic glands as the principal target organs. The secretion of ecdysone in the presence of juvenile hormone in the winged pterygote insects produces a juvenile from; the secretion of ecdysone in the presence of a reduced titer of juvenile hormone initiates metamorphosis, so that at the time of the last larval molt in an endopterygote insect, a pupa is formed instead of a further larva stage. With relatively few exceptions it has been shown that after emergence the corpus allatum again becomes active, and the hormone secreted stimulates the deposition of yolk in the female.

Metamorphosis of the Nervous System

Along with the endocrine system, the nervous system is the controlling center of the body's activities, so that changes in the nervous system are fundamentally concerned with metamorphic changes in other systems. It has already been seen for example, that the nervous system plays a role in the process of differential muscle breakdown. Also, the neurosecretory cells of the brain are virtually the controlling agents of the endocrine system: through regulation of the release of brain hormone molting is controlled; through juvenile hormone maturation is controlled (Schneiderman and Gilbert, 1964). What then happens to the nervous system itself during metamorphosis? As was noted earlier in reference to Power's work, surprisingly little is known about metamorphosis of the insect nervous system at the individual cellular level. Neurometamerphosis has been the subject of study recently by several workers, e.g., pipa and both for Lepidoptera. Many older workers considered the fate of individual nerves; evidence is extensive, although often contradictory, and Wigglesworth (1965) should be consulted for major references on the subject.

For *Drosophila* along (Bodenstenial,1950) there is no detailed account of nervous system changes. What happens to the paired lateral nerves of the larva? Do these disappear in the pupal stage? Are the adult nerves all new developments? How is the relatively longer and narrower ventral thoracico-abdominal nerve mass of the larva converted into the shorter, broader thoracic mass of the adult where the whole abdominal center is reduced to the small posterior region? An answer to some of these questions will be attempted below in discussing metamorphic changes in the cyclorrhaphan nervous system.

Power (1952) drew a graph for the *Drosophila* nervous system development in the pupa, plotting volume increases in the components of the central nerve mass, of cells relative to fibers. He concluded that there was essentially little cell breakdown, but that there was a great increase in neuropile. Clearly, where larval organs break-down their nerve supply is no longer required. Conversely, where new adult organs are formed, a new nerve supply also has to be available. Do the old nerve fibers degenerate? Are the old nerve fibers taken over by other tissues? Do new neurons come into being in the pupa? Do the larval neurons dedifferentiate, their nerve processes degenerating, and later redifferentiate and form new nerve fibers? These are problems touched on but not completely solved for any of the insects in which changes are extreme. The field lies open for investigation at the light and electron microscope levels. Of course, the problem is far more extreme in a case like the Cyclorrhapha than in, say, the Diptera Nematocera such as mosquitoes or crane flies, where most tissues and organs are carried over more or less unchanged from larva to adult. In the mosquito, it is said that central nervous system grows by cell multiplication in the larva, with very slight increase in cell size. Only slight histolysis occurs in the pupa, and in the adult there is again cell multiplication (Trager, 1973). However,even here we will have the large flight muscles developed. How is the new innervation to these developed? How do nerve fibers grow out of and into (sensory) a central nerve mass that is inverted in a thick neural lamella? Breakdown of the connective tissue sheath would seem to be one prerequisite for this (Whitten, 1962a, Ashhurst and Richads, 1964; Pipa and Wooleve, 1965).

General Metamorphic Changes in the Cyclorrhaphan Nervous System

Instead of complete breakdown and reorganization of the peripheral nervous system there is actually continuity in the *basic* framework from larva to adult (Whitten, 1965). At the time of puparium formation there are numerous mitoses among the sheath cells of the lateral abdominal nerves. Similar mitoses are present in the recurrent nerve from hypocerebral ganglion to the midgut junction, and the mitoses are paralleled in the string of cells (mesodermal gonad duct, rudiment) arising posteriorly from the testis and ovary. This burst of mitotic activity is followed by the breakdown of the sheaths, thus allowing cell growth and migration to occur. In the ventral nerve mass, growth occurs along the region and the net result is a dramatic sweeping backward of the abdominal nerves, aided by a general contraction in length of the abdominal nerves themselves (brought about by the shortening of individual sheath cells now that the sheath has gone). In this way the sheath cells, be cell division, growth, and migration, lay down the skeleton of the adult nervous system. This system is basically the same as the larval in general plan, but it differs in the details of individual branching. Differences are most extreme in the thorax where tremendous growth of the sheath cells, from the ventral nerve mass, form the framework of the nerves to and from the flight muscles, legs, wings, and halters. The *external framework* is thus laid down.

Internal to this, individual nerve fibers may degenerate and new fibers regenerate. Evidence so far suggests this and also that there may be dedifferentiation and redifferentiation of fibers of individual neurons. Pipa and Woolever, using electrons microscopy, observed little evidence for "axoplasmic degradation", but the situation in the lepidopteran nerve cord may well be less extreme than in the cyclorraphan system. Muscles breakdown and tracheal histolysis are certainly less extreme in the former group of insects. Early in pupation, at the time of breakdown of the larval neural lamella, there is a concurrent breakdown in dorsal nerves and in the larval abdominal ganglionic trachea- the reduction in volume of the abdominal center is such that these trachea could not even be accommodated in the adult nerve mass. One of the most spectacular observations, which will be discussed in connection with the tracheae *before puparum formation* occurs do not produce pupal cuticular linings. Their fate is already determined at this time. They are empty tracheal tubes, their cells destined to die why these? If apparently not innervated, why do they die and other trachea in the same hormonal milieu survive through to the adult? What changes occur in the abdominal nerve mass that can cause such drastic reductions in respiratory requirements? Late in the pupal stage,_when the processes of growth and differentiation are completed, the new adult nerve sheath is laid down. In the meantime, new adult trachea have been formed, including the air sacs. Their sheaths are formed at the same time as, and are confluent with, the adult neural lamella. These events have been followed in *Sarcophaga bullata, Phormia regina* and *Drosophila melanogaster* and are essentially similar in all three. It may be mentioned that staining for neurosecretory material in sections of the nervous system of *Sarcophaga* has revealed the presence of dorsal neurosecretory cells throughout pupal life. They are always gorged with PAF staining material, which does not imply that they are active but does show that they pass through unchanged from larva to adult.

Nerves and tracheae to organs destined to disappear in the pupa, breakdown; new branches grow out at a time when the innervated organs themselves are still developing What stimulates the sheath cells of the abdominal nerves to undergo mitosis? The number and location of these mitoses determines the future pattern of the abdominal nervous system. How do the nerves at this stage know where future adult structures will develop and require innervation? This brings us into the realm of cell differentiation and pattern formation, problems that will be considered in more detail later. The coordination of the different metamorphic processes is amazing, as is the realization that all these events occur within, in response to, and at speeds determined by light darkness cycles such as those shown by Harker for *Drosophila* (1965a and b).

Hemocyte Activity in Relation to Metamorphosing Tissues

Older works on the hemocytes were largely concerned with the role of blood cells in relationship tissues. More recently, work has centered more on hemocyte identification, origins, and their relative numbers at various stages, together with their role in coagulation. The vast amount of work up to 1962 is reviewed by Jones (1962). Since that time renewed interest has arisen in the part that blood cells may play in 1962, described briefly the fate of blood cells during the life history of the alder fly, *Sialis lutaria.* Here, phagocytes were found to ingest all particles freed in the hemolymph as a result of the cytolysis of the cells of the fat body, and in the gills the epidermal cells are ingested by the blood cells also only after cytolysis. The phagocytes or morula cells, as they are later called, were said to disintegrate immediately before darkening of the cuticle. Ashhurst and Richards (1964b) described adipohemocytes in the vicinity of the nerve cord at the time of breakdown of the sheath in the lepidopteran *Galleria Mellonella.* The participation of these cells in the breakdown of the neural lamella was subsantiated by autoradiographic studies by Shrivastava and Richards (1965) and electron microscopic observations by Pipa and Woolever (1965) on the same species. The latter demonstrated the presence within the adipohemocyte of neural sheath fibrils These blood cells were not found to be involved in formation of the later adult nerve sheath.

In the Diptera Cylorrhapha had been reached by Whitten (1964b) in an investigation of the blood-tissue relationship primarily in *Sarcophaga bullata*, with observations also on *Phormia regina* and *Drosophhila melanogaster*. It was found that phagocytic hemocytes underwent very complex activities at the time of puparium formation. They were responsible for the ingestion of all tissue fragments in the hemocoele; they were also located in such consistent and characteristics positions in the developing pupae as to suggest that they participated in the transfer of materials to the developing tissues, particularly hypodermal cells of the general body surface and internal tracheae and muscle tendons. As with the lepidopteran nerve cord, these blood cells were not found closely associated with the developing adult neural lamella. The blood picture is outlined. The sequence of events for cell type 5 shows nuclear and not cytoplasmic division to form enoromous phagocytic bodies, which then over an extremely short period of time ingest larval fragments, including whole nuclei. These cells correspond to the granular hemocytes of Jones (1962), and in their later gorged form-as"spherules"–they correspond to the "spherule cells" and "spheres of granules" of earlier workers. These cells have recently been followed at light and electron microscope levels (Whitten, 1968) through the pupal to the adult stage. Their continuing close relationship with the developing adult tissues has convincingly shown the importance of the cells not only in removing larval fragments but also in processing and passing on materials to the developing adult tissues.

By the time the adult emerges the blood cells contents are depleted. These cells undoubtedly correspond to the type F hemocytes described by Crossley for *Calliphora* (1964). In a later paper (1965), Crossley described breakdown of type F cells in the midpupa, whereas it will be shown below in a study of co-ordinated development in the fly foot that the corresponding cells in *Sarcophaga* pass through into the adult stage, functioning even at the time of emergence, in the ingestion of selected epidermal cells.

We have now reached the stage in studies on insect hemocytes where it is clear that phagocytic blood cells play as important a role in tissue destruction as phagocytic blood cells do in a metamorphosing amphibian. There is also no doubt what so ever, in the cyclorrhaphan dipteran pupa, that the phagocytic blood cells can and do play an extremely important part in providing" processed larval tissues" as raw materials for the developing adult cells. Blood cells have been variously described as giving rise to adult tissue. Attention may be drawn here, where the possible origin of the cyclorrhaphan adult fat body cells from the phagocytic hemocytes is suggested. This suggestion is not new, having been made by several of the older workers at the turn of the century.

Electron microscopy has substantiated, rather than refuted this suggestion. As with other systems, there is a vast source of information awaiting investigation with modern techniques. There is no evidence to suggest that phagocytic hemocytes attack normal, healthy insect tissues. The beginnings of focal degradation in foot pad tenant hair cells. This occurs before any attack by the hemocytes which only subsequently ingest the pycnotic tenent nuclei and surrounding cytoplasm. Figure parallels closely micrographs of vertebrate tissue in which focal degradation is occurring (e.g., Swift and Hruban, 1964).

Lysosomes

The cyclorrhaphan phagocytic hemocyte provides ideal material for study of lysosomes origin and fate and function. DeDuve's article on the lysosome (1963) describes lysosome activity with respect to the tail resorption and tissue breakdown in the metamorphosing frog. A similar story can be given for lysosome activity in the cyclorrhapphan phagocytic hemocyte. Figure 3.10 shows a typical picture of different stages of degradation of *larval* fragments in the vicinity of the nucleus in a day four (at 25°C) pupa. Acid phosphates has been demonstrated in these cells and in their apparent homogoues in *Calliphora crythrocephala* (Crossley, 1964) This is particularly good material for such studies of lysosomes in that within a time span of 12 days at 25°C these blood cells can be followed at all stages of their various phagocytic activities.

The Reproductive System

Since the reproductive system matures in all insects regardless of whether they undergo metamorphosis or not, considerations of general development of the reproductive system are not directly relevant to the present discussion. However, certain aspects bear on the subject. In the apterygote ametamorphic insects, the reproductive system is laid down at the time of hatching from the egg, and maturation occurs later, although even in the Apterygota, genitalia may be undeveloped at the time of hatching and develop in later instars. In insects with complete metamorphosis, the larvae may hatch with the reproductive system in an embryonic state. Gonads are present but the ducts are limited to the mesodermal elements that extend from the gonad distally, to end posteriorly in a coelomic sac or ampulla, located in different segments in male and female.

During metamorphosis, development of the reproductive system is completed when the genital imaginal rudiment develops both the external genitalia and the ectodermal elements of the internal reproductive system, including median ducts, accessory glands and spermathecae. When growth is completed and ectodermal and mesodermal ducts have met and fused the breakdown of cells and continuity of the ducts to the outside is established. In one of the best-known insects, *Drosophila*, onset of metamorphosis in characterized by immediate and rapid growth of the imaginal discs, including the genital disc. In this same example an anomaly exists in that the ectodermal component is considered to give rise to the entire duct system right to the gonad. This anomaly has recently been doubted in the higher fly *Sarcophaga*, where the mesodermal ducts have been shown to be present in the larva in a typical embryonic state. The ducts themselves represent true examples of "imaginal rudiments" (Whitten, 1965). The onset of papation is characterized by extensive mitoses in this strand of cells.

Subsequent dissolution of the connective tissue "basement membrane" and shortening of the strand (closely similar to events in the peripheral nerves) produces the proximal mesodermal genital duct in both male and female. This meets up and fuses with the ectodermal components derived from the genital imaginal disc. Thus *Sarcophaga* falls into line with other endopterygote insects in which metamorphosis involves passing from the embryonic to the adult condition. It would be surprising if *Drosophila* is really exceptional in this respect. During metamorphosis, extensive changes occur within the reproductive system, the Degree of maturation attained by the time of adult emergence varies from male and female and from species to species. In many cases gametes are fully developed at the time of emergence of the adult insect. Review articles of King (1964) and Telfer (1965) provide interesting and instructive reading on this subject.

The Epidermis in the Metamorphosing Pupa

The epidermis and its derivatives have been left for consideration until last because in many respects this is the most important and interesting cell layer to the insect development biologist. Included here are the imaginal discs which are essentially epidermal sacs containing nerve and tracheal endings along with the hemolymph and contained blood cells. Strictly speaking, the ectodermal region of the reproductive system and of the gut also fall into this category, but they have already been considered briefly along with the rest of these systems. The tracheal system is an invaginated extension of the general epidermis and will be considered first.

The Tracheal System

The tracheal system is composed of two cell types: the tracheal cell and the tracheolar cell. Essentially, the difference between a trachea and tracheole is that the former is formed as a layer of cuticle by the tracheal epithelium, while the tracheole is terminal and formed intracellularly by a single cell. At one time it was thought that trachea possessed the spiral o annular thickenings, whereas the tracheoles did not. However, electron microscopy has shown that tracheoles do also. A physiological distinction made between tracheae and tracheoles is that little gaseous exchange occurs through the tracheal wall, exchange being mainly by way of the tracheoles at the tissue level. It will be seen below that the metamorphosing pupa may be an exception to this generalization. The pupa has respiratory problems that are unique in that a single system laid down at the larval-pupal molt has to tracheate tissues that are destined to disappear during the pupal stage, tissues that persist through from larva to adult, and more significantly, yet other tissue that develop during the pupal stage and are not in existence at the time that the tracheal system will be taken up following a general survey of the development of the system from first instar to adult. Details can be obtained from Buck (1962) and Miller (1964) on the large amount of work that has been done on the tracheal system; however, problems of development are relatively unexplored.

In general, it has been thought that at each molt the tracheal lining is replaced but that the tracheolar lining is retained throughout the life of the insect. However, the situation is much more complex than this, particularly in the metamorphosing insect where drastic changes are occurring in other systems. A schematic drawing of the types of development that can occur in individual tracheae. In general, the insect tracheal system is divided into 10 - pairs of tracheal metameres, occurring in segmentally repeating order and, in the pterygote insects, joined laterally and longitudinally at junction points of *tracheal nodes*. The original segmental pattern can be seen at each molt when the continuous tracheal tube lining "break" at the tracheal nodes, and each segment is removed through its respective spiracle. This occurs even in forms in which the spiracles are physiological closed and not functional as far as respiration is concerned. The degree of change or growth in he tracheal system from instar to instar is determined by the physiological requirements, so that as far as the metamorphosing pupa is concerned, there is a direct correlation between the degree of metamorphic change in, say, the muscular system and that in the tracheal system.

The pattern of development in a generalized dipteran, and members of the nematoceran Bibionidae would follow this closely. Here the larval basic pattern is taken right through from larva to adult, and the only drastic change is in the thorax where the tracheation to the new flight muscles, wings, halters, and legs is superimposed on the larval pattern by growth of new tracheal branches at the end of the last larval instar. In contrast, the sequence of events i . *Drosophila melanogaster* where there is considered change from larva to adult, correlated closely with the drastic changes in other systems and with the virtual reduction of the pupa to an embryonic system. At the larval-pupal molt the posterior half of the larval system does not secret pupal cuticle and is subsequently histolyzed as are the ganglionic trachea.

Later in the pupal stage there is new growth of adult tracheal epithelium in this region, but the last metamere is nerve replaced. Between these two extremes there are all degrees of relative histolysis in the abdominal system, and these are highly consistent within dipteran groups of close affinity. This whole subject is an intriguing and probably as important in terms of the control of cell death as are problems of muscle cell death, particularly as tracheae are generally considered to be noninnervated. Without a nerve supply, why would an individual tracheae or a whole tracheal segment cease to secrete cuticle and the cells subsequently die when its neighbour remains and functions normally? Perhaps the answer lie in physiological changes in the organ that is tracheated and destined to die. However, the converse cannot apply to the development of new tracheal branches to tissues that are not yet developed in the pupa. It has been said that the elaborate tracheal development of the thoracic region of adult systems bears little relationship to that of the earlier larrval system. This is not so. From a survey by the author of the development of the adult system within some 20 families of Diptera, it is evident that the very elaborate system of tracheae or air represents branching of the tracheal epithelium of the earlier basic larval pattern. Superficially, they look very unlike each other and their corresponding larval systems. However, close inspection shows the same basic pattern of transverse and longitudinal elements, with very constant positions of origin of leg, wing, and halter tracheae. No comparable survey is apparently available for other endopterygote orders.

Some Unique Characteristic of Pupal Tracheal Systems

The pupal system of the cyclorrhaphous Diptera, including forms like *Drosophilla, Sarcophaga* and *Calliphora,* is characterized by grape like bunches of tracheoles suspended in the hemocoele. These arise by mitosis of cells of the tracheal system, and after the larval - pupal molt there is often extension of the tracheal cell cytoplasm toward various organs, including the developing epidermis. The pupal stage has been seen to be characterized by the absence of basement membrances : there is no penetration of individual tissues by pupal tracheoles which lie essentially within the hemocoele. Some of these tracheoles. They are particularly interesting in that respiratory exchange is apparently occurring between the tracheole and the hemolymph.

The lining of cytoplasm, particularly on one side, is infinitesimally thin and so is any basement membrane. These pupal tracheoles are seen to move in later pupal stage to come in closer proximity to developing adult organs such as the reproductive system. For example, an early pupal testis or ovary has no tracheation, whereas later pupal gonads have functional pupal tracheoles associated with them, as well as the as yet nonfunctional adult tracheae and tracheoles. The condition in the Diptera Cyclorrhapha is extreme, but the same condition prevails in at least the thorax of more generalized forms where there is extensive development of muscles in the later pupal stages. The situation diagrammatically represents in an organism such as *Drosophila.* In the present case it is the tracheal cell cytoplasm that extends towards the tissues, whereas in the bug *Rhodnius,* Wigglesworth (1959a) described the drawing of tracheoles, by epidermal cell extensions, to areas requiring oxygen. Epidermal cells at quite considerable distances away would send out processes towards tracheoles. "capture them", and draw them to the deprived area. By comparison with other epidermal cell types the tracheal cell shows remarkable unifomity, the only distinction being between tracheal and tracheolar. Whether the cells at the tracheal nodes are structurally different seems never to have been investigated. That they are important physiologically in connection with gradients has been shown by Locke (1964).

Cells of the spiracular region are not included in this tracheal system since they are most easily thought of as belonging to the general body surface. It would be of considerable interest to understanding of cell differentiation to know whether the same tracheal cell ever produces specific larval, as opposed to pupal and pupal as opposed to adult, tracheal cuticle. Whether the air sac cuticle is specifically different in an adult is not known.

Metamorphosis of the General Epidermis

The rest of the organ systems differ form the epidermis in proceeding more gradually from larva to adult except in cases where metamorphic changes are involved. The epidermis is,however, characterized by periodic molting and secretion of new cuticular layers, and often the cuticular layers of different instars are characteristically different from those of others instars. Thus, newly hatched *Thermobia* (firebrat: Thysanaura, Apterygota) that undergoes no metamorphosis during development, is scaleless whereas later preadult and adult instars have scales; a fly pupal cuticle is smooth, while that of the adult consists mainly of small and large bristles and thei sockets. There has been a tendency in recent years to talk of the cuticles of metamorphosing endopterygote insects as "larval", "pupal", or "adult" and to relate the appearance of the specific type to the hormone balance between juvenile hormone and molting hormone at a particular period of development. Thus, the larval cuticle has been set to be produced in the presence of high levels of juvenile hormone and low ecdysone; the pupal cuticle is produced in the presence of high level of ecdysone and lower juvenile hormone; and finally the adult in high ecdysone and virtually no juvenile hormone.

In the case of the non-metamorphosing *Thermobia*, a piece of scaleless cuticle from a newly hatched larva, when implanted into an adult, is found to molt and prematurely produce scaled cuticle. The larva has a high and the adult a low concentration of juvenile hormone (Watson,, 1963). While there are in most Endopterygota considerable differences between larval, pupal, and adult cuticle, one cannot say, for instance, that bristles are confined to adult cuticles. A particular insect type will have characteristic cuticles at different stages, but whether a larval cuticle is ultrastructurally diagnostic has not been shown for any insect. Certainly the tracheal cuticle appears to be essentially similar at all stages, and the epidermal cells secreting this are every bit as exposed to hormone fluctuations as are the cells of the general hypodermis. That changes are not simple responses to different hormone titers has recently been shown by Krishnakumaran and Schneiderman (1664). They questioned whether perhaps the cells of the lepidopteran adult differ from cells of the pupa in response to juvenile hormone, since it was seen in experiments that adult abdomens could be induced to molt again when joined to pupae, but that the second adult cuticle, although distinctly adult, seldom produced scales. A much older adult did produce normal scaled adult cuticle.

A high level of juvenile hormone did not seem to influence the development of the adult cuticle since experimental results were similar with abdomen containing high levels of juvenile hormone (e.g, *cecropia, cynthia*) and others containing low levels (*polyphemus).* Krishnakumaran and Schneiderman quote a comparable phenomenon from the work of Piepho and his students, who showed that when fragments of adult cuticle were implanted into larvae they did not secrete larval cuticle immediately but required several molts before doing so. It may be of interest to observe in this context that the cells of the imaginal discs in the fly (which will be seen below to be determined but not differentiated), secrete first pupal cuticle and subsequently undergo extensive mitoses followed later by differentiation into adult cuticle secreting cells. These disc cells in an adult host continue mitoses indefinitely, but implanted into a metamorphosing larva they first secrete pupal cuticle, then further divide, and finally differentiate and produce adult cuticle.

Metamorphosis of the Epidermal Cells

The degree of changes occurring in individual epidermal cells varies from insect to insect. In the more generalized endopterygote or even the more generalized members of the specialized orders, e.g., the Nematocera of the Diptera, the epidermis passes more or less unchanged from larva to adult. On the other hand, in the higher flies, the Cyclorrhapha, there is said to be a complete replacement of larval epidermis by adult epidermal cells that originate from areas of undifferentiated imaginal cells–those of leg, wing, halter, genital, labial, antennal and eye discs, and paired groups of cells segmentally arranged on the abdominal segments.

The pupal cuticle is secreted by the larval epidermal cells of the abdomen and small areas of the thorax, and the remainder is secreted by the undifferentiated cells of the different imaginal discs. Only in the cyclorrhaphan tracheal system might it be possible to follow individual epidermal cells through from larva to adult, whereas in the Nematocera this should be possible for general epidermal cells also. The metamorphosis of the epidermis may best be looked at now from the viewpoint of "imaginal disc" development, since a tremendous amount of thought provoking and fascinating work has been done on this subject. Following this, the fate of a few individual cells can be traced through from the time of the larval -pupal molt to the time of adult emergence. This will demonstrate the complexities on the one hand and the great potential on the other that the metamorphosing insect epidermal cells have in helping to solve some of the most puzzling of biology's and life's problems–those of nuclear-cytoplasmic relationships and of cell and tissue differentiation. The insect epidermal cell has been the fascinating subject of several articles including that of Stern (1954) and of Wigglesworth (1959b). There have been new developments involving the epidermal cell, that include gradient studies (Locks 1964) and the discovery of giant polytene chromosomes in the dipteran epidermis: these, along with the fact the epidermis is undoubtedly the main "target organ" of the molting hormone ecdysone, give the epidermal cell greater importance than ever.

Metamorphosis of the Imaginal Discs

In the more generalized Diptera the imaginal discs of wings and legs lie beneath the epidermis and are not withdrawn into the body as are those of the specialized Cyclorrhapha. Also, the cavities connecting with the general epidermis are never closed and the eyes do not develop from invaginated buds. Developing pigmented eyes can often be seen, as in the mosquito, through the semitransparent head capsule of a late larva. The imaginal discs of a cyclorrhaphan fly include genital and leg discs, genital and first and second leg discs, wing, halter, labial, eye and antennal. The groups of cells between duct and larval salivary gland, those at the junctions of foregut and midgut and midgut and hind gut, the mesodermal genital duct rudiment of *Sarcophaga*, and the hypodermal discs of the abdominal segments, are also imaginal rudiments.

It is difficult to draw a hard and fast line of distinction between some of these cell groups and those associated with individual systems, such as the tracheal cells that will suddenly at puparium formation undergo intense mitosis to produce the bunches of specialized pupal tracheoles, or the sheath cells of the peripheral nerves, which presumably functioned in the embryo to secrete the larval nerve sheath and now at puparium formation are stimulated into intense mitotic activity to give rise to the cells that form the skeleton of the adult peripheral nerves. Studies on the leg or wing or genital imaginal discs, are essentially studies on epidermal cells line. Embryos of endopterygotes that undergo complete metamorphosis, such as *Drosophila*, possess two cell types. Firstly, there are those cells that are determined to form the larval body and, in these, differentiation immediately sets in. Secondly, there are those cells that are determined but are set apart and remain in an embryonic state until the time of pupation (Hadorn, 1965). Cells of the imaginal discs come into this category. They do divide and the discs grow considerably in size from their earliest appearance to the end of the larval stage, but they remain undifferentiated.

Due largely to the classic earliest works such as those of Bodenstein using transplantation techniques and more recently the fascinating works of Hadorn and his collaborators, some amazing phenomena have come to light with respect to potentialities of imaginal disc cells in the metamophosing insect. It would seem that the cells of these discs are already determined for "leg" or "wing" or "eye" early in embryonic life, as early as the first to fifth hour, at a time when the imaginal discs are not yet visibly set apart within the still apparently homogeneous blastoderm. There is even determination for particular areas of these discs. There is *determination* at this time but no *differentiation.* This occurs only at metamorphosis. As stated by Hadorn, the change from undifferentiated to differentiated state in holometabolous insects is controlled by hormones. The juvenile hormone somehow in the larval instars blocks any differentiation; a drop in the level of juvenile hormone at the onset of metamorphosis results in a "deblocking" of readiness for differentiation present in the discs thoughout larval life. *Perhaps the juvenile hormone acts as a repressor of the sets of genes needed in differentiation* but here Hadorn warns that much more information is needed before the happenings that clear the path for differentiation in a determined cell can be interpreted in molecular terms. Until details are clearer of the responses by particular cells of known fate and history, it would be good to extend such reservations to work on hormone action, particularly with respect to the influence of the balance between juvenile hormone and ecdysone on the production of specific "laval", "pupal", or "adult" cuticle.

The same reservations may be extended to the effects of hormones on chromosomes puffing sequences. The results with larval salivary glands cannot necessarily explain hormone action on tissues that will undergo metamorphosis. The exciting results of Hadorn and co-workers involve the maintenance of imaginal discs within adult hosts, where they increase in size by extensive mitosis but do not differentiate, Removal from the adult and transplantation into a metamorphosing larva brings about cessation of cell division and differentiation into adult structues. Of particular interest is the process of "transdetermination" whereby after some eight or nine transplants (from one adult host to another) the disc when allowed to determination" whereby after some eight or nine transplants (from one adult host to another), the disc when allowed to differentiate in a metamorphosing larval host, differentiates into an adult structure different from the original implant. Thus, a genital disc after many transplants may give rise to leg or wing or thorax. Transdeterminations occur in the probability order of genitalia, antenna, leg, palpus,wing, and thorax the latter occurring in the oldest transfer generations.

Reversals were also found. No explanation can as yet be given for these amazing results, but it is suggested by Hardon that possibly a degree of dilution reached by cell cultures might have a feedback action on gene activation. Instead of the genes that formerly determined the original organ, other genes come into play, resulting in the transdetermination. Some of the transdetermination effects are reminiscent of some Drosophila mutants such as aristapedia were a tarsus is formed on the antenna. However, most of the transdetermination effects are not known as mutants.

Metamorphosis of an Imaginal Disc: The Leg Disc

The type of development undergone by individual elements of an imaginal disc can be seen in the case of the leg. Details have been followed for *Sarcophage bullata*, the flesh fly, but development is essentially similar in other Cyclorrhapha, including *Drosophila.* Event followed at light and electron microscope levels demonstrate the complex changes and coordinated development of the different components: various epidermal cells, tracheoles, and phagocytic blood cells. Some of these events have already been discussed in relation to hemocyte activity during metamorphosis. At this point the main concern will be with the development of individual epidermal cells. The mass of thousands of determined yet undifferentiated epidermal cells (stage 1) first secrete the pupal cuticle of the leg disc. Retraction of the epidermal cells from the pupal cuticle follows and cell division continues (stage 2). Five cells forming the future end of the claw and four future giant cells become recognizable at stage 3 while cell growth and continued mitoses occur among "mother" tenent hair cells whose descendants will occupy the future ventral surface of the foot. Growth and cell division thrust the five claw cells into the tip of forming claw, and the foot pads take shape by progressive growth and median indentation. Dorsally, the two giant cells extend over each foot pad; the nuclei are characterized by giant polytene chromosomes.

The tenent cells give out very fine cytoplasmic processes, while the dorsal cells grow in width as broad, flat cells. The tarsal hair cells at this same time from, like the cells, cytoplasmic processes, varying in size according to whether a large or small hair is to be produced. Stage 4 marks the end of growth and differentiation, yet at this time no cuticle is as yet deposited. Thus, the future shape of a hair is laid down by cell growth (c.f. Lees and Picken, 1945) and is completed by the time cuticle secretion begins. This is of interest in questions of abnormal bristle development in mutant *Drosophila.* Cessation of growth and attainment of cell maturity is followed by a well-spaced sequence of cuticle deposition accompanied in the giant cell by constant and characteristic patterns of chromosome puffing activity. First is secreted an *ecdysial membrance*

(1), followed by a cuticulin layer (2), a layer of dense exocuticle (3), which later becomes darkened and sclerotized, a homogenous inner layer of exocuticle (4), and a narrow layer, of "mesoculticle" (5), each layer is indicated by an arrow. By stage 9 these sequences are completed in giant, tenent and trichogen cells. Cytolysis is now observed within the tenet cell nuclei and cytoplasm. This is seen at the electron microscope level, in which centers of focal degradation can be clearly seen.

The hemocytes, which at the time of first evagination of legs were gorged with larval tissue fragments, have subsequently been digesting them and passing on the degradation products to the ceveloping epidermal cells. These phagocytic blood cells now, at stage 9, embark on a second phase of phagocytosis. By stage 10 the phenotic tenent cell nuclei are nearly all engulfed. In fact, the exact age of the foot can be determined by three means: (a) by close inspection of the extent of the darkening process and actual shade of specific areas, (b) by close investigation of the relative degree of phagocytosis of tenent cell nuclei, and (c) by close inspection of puffing pattern of the chromosomes. All are equally exact and reliable. The tracheoles and giant cells remain unattacked at this time. Endocuticle secretion within the giant cell cytoplasm occurs after this at a time when darkening of the exocuticle is occurring over the ridges of the dorsal cuticle. Endocuticle (layer 6) secretion seems to occur by a combination of intracellular secretion and retraction of cytoplasm (and nuclei and blood cells) to the base of the foot, so that the space formerly occupied by giant and tenent cells is now occupied by endocuticle. The position of blood cells in relation to that or giant nuclei and former position of tenent nuclei.

Even at the light microscope level, association of blood cells with the giant cytoplasm can be seen. The associated becomes more close, and at emergence the giant nuclei and cytoplasm also become ingested by the blood cells. The blood cells finally come to lie as a layer over the inner border of the foot pad cuticle, at the base of the pad. There is no innervation of individual cells in the foot, though possibly a basal stimulus is received that can be transmitted from cell to cell. What determines that cell division should cease in the giant cells and cells of the future claw at a time when mitoses are still occurring among the future tenent cells? What determines the regular pattern and uniform position of the different cuticular layers in the various epidermal cells? That the layers are similar and are deposited at comparable times in tenent, trichogen, and giant cells. The biochemistry must be similar, yet the spatial distribution of the various layers is different, and this is essentially what gives each product its characteristic "differentiation". What initiates cell death among the tenent hair cells? Is it simultaneous or is there a focal point at which the process starts?

Focal degradation begins before any phagocytosis on the part of the blood cells is to be observed. Does this prompt the blood cells to attack the tenent cells and not the giant or the tracheolar cells? Interestingly, the final deposition of endocuticle by the cytoplasm of the tenent cells is carried out in the absence of their nuclei. What of the chromosome activity while these sequences of growth, cuticle secretion, and finally death are occurring in the giant cells? For the first time, we seem to have a system where one should be able to correlate closely chromosome activity and puffing, sequences with the activity of a single cell through growth, various phases of clearly defined activity, and finally cell death. Detailed analysis will be made in the vast amount of information that has accumulated as a result of work carried out on dipteran salivary glands and to a lesser extent on the chromosomes of cells of the Malpighian tubules, gut, or seminal vesicles.

The Malpighian tubules, persist through pupation but are of small dimension while the salivary glands degenerate shortly after pupation. Since epidermal cells are specifically the "target organs" for the molting hormone ecdysone repetition of much of the work carried out on salivary glands should produce particularly interesting results with foot pad cells, particularly work such as that of Karlson and his colleagues (e.g.,1965). It would be exciting to see if and how hormone changes can affect puffing activity and in what way they may alter the biochemistry and ultrascopic structure of the cell. These discussions might seem to have taken us far from considerations of the whole metamorphosing insect, but in fact is visible changes in the epidermal cells and their products that constitute external evidence of metamorphosis. What determines the difference between the various cells of the epidermis of the developing insect? They are variants of the same cell type–the epidermal cell.

When the layers of cuticle that are secreted are identical in two or more variants, what determines the spatial distribution of these cell products, and what determines the cell shape? If one could answer these questions one would be nearer to answering the problems posed by the phenomena of transdetermination of Hadorn and of the problems of patterns and prepatterns discussed at length by Sondhi (1963) and in a broader context by Waddington (1962).

It will be seen from the foregoing necessarily superficial survey of insect metamorphosis that each system in turn has its own facinating problems, which, to name but a few, include general considerations of body change, diurnal, rhythms, problems of cell determination, cell differentiation, cell death, and cell interaction. Many of the problems are common to those found in other metamorphosing animals, as will be seen in succeeding chapters. Nevertheless in the insects alone one cannot but be impressed by the infinite number of problems that the one phenomenon of metamorphosis poses to the inquiring developmental biologist.

Changes in Body Covering

The external covering of insects varies greatly in its consistance. It is generally a tough and flexible skin. The integument of some full-grown insects is almost leathery, but in the majority all the outside tissues are very thick, solid and hard. They look like horn, but their intimate composition is very different. Horn dissolves away, but the integument of insects carbonises and retains its form when exposed to great heat, and their chemical composition is different. This integument, whether it is as thin and flexible as the skin of a silkworm, or as hard and dense as the envelope of a beetle, is always composed of a particular substance, called *chitine.*

The skin is formed of two layers, one deep, soft, and not made up of *chitine,* and the other external and constituted mainly of this substance, to which are added, according to the advanced or retarded condition of the development of the insect, more or less colouring matter, fat, and calcareous salts. The deep layer is the true skin, and the superficial is the epidermis or scaft skin. It is the epidermis which is detached and moulted off during the progress of growth and development, for instance, when silkworms change their skins, during their caterpillar state. Both layers are intimately connected, and the true skin has small glands in it whose tiny ducts transverse the epidermis, and even enter the hairs. The epidermis is composed of an assemblage of very regularly-shaped cells containing colouring matter. If the skin of a caterpillar, a chrysalis, and a butterfly is examined, the marvellously beautiful cells and hairs of the perfect insect can be seen to be modified epidermal cells, whose predecessors were infinitely more simple and less elegant in the immature insect. It is the essential peculiarity of articulate animals to be divided into segments or rings, and these somites, as they are sometimes called, can be traced in the embryo within the egg, in the larva, nymph, or chrysalis, and in the perfect insect.

Huxley distinguished six segments in the head of the young insect, and he has expressed the opinion that less than fur are never found in that part. But after growth has proceeded even for a short time the segments of the head become fused, as it were, in one mass, and are no longer distinguishable. The body segments of many larvae present very slight differences, and have, therefore, a general likeness to each other; but the perfect insect is evidently furnished with three sets of segments, those of the head, body or thorax, and abdomen. The segments of the head of the larva are fused together, but those of the thorax, or chest, and of the abdomen, remain distinct. The head of the larva is readily distinguishable from the rest of the body, but as there are not feet attached to the segments of the thorax and abdomen in hymenopterous and dipterous larvae (bees and flies), no satisfactory distinction can be made between the parts of those insects. However, in most

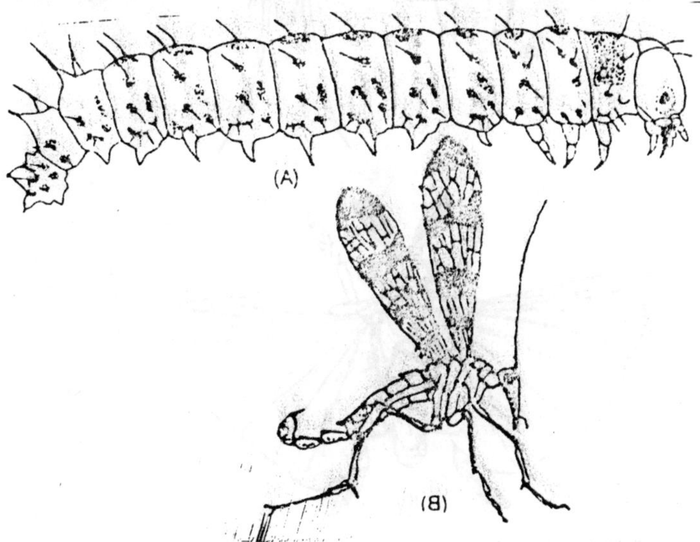

Fig. 6.1. The Scorpionfly, Panorpa–(A) larva and (B) adult (male).

caterpillars and beetle grubs the presence of two classes of legs enables the rings of the chest and abdomen to be numbered and recognised. In its most simple condition, as seen in a great many larvae a segment is homogeneous throughout, and a transverse depression along a line where the tissues are thin separates it from the others. These separations of the segments are shown in the representation of the legless larvae of the dipterous insect. In other instances there are two arches in each of the growing segments, one above and other below, and a membranous species is to be

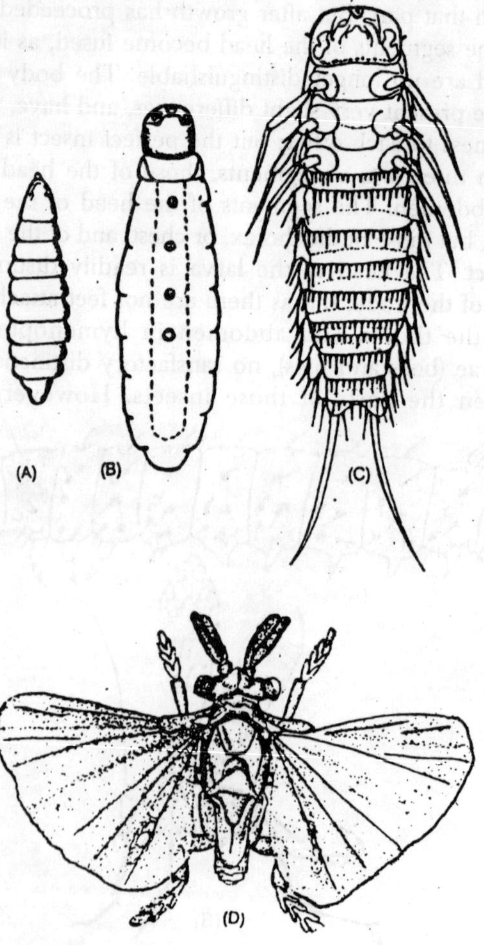

Fig. 6.2. Stylopid–Xenos (A) male larva, (B) adult female; Stylops (C) triangulin larva and (D) male adult.

observed on either side of the body between them. Moreover, each arch is formed of two side halves, and there is often a very distinct line of separation to be noticed between the portions of the arches, and also a mark which occupies this position on the head and along every segment of the body of the grub or caterpillar. This middle or median line is well seen in the immature *Calosoma sycophanta.*

The segments of the abdomen, which are not provided with true legs, appear to be at first nine in number, and there are some larvae which have twelve or more of them. Now, on comparing the segments of the larva, chrysalis, and perfect insect of the same species, it is noticed that the changes in the form of the insect are not brought about by an alteration in the consistence of the integuments only, but by a diminution in the number of the rings or segments.

It is true that this diminution is rather deceptive in its appearance, for there has not been any fading away of the segments, but really a union and a consolidation of several in one or more single and separate pieces. Thus it is not uncommon for the first rings of the abdomen to become united to the last segment of the thorax or body, or for several of them to conjoin and leave very faint traces of their former distinctness. Take as an example a kind

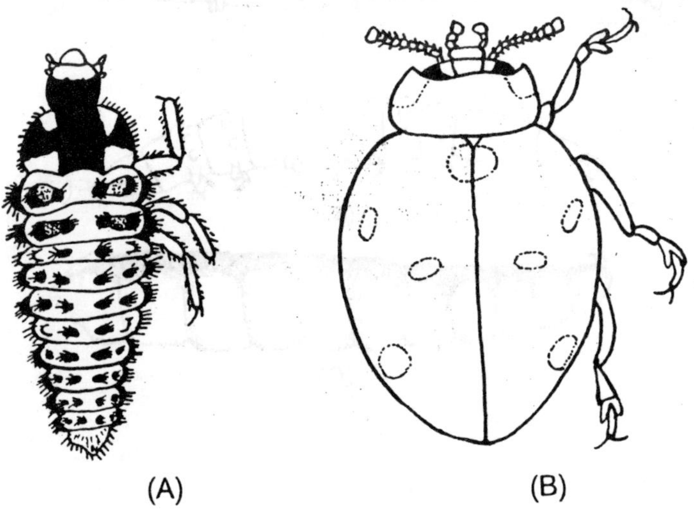

Fig. 6.3. The lady bird beetle, Coccinella–(A) larva and (B) adult.

of lepidopterous insect–a moth–whose metamorphosis, so far as regards the shape of the caterpillar and perfect insect is less than is usual. The large *Attacus pavonia major* is admirably adapted for our purpose. The abdomen of the caterpillar commences at the

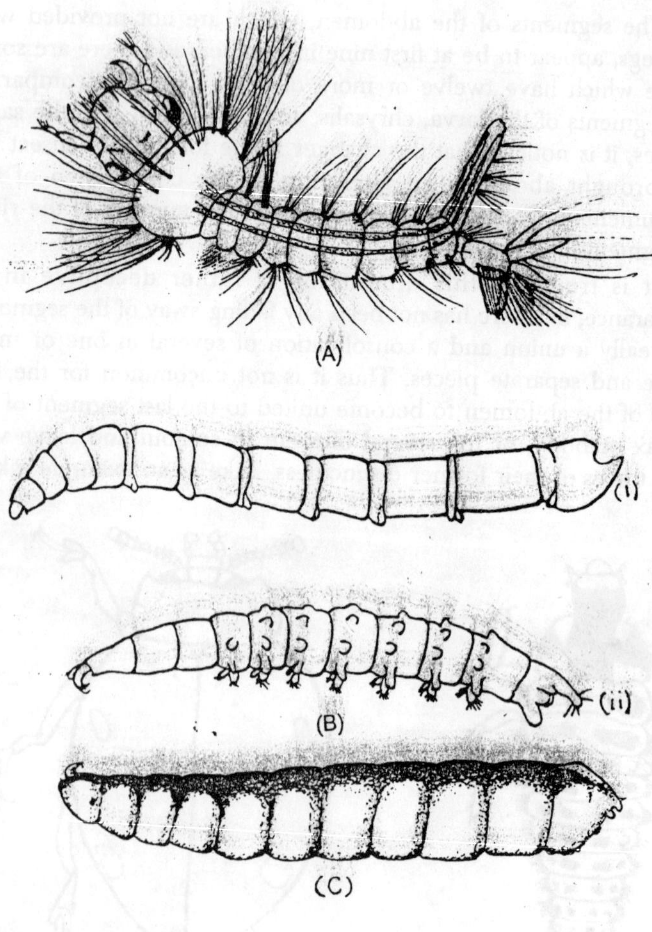

Fig. 6.4. Larvae of Diptera (A) Nematocera, Culex, (B) Brachycera (i) Rhagio, (ii) Tabanus, (C) Cyclorrhapha–Musca.

fourth segment of the body, as in all other insects, and is formed of nine nearly equal rings; the last but one is, however, shorter than the others, and the body ends in a tubercle. Their consistence is the same throughout, and the constituent membrane is almost homogeneous. The third, fourth, fifth, sixth, and the ninth segments of the abdomen have a pair of tubercles furnished with spines on their underneath or ventral surface. These tubercles, which are destined to disappear when the insect passes into the chrysalis condition, act as legs. They are not true legs, but only prolongations of the skin; but they are used as holders and locomotive organs, and under the term of membranous feet or legs are invariably present in caterpillars. Now compare the abdomen of the moth with that of the caterpillar. The number of the segments is no longer the same, and only seven can be counted.

The first has joined itself to the chest-piece or thorax; the last two are rudimentary, and have become closed in at the end of the abdomen, like the slides of a telescope. A closer examination of the inside of the moth proves that certian changes in the length and grouping of the important nervous organs have accompanied this shortening and coalescence of the abdominal segments. The change in the consistence of the integuments is very decided in the chrysalis. It is not at first very evident in the moth, but if the body of one is cleared from the hairs which clothe it, the dorsal (back) and ventral (underneath) portions will be noticed to have

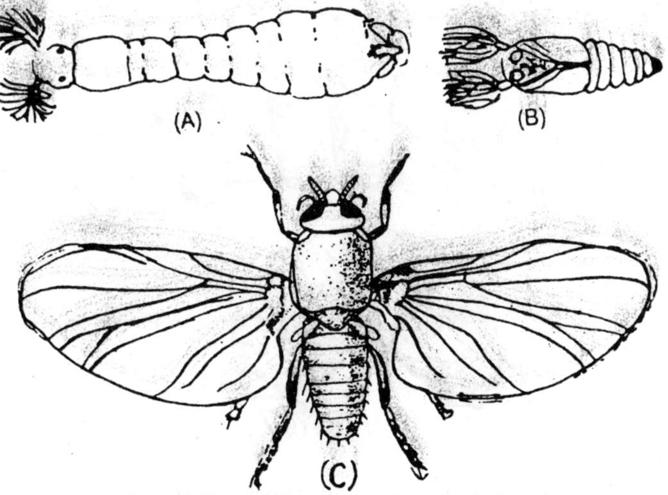

Fig. 6.5. The blackfly, Simulium (A) Larva, (B) Pupa, (C) Adult.

become coriaceous, whilst the lateral parts or the sides have retained their former softness. The vestiges of the two arches of the original segments are thus preserved.

Many insects, instead of having soft skin like the moths and caterpillars, have the integument of the abdomen very hard and strong. Such is the case with most of the larvae of flesh-eating beetles, which come out of the egg much more mature than many others, and whose abdominal segments are nearly covered with solid plates.

Let us examine the larva of a large carnivorous beetle, the *Calosoma sycophanta*. The abdomen consists, as in the caterpillars, of nine distinct segments; ten might be counted if the terminal tubercle were considered to be a ring. Examined from above (dorsal), the rings show a large back piece, divided by a groove in the middle, and on either side a small plate answering to those which are to be seen just above the junction of the legs to the thorax. It may be noticed as a fiant line between the large back plate and the margin of the body, in the engraving. The large cross piece and the smaller side plates form the dorsal arch. In the centre of the lower side of the segment two transverse series of solid plates may be noticed,

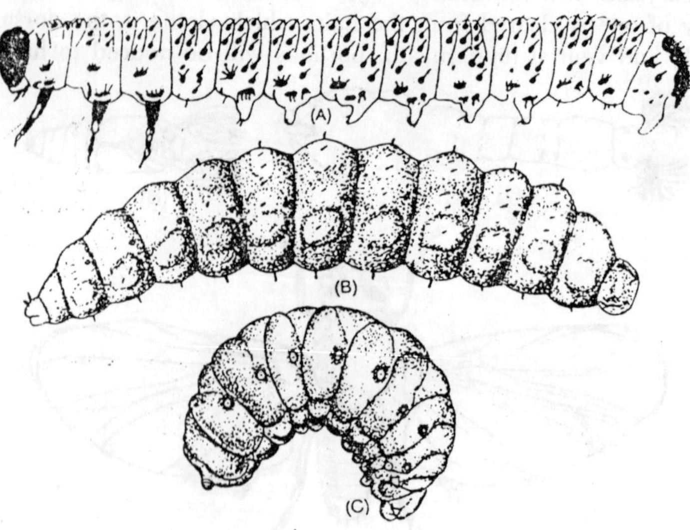

Fig. 6.6. Larvae–(A) Symphyta (Nematus), (B) Apocrita Pimpla (parasitic) and (C) Apis.

the first formed of a single large piece, the second or a range of four very small ones. By their coalescence these separate portions will eventually form the ventral or sternal lamina or plate. The simple arch of the segments of the abdomen of lepodopterous insects or butterflies is formed by a single lamina or plate, commencing from two points of induration; but the more complicated arch of the larva of the beetle results from the union of two series of pieces.

The thorax is that part of the body which gives origin to the legs and wings. In this class of insects the thorax is invariably formed by the three segments which follow the head, and the second and third rings support the wings. The presence of legs and wings involves a considerable development of the segments of the thorax, and particularly of the second and third, for there must be abundance of space within for the passage and attachement of the muscles, which, influenced by the will of the insect, move the legs and the organs of flight. The description of the parts of the thorax is not difficult to understand.

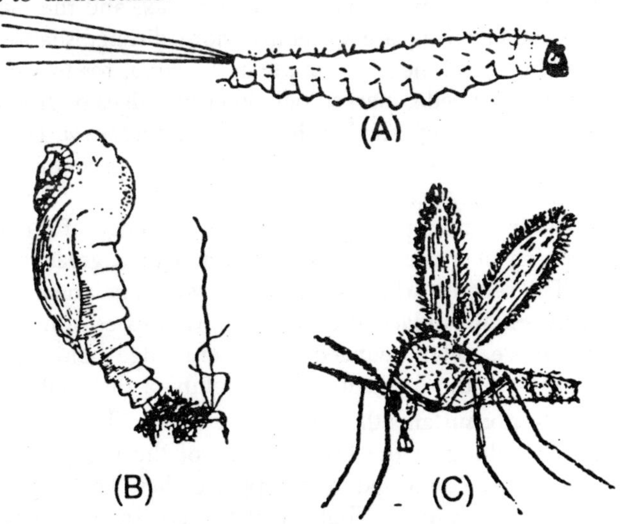

Fig. 6.7. The sandfly, Phlebotomus (A) Larva (B) Pupa and (C) Adult.

The first segment of the thorax which articulates with the back of the head is called the prothorax; the second, to which the first pair of wings is attached, is the mesothorax; and the third, bearing the second pair of wings, is the metathorax. The method of the formation of the abdominal segments is the key to that of those of the thorax. The great development of the mesothorax and metathorax does not usually take place until chrysalis life has progressed.

The larva of the *Calosoma* has its thorax well armoured. Above, the three segments are clothed–as is the case with the abdominal rings–with a double dorsal plate with the small lateal piece on either side. This is the upper arch. The dorsal plate is very large, but it does not exhibit any cross markings or any transverse division, but in the adult beetle four distinct parts are added to the two segments that support the wings. The rings of the under part of the thorax of the same larva are principally membranous. A small plate which is to be observed in the centre of each is the sternal piece. It is very small and rudimentary, but it becomes very large and fully developed in the beetle.

The mesothorax and metathorax become very large in proportion tothe rest of the body in the insects which have wings, and are often so united by growth that they cannot be separated. The mesothorax, which supports the first pair of wings, is almost always better developed than the metathorax, and the constituent parts of the first ring are generally more distinguishable than in the last. The dorsal piece, so simple in the larva, has four transverse divisions in the beetle, marked out either by ridges or grooves.They can be readily distinguished in the *Calosoma sycophanta.* These pieces are intimately united, and have been called precutum, scutum, scutellum, and postscutellum. The first is always very rudimentary, and is a little lamina placed vertically to the edge of the segment, and remains in a membranous condition in a great number of insects. The second, which supports the wings, is larger. The third, often very small, is placed between the wing cases of most beetles. The last piece, often reduced to a mere ridge behind the third piece, assists in the formation of the joints of the wings.

The thorax sustains the locomotive organs. These are the true legs which belong to the lower arches of the segments, and the wings which are connected to the upper arches. The true legs often exist in the young insect, but are then very small and scale-like, and they do not attain a considerable size until adult age is reached. They are from the first in three pairs. Those attached to the prothorax are the anterior legs; those of the mesothorax are the intermediate; and the posterior are in relation with the metathorax.

The legs upon the segments of the abdomen are, as has already been noticed, not persistent; they exist in caterpillars especially,

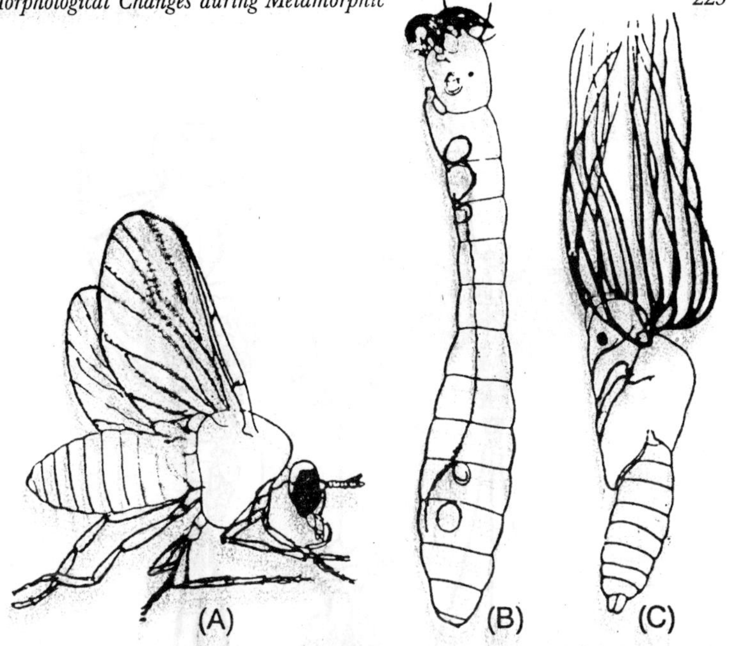

Fig. 6.8. The blackfly, Simulium (A) Adult, (B) Larva and (C) Pupa.

and are really membranous feet, being provided with hairs, spines, and hooks to hold on with, but they are lost during the metamorphosis into the chrysalis condition. The front or true legs are the curious jointed structures so finely hairy and so well provided with suckers, claws, and brushes of hair, and are mainly developed during the period of rest which characterises the chrysalis state.

The distinction between the true legs and the sucker-like feet of the caterpillar is very observable in the engraving of the silkworm larva; the true or scale-like legs are in front, and are not adhering to anything, but the membranous legs support the insect. The joint in the legs of the perfect insect nearest the body is called the hip or coxa, the next the thigh or femur; then the leg or tibia succeeds, and finally the many jointed tarsus with its hooks. It has already been noticed that the larvae of many kinds of insects are entirely destitute of legs and feet. Thus, the grubs of the hornet, *Vespa crabvo,* are legless; and yet the remarkably active limbs of the perfect insect have to be developed with all their muscles and nerves druing metamorphosis; and the apodal or legless condition of the larvae and pupae of the worker bees may be understood by examining the engraving of them.

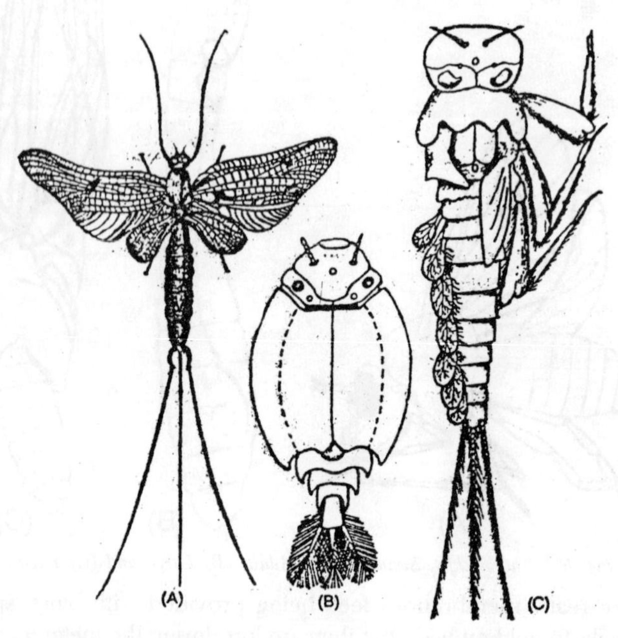

Fig. 6.9. The mayfly–(A) Ephemera adult, (B) Prosopistoma nymph and (C) Heptagenia nymph.

The legs of insects, although true locomotive organs, are often adapted to an infinitude of uses, and they are formed in numerous shapes, and present innumerable special strucutral adaptations, so that their study is really most interesting. Nevertheless, whatever may be the shape of the legs, their uses and functions, or their condition of development, they are always formed upon a definite plan, and have certain anatomical peculiarities. No better idea of the extraordinary development of these locomotive organs in the perfect insect, and of their rudimentary condition in the larva and before metamoprhosis has taken place, can be obtained than by comparing the figures of the larva and adult *Calosoma.*

The membranous feet are not fully developed in some caterpillars, and the "loopers" in particular have rarely more than two instead of five pairs of them. The absence of the central legs produces the curious appearance of these well-known insects. In

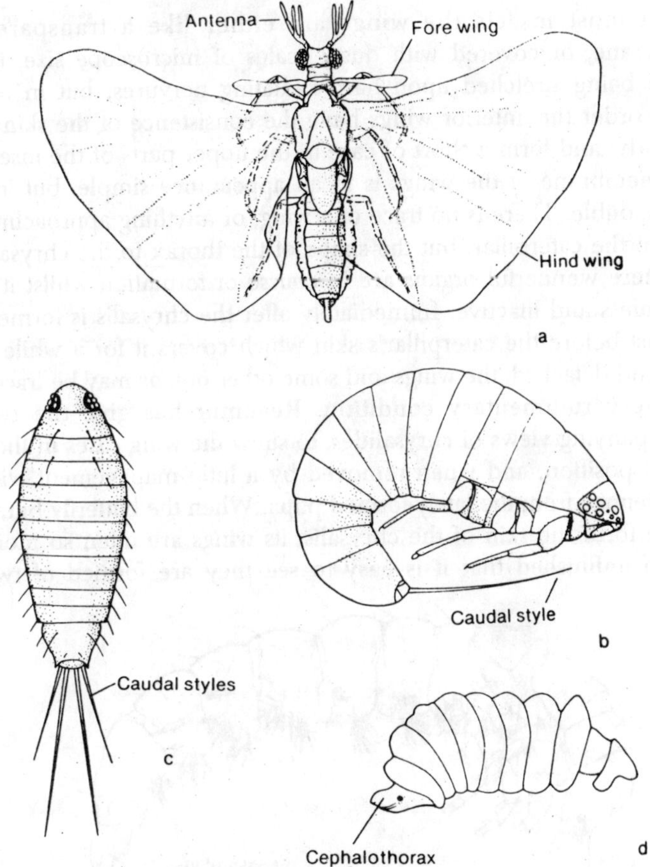

Fig. 6.10. Various instars of Strepsiptera: (a) adult male of Eoxenos laboulbenei (Mengeidae); (b) lateral aspect of first-instar larva or triungulin of Halictophagus tettigometrae (Halictophagidae); (c) dorsal aspect of first instar of Eoxenos (Mengeidae); (d) lateral aspect of third-instar larva of Halictophagus.

the engraving the fore-legs are also shown, but they are very small and pointed, whilst the two pairs of membanous legs are short and distinct.

The wings exist in the great majority of perfect insects. They are four in number, and are in two pairs: one is attached to the mesothorax and the other to the metathorax. One of the largest divisions of the class of insects (the *Diptera*) is supposed to contain species that only have two wings; but the wings of the second pair do exist, although they are in a very rudimentary conditions.

In most insects the wings are either like a transparent membrane, or covered with dusty scales of microscope size, the whole being stretched upon hard radiating nervures; but in one entire order the anterior wings have the consistence of the skin of the body, and form a short of case to the upper parts of the insect. The membrane of the wings is to all appearance simple, but it is really double. There is no trace of a wing, or anything approaching to it, in the caterpillar, but the shape of the thorax to the chrysalis that these wonderful organs are in course of formation whilst it is motionless and inactive. Immediately after the chrysalis is formed, and just before the caterpillar's skin which covers it for a while is burst and detached, the wings and some other organs may be traced in a very rudimentary condition. Reaumur has give the two accompanying views of chrysalides, to show the wing cases in their natural position, and when removed by a little management with the antennae from the lately-formed pupa. When the butterfly bursts for the form the skin of the chrysalis, its wings are often so moist and so unfinished that it is easy to see they are formed of two

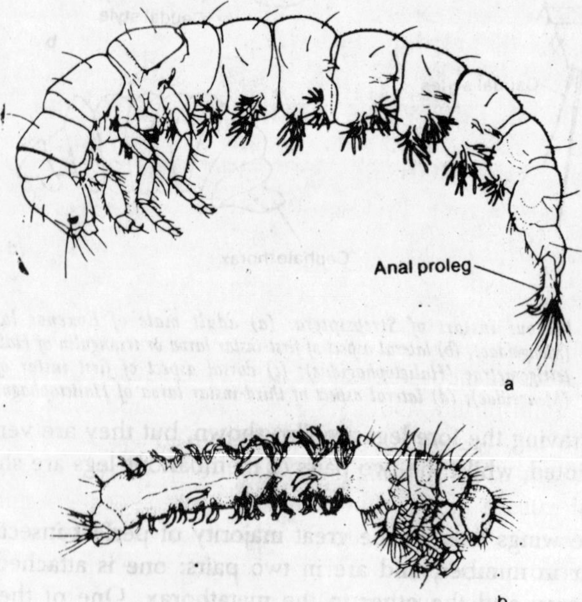

Fig. 6.11. Trichoptera larvae: (a) Rhyacophilidae (Smicridea fusciatella); (b) Limnephilidae (Philarctus quaeris).

membranes, between which run the nervures which enable the insects to move them in its flight.

The heads of insects are composed, like the other portions of the body, of several segments; but, alhough this is a fact, its demonstration has been satisfactorily completed. At the birth of the larva no satisfactory division of the head into segments can be detected; but knowing, as we do, that each pair of appendages–legs and wings–is attached to a particular segment, and observing that the lower part of the head has four pairs of attached organs–the upper lip, the mandibles, the jaws, and the lower lip–the early separation of the head into four segments is fairly inferred. Probably there were six.

There is an evident design in the manifold differences of the jaws and eating apparatus of insects, and the long tubular sucker of the butterfly refers to it quite as much as the horny crushing jaws of many hard-skinned beetles. The change from the leaf-cutting jaws of the caterpillar to the suction tube of the butterfly is a proof of this unity of plan, and that with different instincts, habits, and methods of life there arise different modifications of structures. The examination of the mouth of an eating or mandibulate insect is very instructive. Take, for instance, one of the grasshopper tribe, and examine the different pieces attached around the opening down which the food passes into the stomach.

The mouth is provided with six articulated pieces: a labrum, or the upper lip, two mandibles, two jaws, or maxillae, and an inferior labrum. The curved appendages to the maxillae and the lower labrum are palpi. Now in caterpillars the upper lip is well grown, and there are two strong mandibles working form without inwards, two jaws, and a lower lip. There is thus the clearest resemblance between the mouth of the grasshopper, when it is a perfect insect, and that of the immune caterpillar. Let metamorphosis go on, and the browsing caterpillar becomes a sucking butterfly; though the first, no would think there was the slightest connection between the early and the late condition of the mouth. A slender trunk and two scale-like feelers, or palpi below it are the only apparent structures of the butterfly's mouth. But a carefully examination reveals the fact that the three pairs of mouth pieces are all present, although in a very different condition of form. By removing the fine hairs and scales from the front of the butterfly's head, a small transverse lamina is observable, and its relative

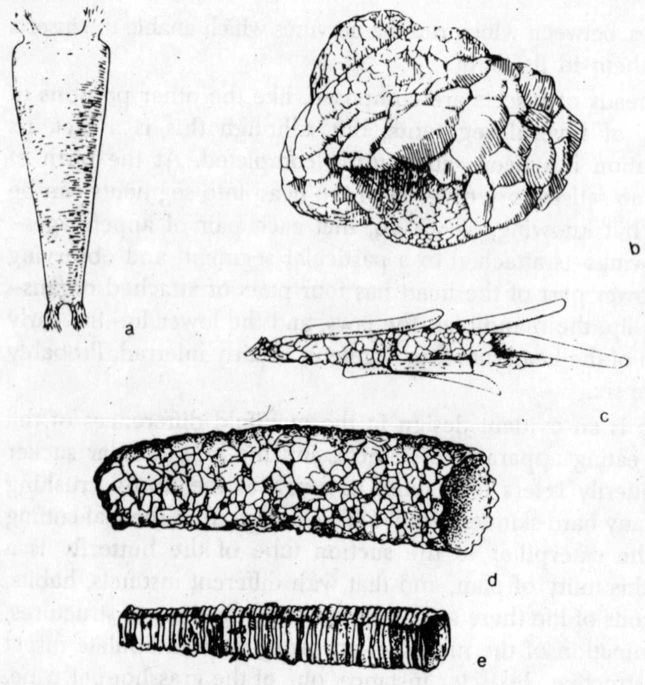

Fig. 6.12. Cases of Trichoptera larvae: (a) Hydroptilidae (Oxyethria serrata); (b) Helicopsychidae (Helicopsyche borealis); (c) Leptoceridae (Mystacides sepulchralis); (d) Limnephilidae (Pseudostenophylax edwardsi); (e) Brachycentridae (Brachycentrus).

position determines it to be the small upper lip, once so large in the caterpillar. On either side and beneath the labrum is a very small piece of skin–the remanant of the once formidable and trenchant mandible. Beneath these is the long trunk, formed by two long and flexible structures, hollow and tubular, and at their base is a pair of small palpi. The flexible tubes are altered inflexible jaws, and the palpi are accessory organs. below all is the inferier labrum, with large palpi attached to it instead of small ones.

The suctorial insects, such as the bees and flies, have transformations of the structures of the mouth quite as wonderful as those just mentioned, and they will be noticed further on; but enough has been said to show that is the nature of the metamorphosis in these organs. The other appendages to the heads of insects are the antennae, commonly called feelers, and the eyes. The eyes are generally developed rather later. Many larvae are blind, and others have immature organs of vision. But in the perfect and adult insects, the eyes are often enormous in size, and most recondite in structure. There are often two kinds of them. Some are very large, and are situated on either side of the head; these are the compound or facetted eyes. The others are found upon the head, above the upper lip, and they are simple eyes or ocelli. The compound eyes are never found in the larva, and yet it is evident that the simple organ of the caterpillar is developed into the wonderful eye of the butterfly, with its ten or fifteen thousand hexagonal lenses. The compound eyes are generally largest in the male insects, and they are frequently magnificently decorated with hairs and with metallic colours. The ocelli are often placed between the compound, and glance like diamonds.

The antennae present every imaginable shape and length, and are situated on different parts of the head in different insects. They are always very small in the larva; they have, however, important function in the perfect, and in which they attain their greatest development. One of the results of the progressive development of the insect is the addition of the organ of hearing to that of touch, which last probably exists in the small antennae of the larvae. Erichson, a German entomologist, and Dr. Braxton Hicks, F.R.S., discovered numerous depression in the antennae of fully developed insects, which are lined with a delicate membrane; and the last-named naturalist found that a small cell, made up of several others, and situated beneath the membrane, was continuous with a nerve which is supplied to the antennae. This may be the organ of hearing or that of smelling; and to which ever sense it may truly belong, it is clearly a new structure, developed after caterpillar life came to close. Having thus briefly noticed some of the modification of the outsides of insects during the metamorphosis, let us consider how the digestive, nervous, and respiratory organs are altered during the progressive evolution of the perfect insect from the caterpillar condition.

The digestive organs in insects are more less tubular, and are continuous with the structures of the mouth. They extent from one end of the body to the other, and are either short and straight or long and convoluted. Certain swellings and contractions of the digestive tube mark the principal division of it, and enable us to distinguish an oesophagus, a stomach, a small and a large intestine. Some glandular appendages, tubular in shape, complete the digestive apparatus; they are the salivary glands, the liver, and urinary tubes. The alimentary canal is formed of several layers; first there is on the outside a delicate, structureless membrane; then beneath it a muscular coating, consisting of fibres arranged across the lengthwise, and which become very dense at both ends of the canal; a mucous membrane is situated beneath this muscular coat, and its inside is covered with multitudes of epithelium cells which have to do with the production of the fluids of the digestive function and to come in contact with the food. After the food has been masticated or sucked in, as the case may be, it passes into the mouth, and then into the gullet or esophagus, being, first of all, mingled with saliva form the glands.

The oesophagus is a passage possessing very dilatable walls, and passes through the thorax in a straight line. In the insects which live on fluid aliments it is usually narrow, but in those which devour more or less solid morsels it is large and has a considerable calibre. There is, in many perfect insects, a considerable enlargement at the back part of the gullet, called the crop. It permit the insect to accumulate and keep a store or nourishing things, without digesting them all at once. This crop is very generally to be distinguished in insects which are great eaters, such as the crickets, grasshoppers, and locusts, as well as in those which make some provision for the larvae which are to come after them. Many have the power of disgorging their food from the crop; and the bee empties and honey which it has collected from its favourite flowers out of this receptacle into the cells of the hive.

Sometime the crops forms a side pocket to the gullet, and becomes almost an appendix. This is the case in the *Lepidoptera* (the butterflies and moths). Occasionally in the *Diptera* (the fly tribe) the sac or crop, when more or less separated, is provided with a long neck which opens into the gullet close to the mouth. When thus formed it has something to do with the suction of food, but the mechanism by which it acts as a sucking-pump is not exactly understood. Usually the stomach follows upon the oesophagus, but a constriction is noticed in most insects at the end of the gullet, which is made up of hard muscular tissue; it is the gizzard, and its inside is thrown into long folds more or less dense, and is often covered with hard growths which triturate the semi-masticated food before it passes into the stomach.

A gizzard only being necessary to herbivorous and general dimensions; sometimes, as in the grasshopper, it is a large and somewhat heart-shaped sac terminating in the form of a tubular canal, and its walls are thick. More frequently it is elongated, and is either smooth or covered with a multitude of such glands, as in the water beetles and other carnivorous kinds. It is the principal organ of digestion, and the food is mixed there with the gastric juice. This liquid, always acid when digestion is going on, possesses the same qualitities as that of the higher animals, and reduces the aliments into a pulp, and finally into chyme. The gastric juice is more freely secretred in some insects than in others, and is most abundant in those which live upon animal food. The glands which produce the juice in these carnivorous insects are very well developed and may be recognized as tiny blunt projections upon the outside of the stomach; they are hollow, and evolve the fluid form their cellular structures. But the glands of the herbivorous insects are not visible on the outside of the stomach, and are contained in its walls instead.

The secretion of the gastric juice is determined by the stomach being more or less filled, just as in man, and when this is the case the little glands give forth their cellular and fluid contents in great abundance. When the stomach is empty, the digestive fluid is very scanty, loses its acidity, and is even occasionally alkaline. Now this repeated secretion of a special acid fluid which continues during the life of the caterpillar–for it is constantly eating, except when molting, or perhaps when in the dark–is discontinued during the chrysalis state, and glandular structures become atrophied and often lost altogether, especially in those perfect insects which do not take food. The intestine follows the stomach, and its commencement is indicated by a constriction and by the attachment of the liver canals; the constriction has a fold internally which prevents the too rapid passage of the food out of the stomach into the intestine. It is very remarkable that the variation in the length and shape of the intestine should not depend upon the food or upon the habits of the insects, and that it should not differ much in these respects in the larva, the chrysalis, and the perfect insect.

The intestinal canal ends in pocket-shaped enlargements or in simple odd-shaped swellings, one of which is shown in the case of *Dytiscus*. The salivary glands are situated on each side of the oesophagus, and look like twisted tubes, or sacs, with cellular walls.

Usually, there are two or three pairs of glands, and they are of two kinds. For instance, in the grasshopper, whose salivary glands are large, they consist of one pair of bunches of bag-like swellings, and of another pair of elongated tubes. Both kinds secrete special fluids, and their admisture is effected in the canal which is common to the two sorts of glands, and the saliva flows straight into the mouth. The saliva is slightly alkaline, and not only lubricates the morsel during deglutition, but assists in the digestion of it also.

In the *Lepidoptera,* or butterfly tribe, the salivary glands are simple elongated tubes, which are largely developed oin the caterpillar, and more or less atrophied in the chrysalis and perfect insect. These long tubes form the web-spinning apparatus in the larva; but after the cocoon is finished, and the first transformation takes place, not only they become small, but their function alters. The small glands of the perfect insect secrete saliva instead of web, and assist in digestion. In the silkworm the oesophagus is short; the stomach, a kind of long cylinder, composes the bulk of the digestive apparatus, and the intestine is remarkably limited in its length. But in the moth the oesophagus is long and has a crop; moreover, the stomach is shorter, whilst the intestine has greatly increased in length, its end being very large and globular. If the chrysalis be examined, the gradual passage of the large stomach of the caterpillar into the small organ of the moth can be traced, and also the formation of the crop and the changes in the glands.

Many larvae get very fat before the chrysalis or pupa condition sets in, and the oily matter collects in the tissue around the digestive organs, and even pushes itself amongst the internal structures of the insect. During the mysterious pupa state this fat disappears, and doubtless it goes to make the beautiful tissues which do not exist in the caterpillar, but which characterise the full growth creature. The differences in the blood and its circulation in the larvae and the perfect insects are not satisfactory determined, but the quantity of the first and the force of the later evidently diminish in the chrysalis.

7

Biochemical Changes during Metamorphosis

The metamorphosis of the higher insects embodies the most profound reorganization of a grown animal that is know. It represents advanced polymorphism, for a single genome determines the development of a differentiated free-living organism which, when fully grown, is largely destroyed and reconstituted in a new form adapted for a totally different life. Such a phenomenon, under the control of hormones, should provide prime material for the study of the regulation of genetic expression and its role in morphogenesis. This challenge has inspired biochemical study of insect metamorphosis for many years. However, since all organ systems of the insect are to greater of lesser extent modified in structure and function during metamorphosis, some change may be expected in almost any biochemical feature that is examined.

Accordingly, the literature of the subject is extensive and diffuse, and reviews of it have tended to give weight to each author's own predilections (Gilbert and Schneiderman, 1961; Karlson and Sekeris, 1964; Agrell, 1964). The contemporary emphasis on molecular genetics and nucleic acid and protein synthesis seems most likely to yield understanding of the fundamental mechanisms in metamorphosis, but investigation of these fields has not yet progressed far with insects. The earlier attention to nutrient reserves and pathways of energy conversion now seems less close to the central question. Nevertheless, these studies have contributed to knowledge of the milieu in which the genes are working and of the nature of the biochemical systems for which they are responsible, and we are not yet in a position to say confidently what is essential and what is not. Therefore, I shall attempt to summarize the present state of understanding in several biochemical areas where enough work has been done to permit some coherent discussion. Other subjects will be arbitrarily omitted; among the more important of these are the development and functions of the silk gland and the ovaries, although much work has been done on both. References to the literature will necessarily be selective, and most examples will be drawn from two insect groups which have received a great share of research attention. The higher Diptera undergo the most complete metamorphosis among insects, replacing most of their organ systems. Some are easy to raise in the laboratory with a short generation time, but the have the disadvantage of small size which has discouraged biochemical work with discrete tissues (except to some extent flight muscle).

In the Lepidoptera, there is far less abandonment of larval cellular structure, but this order includes many species with the experimental advantage of large size, and more work has been done with discrete tissues. Biochemically, there is probably unity in the underlying mechanisms of histolysis and of histogenesis in all insect groups. Processes which are not strictly part of metamorphosis may be relevant e.g., the degeneration and reconstruction of muscles in the larval molting cycle of the bug Rhodnius or in the adult diapause of the Colorado potato beetle. Metamorphosis is not wholly confined to the transformation from larva to adult, as there is often a small degree of change in form at each larval molt.

Use and Interconversion of Reserves

During their last larval state, in preparation for metamorphosis, holometabolous insects accumulate substantial nutrient reserves. These are chiefly deposited as fat and glycogen in the fat body, a cellular tissue which is a center both for storage and for intermediary metabolism and which proliferates greatly during the last larval instar (Kilby, 1963). In the larva of the wax moth, *Galleria mellonella,* this preparation for metamorphosis can be detected early in the last instar, which differs metabolically from the preceding larval stages (Jandaet al., 1966). Peas values for glycogen content are often in the neighbourhood of 10 per cent (but sometimes as high as 30 per cent) of total body dry weight, and the lipid reserves are generally greater, values of 20 to 50 per cent of dry weight before metamorphosis being common. Analytically studies on changes and conversions in the insect pupa have been stimulated by the fact that it is a closed system which exchanges only gases with the environment. A body of the early work is summarized and discussed in an admirable review by Needham (1929), but few conclusions can be drawn from it because of inadequacies in the analyses.

In particular, trehalose was not recognized as an ubiquitous sugar in insects until 1957, and the extent to which the chitin and protein of the cuticle are resorbed in the molt and reutilized was generally underestimated or ignored. References to more recent studies can be found in several reviews (Wigglesworth,1965; Gilbert,1967; Wyatt,1967a). Among the most comprehensive data on a single species are those recently obtained by Birt and colleagues on the sheep blowfly, *Lucilia cuprina.* Several conclusions may be drawn.

The principal source of energy is fat, the content of which at adult emergence is about one third of that in the mature larva. The rate of fat oxidation is more rapid at the beginning and the end of metamorphosis than in the middle, in accord with the changing rate of respiration. At the beginning of metamorphosis before the pupal molt, there is a fall in total protein and in free amino acids, which are partly oxidized and partly converted to carbohydrate and fat. The utilization of protein leads to accumulation of much uric acid in the early fly pupa (RussoCaia, 1960). There is also a fall, before the pupal molt, in chitin as the inner layers of the larval cuticle are digested and resorbed; some of this is apparently deposited as glycogen, which is later reconverted to chitin in the growing adult cuticle. Thus, when chitin is included, the total carbohydrate content undergoes only a slight net fall throughout metamorphosis. Since files depend upon carbohydrate for flight energy, its conservation during their development is advantageous.

A slight temporary rise in total carbohydrate which is observed just after pupation is attributable to glyconeogenesis from amino acids. These conclusions differ in several respects from those from earlier, less complete analyses on related specise, which had suggested synthesis of glycongen at the expense of fat, followed by substantial oxidation of carbohydrate during metamorphosis. The net conversion of fat to carbohydrate is not known to be possible in insects or vertebrate animals. Analyses on metamorphosing Lepidoptera, such as the oriental silkworm, *Bombyx mori* (Niemierko et al. 1956; Zaluska,1959), and the American saturniid silk moth, *Hyalophora cecropia* (Gilbert and Schneiderman, 1961; Bade and Wyatt,1962; Domroese and Gilbert 1964), present a generally similar picture. Again, fat is an important energy source. However, the extent of net oxidation of carbohydrate in passing from caterpillar to moth is greater than in the case of the fly, perhaps because moths use fat as their fuel for flight and so its conservation rather than that of glygogen, is advantageous.

There is a further adaptive difference between the sexes in that male moths, which are more ambitious in flight, conserve more lipid than do the relatively sedentary females of a species. As in the blowflies, there is a phase of glycogen deposition about the time of pupation,, which is attributable to syntesis from chitin and protein. A large proportion of the cuticle is resorbed and reutilized in each molt; thus, in the larval-pupal and pupal-adult milts of the Cecropia silk moth, the resorbed fraction is more than 80 per cent of the cuticular chitin. Evidence concerning the balance of metabolism in insects also often has been sought by measurement of respiratory quotients. Such measurements, especially with pupae, however, are subject to a number of pitfalls(e.g., release of CO_2 in bursts, trapping of CO_2 in nongaseous forms, disturbance of metabolism by accumulated CO_2), and in general more direct methods of study are preferable. The more reliable estimates of respiratory quotients are in reasonable agreement with analytical results. Along with the quantitative changes in body composition are certain cases of apparently qualitative change(Wyatt, 1967a).

In blowflies, e.g., trehalose cannot be detected in the feeding larva, but makes its first appearance just before pupation and then increases to play an important role in the flight metabolism of the adult. In pupae of the Cecropia silkworm, which pass the winter in diapause, most of the glycogen is converted to glycerol, and part of this is reconverted to glycogen at the beginning of adult development in the spring. But this is an adaptation to survival in cold weather rather than a process characteristic of metamorphosis, for glycerol is accumulated by many insects which survive cold winters, and the life-cycle stage in which it appears is always that which passes the winter, be this the embryo, larva, pupa, or adult.

In short, the use and interconversions of reserves in the metamorphosis of different species is variously adapted to their respective needs. Little is know about the regulation of these conversions.

A recent review on metabolic control mechanisms in insects cited much evidence for the existence of control but few instances where enough is known to justify designation as a mechanism (Harvery and Haskell, 1966). Fat and carbohydrate metabolism, like other aspects of insect development, must be more or less directly under control by hormones, and cases can be cited in which metabolic change follows change in hormonal state (Wyatt, 1967a). Certain rapid adaptive changes can result from hormonally released activation of preexisting enzymes, such as glycogen phosphorylase. The longer term changes in metamorphosis are presumably associated with changing enzyme levels, but there exists little satisfactory biochemical evidence on this point. One example is the elevation of chitinase, a participant in digestion of the cuticle, at each molt in the silkworm, *Bombyx Mori* (Jeuniaux, 1961).

Respiratory Metabolism

Respiration and Energy Supply During Metamorphosis

One of the earliest and most frequently recorded metabolic observation concerning insect metamorphosis is that the rate of respiration changes in a manner that can be described as a U-shaped curve. In the earlier papers, the significance of this general course of respiration was much discussed. The first suggestion was that it reflected successive histolysis and histogenesis, the rate of respiration changing in accord with the amount of organized tissue present at any time, but it then became clear that the histological picture often did not support this simple view. When the activities of respiratory enzymes, such as cytochrome oxidase and dehydrogenases, were measured, a somewhat better correlation with respiratory rate was found (Agrell, 1949; Sacktor, 1951). No more than a rough agreement between respiratory rate and activities can be expected from such measurements, however, for the course of change during metamorphosis varies for different enzymes and different tissues. In any case, it is now clear that tissue respiration is generally not limited by the levels of respiratory enzymes.

The capacity for respiratory depends on the content of mitochondria and their electron transport system, which is relatively low in tissues such as fat body and high in others such as muscles.

But the actual respiration is generally substantially less than this capacity, as a result of respiratory control by the coupled phosphorylating system, which, is subject to the levels of ADP, ATP, and inorganic phosphate in the cell (Harvey and Haskell, 1966). Thus the rate of respiration adjusts itself to supply the demand, in the form of high-energy phosphate, of the work done by the tissue, which includes biosynthesis, active transport, muscular activity, and so on. In the pupa and the pharate adult insect, where there is little muscular activity, one may expect protein syntesis to be a major energy user, and indeed the rates of respiration often show some correlation with those of protein synthesis. Some evidence on the balance of energy metabolism in insect metamorphosis can be obtained by analysis of metabolite and coenzyme levels.

It was suggested as early as 1893 by Bataillon, on the basis of crude carbohydrate analyses, that the metabolism of the silkworm pupa might somehow be anaerobic, and similar ideas have recurred in various forms (Wyatt, 1963). Several compounds are now recognized as products of experimentally induced anaerobic glycolysis in insects:lactate, alanine, pyruvate, and glycerol-1-phosphate(Gilmour,1965), but recent analyses show none of these accumulating during metamorphosis of the blowfly(Crompton and Birt, 1967). There is, in fact, much more lactate in the actively respiring larva than in the pupa. In addition, analyses of pyridine nucleotide coenzymes show that the ratio NADH/NAD is somewhat lower in midpupal life than in the active larval and adult stages, whereas anaerobic mechanisms would be expected to lead, on the contrary, to accumulation of the reduced form (Birt, 1966). Thus, there is no evidence for anaerobic processes in the blow-fly pupa. Indeed, it would be surprising to find them, for the supply of oxygen is ample to support the depressed rate of metabolism, and aerobic processes give more efficient use of reserves which the pupa needs to conserve for the construction and initial activity of the imago. The levels of "high-energy" and related phosphate compounds are also of interest.

In the blowfly, ATP and the phosphagen, arginine phosphate, show a decrease in the midpupa, and inorganic phosphate shows an almost reciprocal increase. The doubtless corresponds to changes in the body content of tissues which are characteristically rich in high-energy phosphate, notably muscle. The rise in AMP presumably represents degradation of ATP and ADP from such tissues. That the ATP/ADP ratio remains high through metamorphosis, and that phosphagen is never exhausted, show that the ATP-generating processes in the pupa are at all times fully able to keep up with the energy demand.

Respiratory Metabolism and Diapause

When the pupal stage enters the resting state of diapause, the depression of metabolism is extreme and prolonged. This is the case with *Hyalophora cecropia* and some related saturniid moths, which have received much attention directed toward testing the relationships between respiration, respiratory enzymes, tissues growth, and hormone action (reviewed by Shappirio, 1960; Harvey, 1962; Wyatt, 1963: Harvey and Haskell, 1966). Since diapause is terminated and adult development is initiated by the prothoracic gland hormone, ecdysone, this system has been used for analyzing its action. A provocative observation was that respiration of pupae in diapause is highly resistant to the cytochrome oxidase inhibitors carbon monoxide and cyanide, a fact previously noted for the embryonic diapause of the grasshopper, *Melanoplus differentialts*. In spectroscopic and enzymic assays, the cytochrome content of tissues in diapause was found to be much reduced–cytochrome oxidase is present at low levels, but cytochromes *b* and *c* could not be detected at all. Accordingly, it appeared possible that terminal electron transport in diapause might proceed by some pathway other than the ususl cytochrome system and that resumption of growth might depend upon the resynthesis of cytochromes which in turn might be closely linked to the action of the hormone.

Further experiments showed, however, that when appropriately tested, diapause respiration is indeed sensitive to cyanide and carbon monoxide and since the latter inhibition can be reserved by light, participation of cytochrome oxidase is indicaed. Apparently, the tissues contain substantially more cytochrome oxidase than is needed to support the normal rate of electron transport so that the oxidase can be largely inhibited with no effect on net pupal respiration. When the demand upon this enzyme is increase, however, either by lowering the oxygen tension or by enhancing the rate of respiration (e.g., with 2,4-dinitrophenol), inhibition of the pathway becomes demonstrable. This conclusions raises the question: How is diapause respiration held to its low rate? Since cytochromes *b* and *c* were not detected spectrtscopically, it seemed at first that a limiting step might exist here. Oxygen uptake by the pupae, however, is strongly stimulated by dinitrophenol, which shows that the capacity of the electron transport enzymes exceeds their normal load and indicates respiratory control via the coupled phosphorylation system, as is usual in active tissues.

In addition, analyses of adenine nucleotides in diapause and developing tissues show that the ATP. ADP ratio changes little, even though the turnover rate is much higher in development, so that a nice co-ordination between phosphorylation and dephosphorylation is indicated, just as in the nondiapausing blowfly pupa discussed above. The ATP level in diapause tissues is high enough so that deficiency of metabolic energy can scarcely be responsible for restraint of growth. The suggestion has also arisen that diapause pupae possess partially anaerobic metabolism as a consequence of restricted electron transport. This seemed to be supported by the gradual conversion of glycogen to glycerol in the Cecropia pupa. It now appears, however, that the characteristic glycolytic products glycerol-1-phosphate, lactate, and pyruvate are not elevated in the normal diapause pupa, though they are produced in experimental anaerobiosis. In addition, the NADH/NAD ratio is not elevated in diapause.

The production of glycerol, therefore, is probably subject to special regulatory mechanisms as an adaptation to cold, and the diapause silkmoth pupa is no more anaerobic than is the blowfly pupa. The present indication from these extended studies on the extremely suppressed metabolism of diapause is that the pathways do not differ qualitatively from those of other tissues, but there is a co-ordinated decrease in activities. The breakdown of the cytochrome system is impressive, yet the actual rate of respiration appears to be limited by energy demand which, in the pupa, must largely represent biosynthesis of protein and other tissue constituents. Furthermore, resynthesis of cytochromes, though necessary, is not sufficient to release resumption of development, for injury to diapause pupae is followed by the former but not the latter. Certain problems remain: e.g., cytochromes *b* and *c* have not been detected in tissues other than muscle of the diapausing Cecropia pupa, and the evidence for their participation in electron transport remains indirect.

The measurements of respiration all have been made with whole pupae, and it is not known how much is contributed by muscle, epidermis, fat body, and other tissues whose activities and enzyme complements are quite distinct.

Respiration and the Corpus Allatum Hormone

In addition to the effect of ecdysone in increasing respiration when diapause is broken, enhanced respiration is often attendant upon the action of the hormone of the corpora allata. Increased respiration has been noted both during the "juvenile" action of this hormone, when corpora allata are implanted into larval insects, and during its gonadotrophic action in the adult. When corpora allata were implanted into *Galleria* larvae of different ages, however, and development was modified to various extents, the effect on respiration was always correlated with the degree of morphogenetic change. It appears that this hormone, like ecdysone, may influence respiratory metabolism indirectly by inducing growth and structural change in the tissues (Sehnal and Slama, 1966).

Development of Flight Muscle

A picture of the co-ordinated synthesis of enzymes in the construction of a specialized tissue has been gained from studies on the biochemical development of flight muscle. The indirect flight muscles of insects are of the fibrillar type. Between the bundles of large fibrils are packed rows or giant mitochondria ("sarcosomes") which may be 1 to 3 μ in diameter and make up 30 to 40 per cent of the weight of the tissue. These are directly supplied with oxygen by tracheoles. The flight muscle has almost no lactated dehydrogenase and in its normal activity is totally aerobic. The contents of soluble NAD-linked α-glycerophosphate dehydrogenase and of mitochondrial flavoprotein α-glycerophosphate oxidase are usually high and provide a shuttle for the oxidation of extramitochondrial NADH by passage of glycerol-1-phosphate and dihydroxyacetone phosphate in and out of the mitochondria (Sacktor, 1965).

Flightmuscle is thus constructed for sustained aerobic activity with exceptionally high energy output. Insect leg muscle, such as the jumping muscles of the locust, on the other hand, has a smaller content of mitochondria and possesses lactate dehydrogenase in adaptation for sporadic violent activity followed by rest. The most comprehensive picture of the structural and biochemical development of flight muscle is that due to the work of Bucher and his colleagues on the locust, *Locusta migratoria* (Bucher, 1965). The muscle develops during a period of rather more than two weeks starting from a small anlage, or precursor muscle, that is present just after the fourth molt. Even this precursor muscle resembles flight muscle in having only a low content of lactate dehydrogenase. Four phases of growth are described: (1) larval growth, with increase in size (but not number of nuclei) with little differentiation; (2) a phase about the time of the final molt when the larval tracheal supply breaks down and new tracheoles penetrate into the muscle and a transient elevation of lactate dehydrogenase may indicate some temporary anaerobiosis; (3) a phase of differentiation, during the first days of adult life, in which the characteristic enzyme pattern of flight muscle is established by elevation of mitochondrial enzymes and degeneration of lacta*t*e dehydrogenase and glucose-6-phosphate dehydrogenase; (4) finally, further growth, with approximate doubling of all components. Thus, most of the growth of the mitochondria and synthesis of the eventual enzymes takes place after emergence of the adult insect, and the locust cannot fly until this maturation is completed.

There are "constant-proportion groups" of enzymes, so that the mitochondrial oxidative enzymes (represented by α-glycerophosphate dehydrogenase) increase co-ordinately with one another and, similarly, a group of soluble glycolytic enzymes represented by glyceraldehydes phosphate dehydrogenase) retain an approximately constant mutual relationship. The existence of such constant proportionality within groups of enzymes, despite wide variations in absolute levels, extends to other tissues than flight muscle and presumably indicates some common regulation of their synthesis. Although the flight muscle of the locust is somewhat less specialized than that of holometabolous insects such as flies and bees, the biochemical development of the latter is probably essentially similar.

A greater proportion of the enzymes syntesis may occur before eclosin than in the locust, but there is always period of maturation, involving further mitochondrial growth and enzyme synthesis, in the emerged adult. In saturniid silk moths, cytochrome *c* first appear appears in measurable amounts in the flight muscle on about the sixteenth day of the 21-day period of adult development, and its synthesis thereafter is very rapid (Chan and Margoliash, 1966). In the blowfly, *Phormia,*the cytochrome content rises about fourfold during the first week of adult life, which reflects an increase in the size of the mitochondria while their number remains constant (Levenbook and Williams, 1956). Growth of the flight muscle mitochondria is also reflected in a 50 per cent increase in phospholipid content of the thorax of *Lucilia,* the distribution of enzymes and protein in particles of various sizes has been determined, and inasmuch as the specific activities of several respiratory enzymes relative to mitochondrial protein change greatly during development of the fly, it is proposed that the respiratory enzymes and the nonrespiratory, structural protein follow different courses of synthesis (Lennie and Birt, 1967). This is quite likely, since it appears that the nonextractable, structural protein of mitochondria may be synthesized within them, whereas synthesis of the extractable enzymes is a function of extramitochondrial ribosomes. In the honeybee, the mean volume of a flight-muscle sarcosome increases from 0.015 μ^3 on the day of emergence and to 0.52μ^3 in the 20-day-old adult. The increase in content of several cytochromes, determined by the sensitive technique of spectroscopy at the temperature of liquid nitrogen.

It is evident that their synthesis is by no means synchronous. Cytochrome b_5, which occurs in the microsomal fraction and is believed to participate in biosynthese, is found in the pupa and declines sharply in adult development. Among the mitochondrial cytochromes, although the general pattern is similar, there are distinct differences in time of first detectability and course of synthesis, so it is clear that constant mutual proportionality is not a universal rule among these electron carriers.

Sclerotization of the Cuticle

A characteristic feature of insects is their tough external cuticle. This is secreted at each molt by the epidermis and consists chiefly of protein and chitin, plus smaller amounts of specialized lipoidal substances. Unlike the cuticle of Crustacea, which contains inorganic deposits, it is hardened by organic tanning or cross-linking of the protein in a process known as sclerotization. It has been accepted for many years that sclerotization involves quinines derived from tyrosine, but the chemical nature of the tanning agent was established only recently, and the details of its reactions with the protein are still obscure. The biochemistry of sclerotization of the blowfly puparium has recently been intensively studied by Karlson, Sekeris, and coworkers because of its connection with the action of the hormone, ecdysone. This subject is reviewed by Cottrell(1964), Brunet(1965), and Sekeris and Karlson(1966). When the blowfly maggot is full grown and ready to pupate, its soft, white cuticle is inflated into a barrel shape and then becomes hard and dark brown, as the puparium, before the insect molts to a pupa within it. The principal tanning agent has been identified as N-acetyldopamine its acetylation has the chemical role of preventing cyclization into an indole, and when the phenolic ring is further, oxidized to an o-quinone, it is reactive with protein amino groups.

In the early third-instar larva, metabolism of tyrosine is largely by a degradative pathway that starts with transamination. During the instar, the level of throsine mounts until it is one of the most abundant free amino acids in the hemolymph. Then, at the time of puparium formation, it is largely used in the production of N-acetyldopamine by the enzyme-catalyzed steps shown. The switch in metabolic fate of tyrosine is the result of changed enzyme activities: tyrosine transaminase apparently decrease, prephenol oxidase accumulates in the hemolymph and its activator protein builds up in the integument, and a specific dopa decarboxylase is synthesized. The acetylase,on the other hand, is present throughout the instar. The decarboxylase, whose activity shows the greatest changes, is believed to be the controlling enzyme. The whole process is set off by the increasing level of ecdysone at this stage of the life cycle; it can be prevented by ligation and induced by injection of the hormone, and this forms the basis of the usual assay for ecdysone.

Because of the close correlation between production of dopa decarboxylase and ecdysone titer, and other evidence to be discussed below, it has been proposed that the hormone "induces" this enzyme by an effect on nucleic acid metabolism. The tanning of the dipteran puparium just before the pupal molt is a special case in insect ontogeny, as sclerotization commonly affects newly deposited cuticle just after ecdysis. In several such case recently examined in Karlson's laboratory (the, blowfly adult, the locust, and the mealworm, *Tenebrio molitor*) the mechanism appeared to be similar, for dopa decarboxylase was active, and tyrosine was converted to N-acetlyldopamine. Close to times of sclerotization.

In the blowfly adult, a further source of tanning agent is apparently the hydrolysis of N-acetyldopamine –4-0-β-glucoside, in which form excess N-acetyldopamine is conserved after puparium formation. After the adult molt, ecdysone titer is low, but there is now recognized another hormone specifically associated with cuticular hardening and designated bursicon(Cottrell, 1964; Fraenkel, Hsiao and Seligman, 1966). It is required for sclerotization of the cuticles of newly emerged flies and newly molted cockroaches and will effect the darkening of discs of colourless integument from the latter in vitro. Little is known about the biochemistry of its action; it has a molecular weight of about 40,000 and possibly may be itself an enzyme. It appears to act at a later stage than the production of N-acetyldopamine. Thus, ecdysone and bursicon could have complementary roles in stimulating the production and use, respectively, of the tanning quinines.

Pigment Production

Insects possess many kinds of pigment, including the ubiquitous and refractory melanin, and creation of the characteristic adult colour pattern is often one of the most striking features of metamorphosis. In the period butterflies the wing patterns are due chiefly to pteridines, and the several pteridines of *Colias eurytheme* wings follows characteristic courses of synthesis, beginning on the third day of pupal life (Watt, 1967). But little is yet known about the regulation of such processes. One instance in which there is a relation of hormone action involves certain ommochromes. These are phenoxazone compounds derived from tryptophan via kynurenine, which occur widely in the eyes, integument, and internal organs of insects. They are of historic interest because it was experiments with certain eye-colour mutants of *Drosophila* and *Ephestia* blocked in tryptophan metabolism that some 30 years ago first demonstrated the association of specific genes with successive steps in a metabolic sequence.

The structure and distribution of ommochromes was worked out in a monumental series of investigations in Butenand't laboratory. In the hawk moth, *Dicranura vinula*, the green caterpillar develops a characteristic red-brown pattern six days before pupation as a result of the deposition of xanthommatin and rhodommatin. Ligation prevents this change in the abdomen, and it can them be induced by injection of small doses of ecdysone. The enzymic basis of the change has not been established, but the hemolymph levels of tryptophan and 3-hydroxykynurenine rise sharply at the time of pigment synthesis, and it is proposed that ommochrome formation may serve in the removal of tryptophan liberated by proteolysis (Buckmann, Willig and Linzen, 1966).

Biosynthesis and Degradation of Macromolecules

The Proteins of Larval and Adult Insects

A question of fundamental importance is to what extent to protein constitution of an adult insect differs from that of a larva of the same species. No all-out attack on this problem has been described, but there has recently been a good deal of skirmishing with it. Most attention has been given to insect blood (hemolymph), because this protein solution is easy to obtain and it undergoes large quantitative changes during ontogeny (Wyatt, 1961). The usual pattern with respect to protein is a relatively low level often 1 to 2 per cent), during the earlier larval stages, then a rapid rise just before pupation (often 6 to 8 per cent, but 18 and 20 per cent have been recorded), then there is some decline in the pupa and a further fall during the later part of adult development.

Qualitatively, a number of analyses by electrophoresis on filter paper (summarized by Wyatt, 1961) gave too little resolution to be of much significance. More recently, starch gel and acrylamide gel electrophoresis have been applied to the hemolymph of a number of species, and between 10 and 20 bands can generally be recognized. These have no correspondence with the components of mammalian serum and differ characteristically with the species. There is often a yellow or green chromoprotein band and a dense band of low mobility which may represent the major hemolymph component of very high molecular weight (more than 500,000) which has been detected in several insects by other methods. In the transformation from larva to adult, most of the electrophoretic components generally persist, but there are certain qualitative differences. Thus, in the tent caterpillar *Malacosoma americanum*, out of 17 bands detected in the full grown larva and early pupa, two to five were apparently missing in young larvae and in adult moths, and the adult did not a possess any obvious new components (Loughton and West, 1965). When blood from fourth-instar tomato hornworms was concentrated fourfold before electrophoresis, several otherwise undetected components were revealed, and in this insect the adults apparently lack some components characteristic of larvae and contain several fresh ones (Hudson, 1966).

A study on the blowfly, *Phormia regina*, in which polyacrylamide gel was used with equal amounts of protein from each developmental stage, led to the conclusion that "the newly emerged adult fly has an entirely different picture" from the larva and pupa (Chen and Levenbook, 1966a), though inspection of the published diagrams suggests that about the half of the bands in the adult pattern may correspond to larval components. Bands observed in zone electrophoresis are, of course, rarely pure proteins, nor is their mobility an adequate criterion of identity. The technique of immunodiffusion, in which antigen and antibody diffuse toward one another in an agar gel, can assist with establishing identity, although when applied to whole hemolymph, it has generally revealed fewer components than has electrophoresis.

In an immunological study on Cecropia silkworm hemolymph, Telfer and Williams (1953) observed nine distinct antigens and followed six of the quantitatively from the fifth-instar larva to the adult. Each followed its own course of variation, which shows that their levels are individually regulated. One protein found in female pharate adults was present only as traces in the male (Telfer, 1954). But theappearance of uniquely adult antigens, distinct from those of the larva, is not reported. By a technique of immunoelectrophoresis, which should combine high resolution with definitive identification, Terando and Feir (1967) detected 11 antigens in hemolymph of the milkweed bug, *Oncopeltus fasciatus*, and finding no evidence for state-specific antigens, arrived at the conclusion that nymphs and adults of both sexes showed only quantitative differences; but it should be noted that this concerns a hemimetabolous insect. Laufer(1963) has applied both electrophoretic and immunological methods to the examination of hemolymph and tissue protein of the Cynthia silk moth and several other species and concludes that there exist both stage-specific and tissue-specific antigens.

Enzyme activity (esterases and dehydrogenases) was found associated with a number of the protein zones, but as the enzyme assay is very sensitive it may reveal proteins other than those made visible by staining. Extracts of the cellular tissues would be expected to be more representative than hemolymph of the overall protein makeup of an insect. Butler and Leone (1966) have prepared whole-insect extracts of several developmental stages of the beetle mealworm) *Tenebrio molitor* and examined these by several types of electrophoretic and serological test. By immunoelectrophoresis, 10 precipitin zones were obtained; in several zones. Larval and adult protein were distinguished serologically despite similar electrophoretic mobility. These authors conclude that during ontogeny there are gradual and sequential qualitative and quantitative changes in protein complement. The electrophoretic pattern of basic proteins extracted from an acetone powder with acid also changes during blowfly metamorphosis. From these various studies, it is very difficult to feel confident in any conclusions as to the extent of changeover from a larval set of proteins to an adult set in metamorphosis. For the enzymic requirements for their metabolism, insect cells, like other cells, must contain at least several hundreds of distinct proteins. Yet the methods that have been used have detected at most 20.

These are no doubt the most abundant or the most antigenic components but are not necessariiy the most significant in ontogeny. In only a few studies has there been any attempt at quantitation, and in view of the limited sensitivity of the detection methods, it is very difficult to distinguish a qualitative from a quantitative change. There is suggestive evidence (with respect to hemolymph proteins) that, as might be expected, some correlation exists between the completeness of metamorphosis in an insect group and the extent of change in its proteins. Thus, a bug shows no qualitative change, a moth changes a small proportion of its complement, and lies exhibit extensive change. In all cases, however, proportion of the demonstrable protein components are found at all stages of development. A quite different, and more demanding, approach to the question of identity of larval and adult proteins is to isolate homologous proteins from each stage and compare them chemically. This has been done with tropomyosin, which was obtained crystalline and apparently pure from both larvae and adults of the blowfly, *Phormia regina* (Kominz et al., 1962).

The two preparations were very similar but showed slight differences in amino acid and peptide composition and physical properties. These small differences cound conceivably result from impurity or secondary modifications of the protein rather than genetic difference, and thus the question of genetic identity is unfortunately unresolved. Another indication of similarity of larval and adult proteins, though not conclusive evidence of identity, comes from low-temperature spectra of mitochondria from larvae and adult thoraces of *Drosophila;* the two show no significant differences, while both differ characteristically from mammalian mitochondria.

Free Amino Acids and Peptides

A peculiar feature of insects that always has to be taken into account in a discussion of their protein metabolism is their high level of free amino acids. This is hghest in the bolood, and the tissue pool is generally somewhat less concentrated and differently constituted with respect to individual amino acids. In the more advanced insects, the hemolymph amino acid level is commonly in the range 5 to 20 mg/ml-many times that mammalian plasma. In addition to the amino acids of proteins, some les usual ones have been identified from certain insects, including β-alanine, tyrosine-0-phosphate, methionine sulfoxide, D-alanine, D-serine and others. The literature, of this subject, as well as other aspects of amino acid and protein metabolism in insect development, is reviewed by Chen (1966). The significance of the high amino acid level is not well understood, although it provides a pool for protein synthesis and appears to have an osmotic role.

During metamorphosis, while individual amino acids change in concentration, the variation in total free amino acid pool is remarkably small when one considers the extent of tissue reconstruction that is taking place. Along with free amino acids, insects also exhibit a variable content of peptides. In *Drosophila* larvae, intensive chromatographic analysis led to the conclusion that more than 600 peptides were present, and these incorporated injected C^{14} glutamate much more rapidly than did the total body protein (Mitchell and Simmons, 1962). In *Phormia*, a group of peptided is transiently abundant during the early third instar; these are mostly composed of only tow amino acids one of which, in each that was analyzed, was either lysine or histidine. Their metabolism and significance is unkonw.

The Biochemistry of Histolysis

It repeatedly has been observed since the early histological studies on insect metamorphosis that at certain developmental stages the fat body, and to a lesser extent other tissues, become loaded win protein granules of "albuminoid spheres". These have been regarded as a form of protein reserve and appeared to constitute an exception to the generalization from mammalian biochemistry that animal tissues have no specifically storage proteins. Their probable nature is now becoming apparent with the use of the electron microscope and with growing understanding of the nature and function of lysosomes.

In Lepidoptera, such as the skipper *Calpodes ethlius*, protein granules appear in the fat body about one day before pupation, at the time when the larval cuticle is being resorbed. Since injected radioactive amino acid showed little incorporation into these granules, whereas injected plant peroxidase (detectable histochemically) appeared specifically in them, it was inferred that they contain protein sequestered from the hemolymph and not newly synthesized. Similar uptake occurs in other tissues, particularly the epidermis and especially at times in the molting cycle when there is active protein synthesis.

The granules are surrounded by membranes and are believed to transform into multivesicular bodies in which digestion takes place, presumably by addition of lysosomal enzymes, as a source of amino acids for new protein synthesis. Cell organelles such as mitochondria due for destruction, are believed to be isolated in a similar way. The uptake of hemolymph proteins into fat body, gut , and to some extent other tissues of Lepidoptera is independently demonstrated by immunodiffustion test. Which show several blood antigens to be present in them during the prepupal and early pupal stages. As an example of programmed histolysis, have studies the breakdown of the abdominal intersegmental muscles of saturniid silk moths, whose function is to pump blood into the wings of the newly emerged moth and which lyse when this role is completed. The trigger for lysis when this role is completed. The trigger for lysis is apparently the cessation of nervous motor stimuli to the muscle.

Lysososomelike bodies are observed to undergo change at this time. Proteolytic activity of catheptic type (pH optimum 4.0) builds up is these muscles at first gradually, then rapidly, during the 3 weeks of pharate adult development, and is associated with lysosomelise particles. It can be activated and partially released from the particles by detergent treatment, but in the emerged moth the activity and the proportion of soluble enzyme are naturally enhanced. Similar catheptic activity is found also in the far body, but not in thoracic muscle (which, of course, is not destined for dissolution). The intersegmental muscles and fat body also contain an acid phosphatase which shows greatest activity in advanced adult development. These enzymes are considered to be directly responsible for the destruction of the intersegmental muscles.

It is remarkable that, despite the dominant place of histolysis in insect metamorphosis, there is almost no other information on the enzymic basis of the process. Not much can be concluded from several reports of hydrolytic activities without evidence as to their source in tissues or cytological fractions.

Rate of Protein Synthesis and Turnover

Linked to the question of changing protein composition of an insect during metamorphosis is that of the rates and extent of protein degradation and synthesis. A number of measurements have been reported of the incorporation of injected amino acid into the proteins of hemolymph or tissues of insects at selected stages. It is found that incorporation rates rise sharply when adult development is initiated in diapausing silk moth pupae, and that fat body qualifies kinetically as a source of hemolymph proteins. Similar results have been obtained with isolated insect tissues incubated in vitro, a technique which avoids complications due to the blood amino acid pool. But the systematic quantitative study of protein metabolism during metamorphosis seems to be confined to the recent work of two laboratories. Using *Drosophila melanogaster*, Boyd and Mitchell (1966) fed C^{14} amino acids to larvae, then after appropriate times for incorporation, extracted body fluid (chiefly hemolymph), and separated radioactive protein fractions from it by acrylamide gel electrophoresis. Then, their turnover was measured by reinjecting into larvae or pupae the isolated radioactive protein fractions and, after incubation, subjecting hemolymph samples to the same kind of analysis.

The redistribution of activity was exstensive and varied in different fractions: One was stable in the larva, but after pupation had a half-life of 13 hours; another was too stable for the half-life to be measured. Specificity is also indicated by the fact that injected mammalian hemoglobin did not turn over. When enough nonradioactive valine was injected to increase the blood valine pool about thirty-fivefold, the redistribution of activity from valine-labeled protein during 42 hours was cut from 56 to 39 per cent, which suggests that free valine, but not the valine of the hemolymph, may be an intermediate in the interconversion of hemolymph proteins. The authors conclude that hemolymph proteins "may be an important channel through which the massive protein turnover of the metamorphosing insect passes". Levenbook and colleagues have examined the question of protein synthesis and turnover in *Phormira regina.*

The hemolymph protein is shown to be synthesized chiefly in the larva; at pupation there is a rapid fall in blood protein content, and there is little synthesis from injected amino acids or change in level thereafter. Radioactive hemolymph protein was prepared from C^{14}-labeled larvae, and after injection into larvae, pupae, and pharate, adults, its fate was followed. Very little was converted to $C^{14}O^{2}$, although injected free amino acids were rapidly oxidized, and conversion of hemolymph protein to free amino acids was also slow.

It is inferred, in contrast to the conclusions of Boyd and Mitchell, that the hemolymph proteins are relatively stable. When mature larvae were injected with labeled hemolymph protein and kept until they emerged as flies, however, about there quarters of the recovered radioactivity was found in protein in the tissues. Since the blood free amino acids clearly cannot be intermediates, some from of direct transformation of blood proteins into tissue proteins is suggested. The data seem only to rule out *hemolymph* free amino acids as intermediates and do not oppose degradation to free amino acids which remain within the tissues and are rapidly used in resynthesis. In a related study, Dinamarca and Levenbook (1966) examined the dynamic state of the free amino acid pool and the rate of protein synthesis from it in *phormia regina* at different developmental stages. C^{14}-alanine and C^{14}-lysine were injected, and their conversion to C^{14}-o_2, total disappearance from the pool, and incorporation into protein were measured, as well as the pool size for each amino acid in each insect used. Corrected values for the rate of incorporation into protein could then be calculated and were found a U-shaped curve, closely similar for the two amino acids, except in the adult.

The importance of the procedure became clear, for lysine always yielded more of the injected dose into protein than did alanine, but this was a consequence of the more rapid turn-over of alanine. With the assumption of a single homogeneous pool of each amino acid in the insect, the further calculation could be made of the total amount of new protein synthesis during metamorphosis, and this came out at the surprisingly small value of 0.16 mg, or about 2 per cent of the total protein of the insect. As Dinamarca and Levenbook recognize, the assumption is a great oversimplification; nevertheless, they suggest that the total amount of new protein synthesis during metamorphosis is less than has been generally assumed. The chief conclusions from these studies are, perhaps, to confirm that the hemolymph amino acids (which make up more than half of the total pool) are not in free exchange with those in the tissues and to indicate that the greater part of the protein turnover involves only local intracellular pools. This would be in accord with the cytological evidence that proteins are engulfed from the blood and digested within various tissues that are also synthesizing protein. Calculations based on the total pool of the animal might then fail to give any indication of the true amount of protein synthesis. It seems that still more refined techniques, perhaps involving isolation of specific proteins from individual tissues, are required.

Enzymes Associated with Protein Synthesis

The amino acid activating enzymes, responsible for the first step in protein synthesis, have been assayed and to some extent characterized in different developmental stages of the blowfly, *Lucilia cuprin,* and the fat body of *Sarcophaga bullata.* Activity was found for almost all of the protein amino acids, and where it was not found, this may have been due to inactivation. The total soluble enzyme activity in both series of experiments was minimal in midpupal life and rose to a maximum just before emergence of the adult fly, thus paralleling the changes in rate of protein synthesis in a blowfly pupa.

In *Lucilia*, during the early pupal stage, some amino acid activating enzymes, particularly for tyrosine, are bound to a class of cell particle, and it has been suggested that these may have some role in storage of tyrosine for use in cutincular synthesis and tanning. The activity of glutamate-aspartate transaminase in the housefly declines from a high value at pupation to a minimum just before adult emergence and then rises steeply. A relation to protein synthesis is suggested, but the curve appears more closely correlated with that to be expected of protein degradation. In the Cecropia silk moth, however, this enzyme rises strikingly in thoracic muscle during late adult development, which is certainly an active site of protein synthesis.

Ribonucleic Acid and the Machinery of Protein Synthesis

Increasing attention has been directed recently to the nucleic acids as determinants in the specificity and possible regulatiors in the rate of protein synthesis, but there is not yet a great deal of information relevant to insect metamorphosis. Measurements of total RNA during metamorphosis of several insect species reveal little change, which is consistent with the histological observation that histolysis and histogenesis are concurrent rather than consecutive, so that the total amount of cellular tissue does not change much. Some changes reported in RNA fractions differing in extractability with phenol from *Calliphora evythrocephala* are very difficult to interpret. When a single tissue is examined, more significant changes may be seen. Thus, in the fat body of the saturniid silkworm,

Samia Cynthia ricini, during pupation there is a great decrease in cytoplasmic RNA, which is shown by P^{32} incorporation to have been synthesized early in the fifth larval instar.

In the Cecropia silk moth, when adult development begins after the pupal diapause there is no discernible net increase in fat body RNA, but there is an increase in turnover, associated with renewed activity in protein synthesis. Other recent studies on RNA in insect metamorphosis have concentrated on particular developmental processes related to hormone action, Karlson, Sekeris and colleagues have examined changes in fully grown *Calliphora* larvae, at the stage when formation and tanning of the puparium follow upon the increasing titer of ecdysone. After injection of ecdysone into larvae taken about two days before pupation, the incorporation of P^{32} into RNA is elevated. With 1.5-hour pulse corporation, the maximal stimulation of 80 per cent over controls was observed at 4 to 7 hours after giving has hormone. A greater effect was observed in RNA from the cpidermis, a target tissue for ecdysone, and in RNA prepared with hot phenol from epidermal nuclei there was some stimulation within 1 hour of giving the hormone.

The nuclear RNA stimulated amino acid incorporation in microsomal preparations from rat liver or *Calliphora* epidemis. It has been reported from the same laboratory that ecdysone added to isolated epidermal nuclei stimulates their incorporation of C^{14}-uracil into RNA and that RNA prepared from pupating larvae when added to rat liver microsomal preparations acts as a template in the synthesis of dopa decarboxylase, but fuller data are required before one can interpret these observations with confidence. The results obtained with this system have been presented in support of the view that ecdysone and other hormones act by directly causing derepression of DNA in the cell nucleus and release of specific messengers. The evidence, however, does not eliminate the presence of further links in the chain of hormone action. The synthesis of RNA under the stimulus of ecdysone has also been investigated in the pupal wing epidermis of *Hyalophora cecropia*, during the initiation of adult development after diapause. During natural development, in pupae incubated after a period of chilling, the synthesis of RNA begins well before morphological change, and by the second day of visible adult development the content of RNA in the wing tissue is about fourfold that in diapause. After initiation of development by injection of ecdysone into pupae whose endogenous endocrine sequence is blocked by removal of the brain, stimulation of RNA synthesis is detectable at 8 to 10 hours, intense (about sixfold) at 24 hours, and falls off thereafter. Some effect on protein synthesis begins to appear at about the same time, but the rise is more gradual and more prolonged.

The RNA produced under the influence of ecdysone has been characterized in several ways. Its base composition (by pulse labeling with P^{32}) is intermediate between the bulk RNA and the DNA of the species and could be accounted for by synthesis of mixture of ribosomal RNA and DNA-like RNA. Fractionation of the RNA, after pulse labeling with uridine, showed incorporation in the 4s,17s and approximately 30s regions(corresponding to transfer, ribosomal, and ribosomal precursor RNA, respectively) and also in the residue from cold phenol extraction, which would contain nuclear RNA. The extent of stimulation by ecdysome was rather similar in each of these fractions. When assayed for stimulation of amino acid incorporation in an E. coli ribosomal system, the RNA produced early in Cecropia wing development was found to be highly active, but here are reasons for believing that such stimulation in a heterologous system may not detectable in the epidermis in diapause, become abundant early in development. It is concluded from this work that new RNA synthesis forms an important early part of the action of ecdysone and contributes to building the stage for increased protein synthesis. But the connection between hormone action and RNA synthesis, and the question of specificity in template RNA production, require further study. The picture in a tissue being awakened from diapause may, of course, differ from that in a tissue which, when already metabolically active, is influenced by ecdysone. But the observations on the Cecropia wing epidermis are consistent with those on the action of various developmental hormones in vertebrate material.

The juvenile hormone of the corpora allata, acting in the presence of ecdysone, permits growth and molting in insects but modifies their course, so that the characteristics of immature stages are retained. A plausible basis for this would be a regulatory effect upon RNA synthesis so that larval-specific templates are produced and adult templates repressed, but there is as yet no biochemical evidence for this. When juvenile hormone is injected into *Antheraea polyphemus* pupae at the beginning of adult development, the rate of RNA synthesis is increased, but this may be a result of stimulated output of ecdysone from the prothoracic gland rather than a direct effect on other tissues. A provocative observation for which no interpretation cab be offered concerns RNA-methylating enzymes.

In *Tenebrio molitor*, the enzymes active in adding methyl groups to E. *coli* tRNA are more active(with a characteristic day-to day pattern) in the early pupa than in the larva or pharate adult, thus portraying the approximate reverse of the U-shaped curve seen in various other metabolic activities. Another intriguing observation concerns the composition of insect ribosomal RNA. Analyses of ribosomal RNA from bacteria, yeast, and vertebrates invariably have found guanine as the most abundant base and the ratio of guanylic plus cytidylic acids to adenylic plus uridylic (G+C/A+U) to be greater than 1, 0. Insect ribosomal RNA, however, varies widely, the G + C/A + U ratio ranging, in several species examined, from 0.70 to 1.20 or higher.

A correlation was noted between ribosomal RNA composition and the rate of growth and rate of growth and development in a number of species, and it was suggested that these may be related because of differences in heat stability of the RNA (Mednikov, 1965). Inspection of the data, however, suggests that there is also a correlation with polygeny. The more primitive insects (Orthoptera) tend to have high G + C, Coleoptera and Lepidoptera are intermediate, and the most specialized group (Diptera) have RNA distinguished by low G + C content. Studies on the properties of ribosomes with such differently constituted RNA would be of interest. Understanding of the control mechanisms of rate and specificity of protein synthesis in insect metamorphosis will undoubtedly depend on the development and study of cell-free amino acid incorporating systems of insect origin. In such a system for *Tenebrio molitor*, the ratio of incorporation of tyrosine to that of leucine rises greatly during the last two days of adult development, which is believed to reflect the high tyrosine content of the cuticular proteins then being synthesized.

Deoxyribonucleic Acid

When one considers the pivotal position of DNA in determining the characters of organisms, and wealth of recent critical and ingenious research on this nucleic acid in various living systems, the paucity of information of DNA in insect metamorphosis is remarkable. For very few insect species do we even have reliable estimates of the changing quantity of DNA during development. Nucleic acid determinations with insect material require special attention to technique because of various complication that may arise, e.g., difficulty in securing quantitative extraction, the presence of compounds which interfere in colorimetric reactions, and uric acid which interferes in the ultraviolet range. In *Lucilia cuprina*, the total body DNA falls to about one half just before pupation and then rises rather steadily during the pupal stage and adult development, presumably reflecting breakdown of larval cells, followed by cell multiplication in the imaginal *anlagen*.

In the mosquito, on the other hand, there is apparently little change in DNA during metamorphosis, although in larvae there is the unusual situation that up to 30 per cent of the total DNA is found in the soluble supernatant fraction of homogenates. Patterns of DNA synthesis in different tissues are readily surveyed by autoradiography after incorporation of H^3-thymidine, and this technique has been applied to the development of saturiniid silk moths. ?In the larval molting cycle there is a burst of synthesis in most tissues centered about 3 days before the molt, but in the hemocytes and midgut, synthesis continues at a reduced rate at other times, and in the imaginal wing buds synthesis is apparently unrelated to the molt cycle. Shortly after pupation, in diapausing species such as *Hyalophora cecropia*, DNA synthesis stops in all tissues except the hemolytic and spermatogonia, in which a significant rate of incorporation continues. In a species without diapause(*Samia Cynthia ricini*), incorporation in most tissues comes almost to a stop at the time of pupation, but then resumes within a day with the onset of adult development.

At the conclusion of a pupal diapause, DNA synthesis appears to be correlated with the stimulation of growth by ecdysone. Adult development of moths is blocked by misogynic C, an inhibitor of Dna synthesis, and it is clear that DNA replication is an essential part of ecdysone-stimulated development. The cecropia wing epidermis developing after the pupal diapause has also been examined by biochemical methods (Wyatt, 1967b). When incorporation of thymidine is measured at intervals after injection of ecdysone, clear stimulation of DNA synthesis is not detected until about 48 hours-much later than the effect upon RNA. Thus, the stimulation of DNA synthesis seems to be separated from the earliest effects of the hormone.

During the 21-days period of natural development to the moth, two cycles of DNA replication centered at about the second and sixth days can be detected in the wing tissue. In a quest for the enzymic basis of the cessation of DNA synthesis in diapause, the crucial enzymes thymidine knase and thymidylate kinase were assayed in silk moth pupal wing epidermis. Both increased manifold at the initiation of adult development, yet diapause tissue did contain appreciable activity, so that the lack of these enzymes could not be responsible for the total cessation of synthesis.

Inorganic Ions

One further subject which must be mentioned is the remarkable inorganic ion content of many insects. In contrast to the usual composition of body fluids of vertebrates and most other animals, the higher insects, particularly phytophagous species, contain much potassium and little sodium, so that the Na^+ :K^+ ratio is far below unity. Further, the level of magnesium is often extraordinarily high (often up to 30 mM and occasionally to 100 mM, though part of this is chemically complexed). This situation is regarded as an adaptation to a vegetarian diet high in potassium and magnesium. Normal ion levels are maintained with the aid of active transport mechanisms, including the transport of k^+ through the gut wall.

In metamorphosis, the inorganic levels may be changed Thus, in the silkworm, *Bombyx mori*, falls in sodium and magnesium at pupation have been reported, though the data of different authors are not consistent. Since the ionic proportions found in different insect species vary greatly, their having an important role in metabolic regulation is perhaps unlikely. Little is known, however, about the distribution of ions between hemolymph and intracellular fluids in insects, and it is interesting that an effect of ion levels on protein production by blowfly larval fat body in vitro recently has been reported. The matter deserves further attention, in view of the indications that potassium can have roles in the regulation of gene activity in insects and in protein synthesis in various living systems.

8

DEVELOPMENT OF FRUIT FLY

The study of development of fruit fly was radically changed by the advances made in the genetics in the early 1900. With the growing perception that gene action underlies most embryological events, biologists began to seek genetic answers to fundamental development questions. *Drosophila* has played a significant role in this search, and the importance of this small fly for developmental studies continues to increase. Why *Drosophila*? Fruit flies are popular research subjects because their unique biological characteristics offer important advantages for experimental studies. *Drosophila* are small, rapidly breeding animals that can be raised in the laboratory in small glass vials on an inexpensive, easily prepared medium. Because of their rapid life cycle (14 days), many generations can be raised in a short time.

In addition, at each stage in their development. *Drosophila* have a complex body structure with many internal and external structure that can serve as morphological markers in development studies. Thousands of mutations have been isolated in *Drosophila*, including many that affect development, and powerful techniques are available for isolating new mutations that have defects in specific genes. *Drosophila* also offer molecular biologists a number of unique advantages. They can be grown in large pupulations, yielding gram quantities of material for biochemical analysis. In addition, it is relatively easy to move form a genetic mutation to cloning and characterization of the gene containing that lesion. Unlike any other organism, genes in *Drosophila* can be isolated based solely on their physical location in the chromosome, independent of their function.

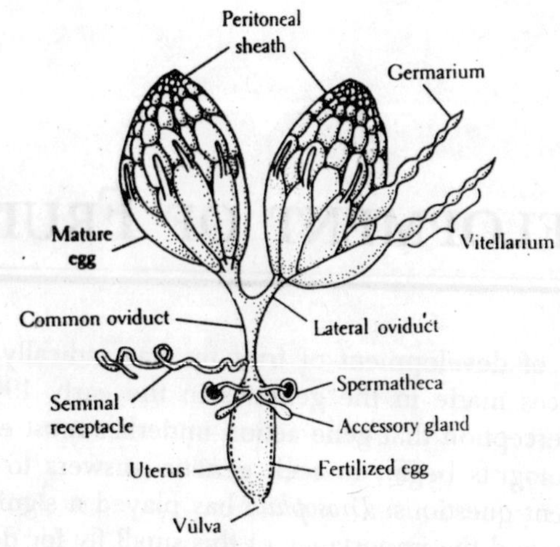

Fig. 8.1. The reproductive system of an adult female Drosophila.

Furthermore, genes can be reintroduced into the *Drosophila* genome and still show correct satial and temporal expression. Because of the ease with which *Drosophila* can be studied both genetically and biochemically, this organism is now the focus of intensive investigation. As a result, we know more about the molecular basis of development in *Drosophila* than in any other multicellular organism. In this chapter we shall review our current understanding of this complex process.

Developmental and Genetic Features of the Fly

The Drosophila Life Cycle Consists of Four Stages

The life cycle of *Drosophila* can be divided into four distinctly different stages: *embryonic, larval, pupal* and *adult* (or *imago*). During embryogenesis (0 to 24 hours after egg laying) the zygote develops from a single, large cell into a motile larva. The larvae pass through three stages, or *instars,* which are separated by molts. Larvae are highly specialized for feeding and increase greatly in size, especially in the third instar. At about 120 hours after egg laying, larvae stop eating and undergo *pupation,* during which the external larval cuticle expands, hardens, and becomes the tan-coloured pupal case, or

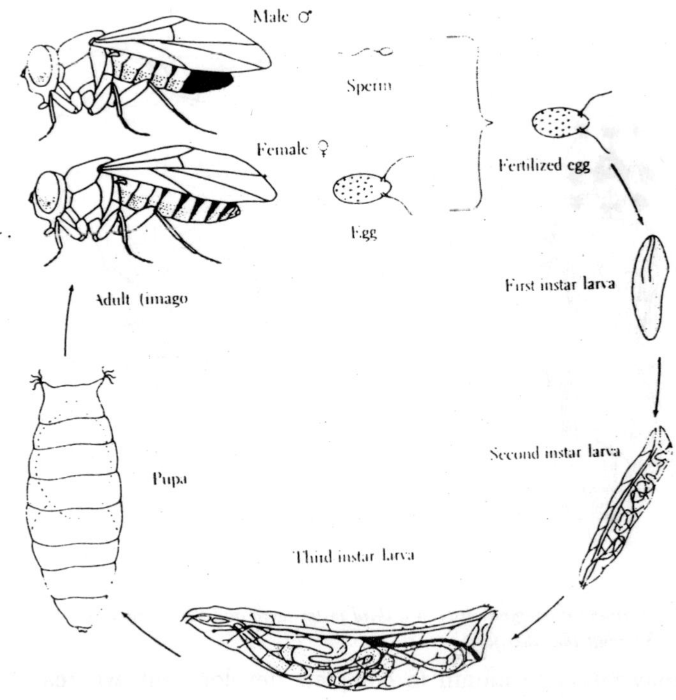

Fig. 8.2. The life cycle of Drosophila melanogaster.

puparium, within which ***metamorphosis*** occurs. Metamorphosis takes four days, during which time most larval tissues degenerate and new adult tissues form.

The precursor cells of the adult cuticular structures (eyes, legs, wings, etc.) are determined early in embryogenesis and are organized into discrete clusters called *imaginal discs.* There are nine pairs of imaginal discs in the *Drosophila* larva, along with a single genital disc, each of which occupies a characteristic position and forms one particular portion of the adult during metamorphosis. About ten days after egg laying, metamorphosis is complete, and the new adult burst open the puparium and emerges, a process called *eclosion.* Within two days the adults mate and the females begin to lay eggs. As with all insects, the time required to complete development depends on the surrounding temperature. Thus, *Drosophila* reared at

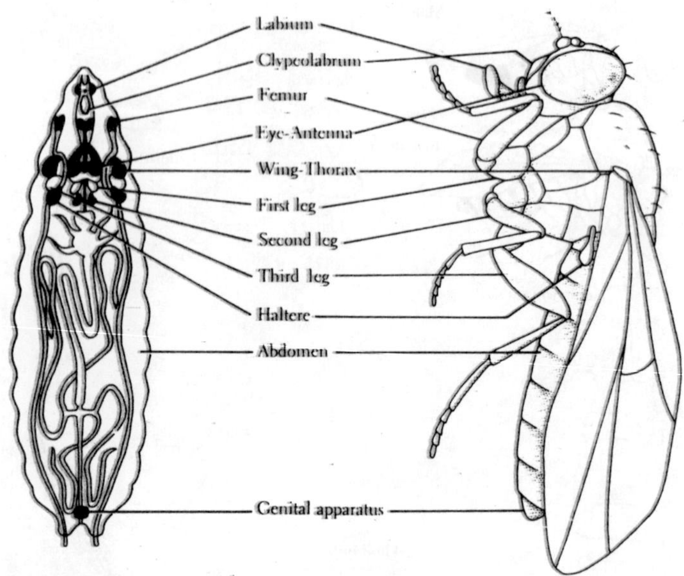

Fig. 8.3. Locations of imaginal discs in a third instar larva and the corresponding parts of the adult that they form.

15°C may take one month to complete development, whereas at 29°C development is complete in ten days.

Molting and Metamorphosis are Controlled by Hormones

The dominant feature of larval growth in the process of molting *Drosophila* undergo two *larval-to-larval* (L-L) molts between each instar, during metamorphosis there are two additional molts, the *larval-to-pupal* (L-P) and *pupal-to-adult* (P-A) molts. In each molt, the epidermal cells (either larval or adult) secrete a new cuticle, and the old external cuticle is shed. Molting in insects is controlled by hormone that regulate both the initiation of the molting process and the outcome (i.e. whether the molt is L-L, L-P, or P-A).

The regulatory hormones are *prothoracicotrophic hormone* (PTTH), *ecdysone,* and *juvenile hormone.* In *Drosophila,* as in all insects, the first step in molting is the secretion of PTTH by neurosecretory cells in response to external signals. Once secreted, PTTH stimulates the production of the steroid hormone ecdysone, the actual molting hormone. This production rises to a peak level before each molt. Finally, the nature of the molt is controlled by juvenile hormone. Juvenile hormone is so named because its presence ensures that

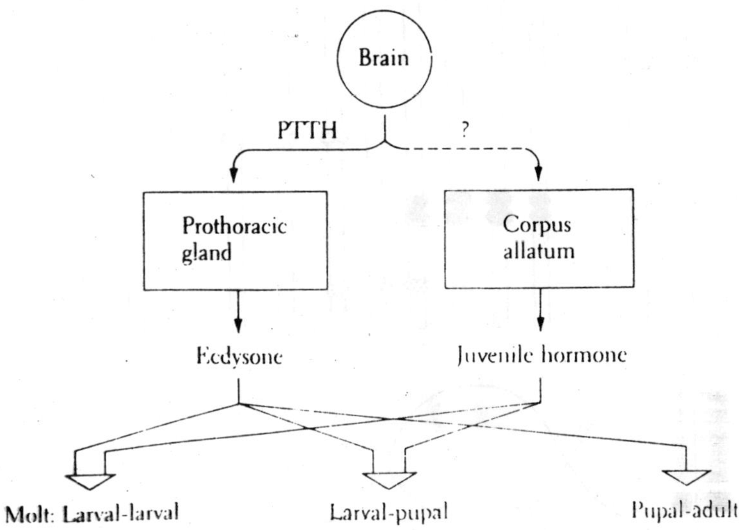

Fig. 8.4. The hormonal control of molting in Drosophila.

the molt will result in another larval stage. Prior to an L-L molt, the juvenile hormone level rises along with the ecdysone, whereas prior to L-P and P-A molts, only ecdysone is present. Michael Ashburner in Cambridge has demonstrated that during metamorphosis ecdysone triggers the formation of specific puffs in the polytene chromosomes of the salivary gland.

Recall from that puffing represents a direct visual manifestation of the transcription of specified genes. Ashburner found that about six puffs are rapidly induced in direct response to ecdysone. These puffs appear to encode regulatory proteins that act to repress their own expression and induce the formation of more than 100 new puffs, which in turn are believed to encode the proteins required for metamorphosis. Ecdysone thus triggers a hierarchical casade of genetic activity; its final effect on gene expression is mediated and amplified by a small family of regulatory proteins. The regulatory genes that are induced directly by ecdysone are currently being studied at the molecular level. Although at present little is known about their mechanism of action, we do know that these genes comprise unusually long transcription units, similar to other key development regulatory genes that will be discussed later in this chapter.

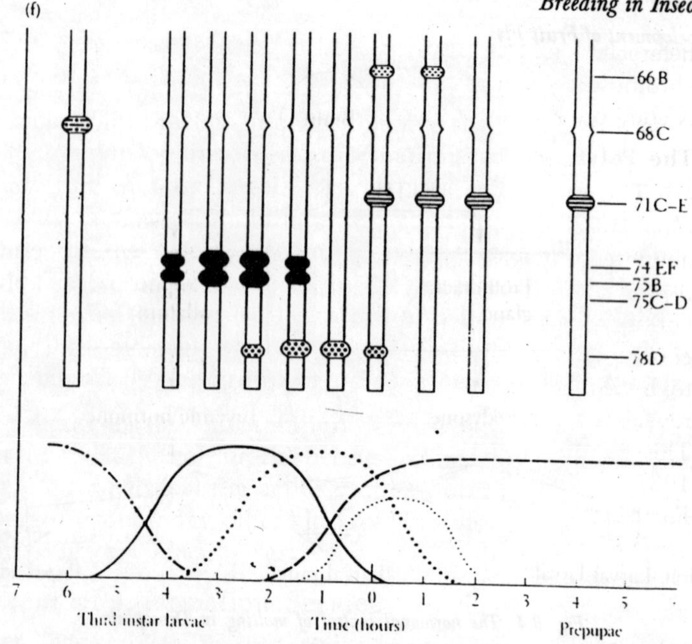

Fig. 8.5. Changes in the puffing pattern of the polytene chromosomes in response to ecdysone. The puffs at positions 74EF and 75B are representative "early" puffs, which arise in direct response to ecdysone. The puffs at positions 71C-E and 78D are "late" puffs, which arise in response to the early puffs products. This puffing pattern is in schematic form.

The Drosophila Genome is Very Small for a Higher Eukaryote

The genome of *Drosophila* contains four chromosomes. The first, or X, chromosome is *telocentric* (the centromere lies at one end) and contain approximately 39,000 kbp of DNA, of which one-third is heterochromatin; this DNA consists largely of repetitive satellite sequences, is not transcribed, and replicates late. The rest of the chromosome, the Y, consists largely of heterochromatin; the few euchromatin. The other sex chromosome, the Y, consists largely of heterochromatin; the few genes that it does contain are necessary for male fertility.

Females have two X chromosomes, and males one X and one Y. Similar to nematodes and unlike humans, sex in *Drosophila* is not controlled by "male-factors" on the Y chromosome but rather by the ratio of X chromosomes to autosomes. This is a fact of great experimental significance, as we shall see later. The second and third chromosomes are *metacentric* (centromere in the middle), and each contain 60,000 kbp of DNA, 17 percent of which is heterochromatin, mostly localized around the centromere. The fourth chromosome is very small, containing only 6000 kbp of DNA and is composed mainly of heterochromatin.

The Polytene Chromosomes are a Valuable Tool

The salivary glands of *Drosophila* larvae contain *polytene chromosomes* in their nuclei. These giant chromosomes result from about ten rounds of replication of euchromatic DNA, the products of which remain tightly pressed together along their length. Polytene chromosomes therefore represent giant versions of their diploid counterparts and allow the interphase chromosomes of *Drosophila* to be directly visualized by light microscopy. Such observation reveals a characteristic array of bands along the chromosome arms. This banding pattern was carefully recorded by C.B. Bridges in 1935, who produced physical maps of each chromosome. Furthermore, the polytene chromosomes are colinear with the genetic map and allow mutations caused by large chromosomal rearrangements to be mapped with great precision. In addition, as described earlier, in *situ* hybridization can be used to determine the site on the polytene chromosomes from which a cloned segment of *Drosophila* DNA is derived. The polytene chromosomes, therefore, provide a valuable reference point for both genetic and molecular analysis of gene function in *Drosophila.*

Genetic Analysis in Drosophila Uses Chromosomes with Deletions and Rearrangements

Mutations are the foundations of *Drosophila* studies and can be divided into two general classes. In the first class are *chromosomal aberrations*, physical changes in large portions of a chromosome (deletions, duplications, or rearrangement) that can after many genes. These mutations can be produced by X irradiation. The second class consists of *point mutations,* base changes within genes that can be induced by chemical mutagens such as ethyl methane sulfonate. The most extreme form of chromosomal aberration, *balancer chromosomes*, contain multiple inversion that effectively prevent these chromosome from recombining with their unrearranged homologues. This is of significant experimental value because the genetic properties of any chromosome bearing a new mutation can be preserved simply by maintaining it in heterozygous combination with a balancer chromosome. Chromosal deletions can also be used to determine the locations of point mutations; in this technique of *deletion mapping*, a fly that has a chromosome containing a deletion is mated with another fly carrying the point mutation of interest. If the point mutation is located within the portion of the chromosome that has been deleted, then the deletion-carrying offspring will show a mutant phenotype.

If the mutation is not located in this region, then the deletion chromosome will complement the mutation and the offspring will be wild-type. A basic assumption made in mutation studies is that the function of a gene can be deducted from the difference between the normal and mutant phenotype. Mutations in the *yellow* gene, for example, result in a lighter-than-normal body colour. This suggests that the normal form of the gene is needed to produce body pigment, thus to identify the genes that control a particular aspect of growth or development, mutations are collected that perturb that process. The mutations discussed in this chapter can be placed into one of several categories, based on their phenotype effects: ***matenal effect mutations, lethal mutations,*** and ***visible mutations.*** Maternal effect mutations affect the phenotype of the offspring by altering egg formation. These are studied to discover to what extent and in what manner maternal gene products stored in the egg contribute to normal embryogenesis. Lethal mutations are the most frequently induced mutations.

The stages in the life cycle at which lethal mutants die indicates when specific genes must function for the fly to survive. Conditional lethal mutations, such as temperature-sensitive mutations, are clearly of the most value, since their lethality can be suppressed under appropriate conditions. Visible mutations are often the easiest to study because their effect can be observed directly in the adult. In particular, visible mutations that affect the adult cuticle, such as *yellow*, are often used as genetic markers in studies of the development fate of cells.

Molecular Cloning of Drosophila Genes Involves Chromosome Walking

The *Drosophila* genomes is 1.4 x 10^6 bp in length, 30 times the size of the *E.coli* genome, twice the size of the *Caenorthabditis elegns* genome, and 5 percent in the size of the human genome. The relatively small size of the *Drosophila* genome simplifies the isolation first step in isolating a *Drosophila* gene is to cytogenetically map the gene in the polytene chromosomes by determining the location of appropriate mutations as described above. The investigator can then test a library of cloned *Drosophila* DNA segments to determine

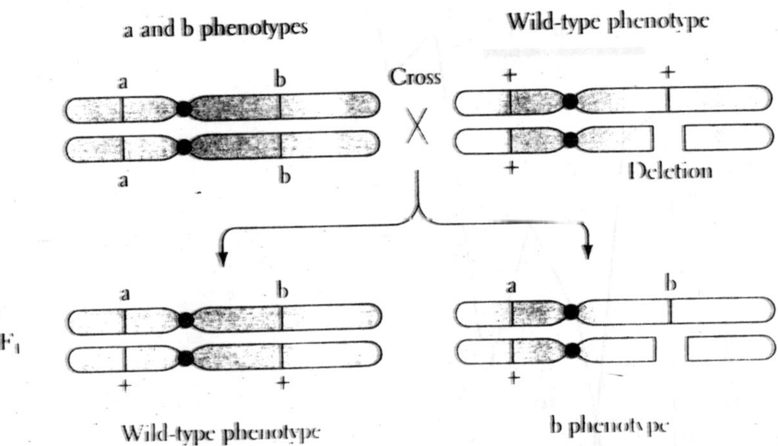

Fig. 8.6. Deletion mapping of point mutations. A fly carrying a chromosome homozygous for two different recessive point mutations (a and b) is crossed with a fly carrying a wild-type chromosome (a^+ and b^+) and a chromosome with a deletion. Since the b mutation is within the region deleted from the chromosome, some of the offspring will show the b phenotype.

whether any of these DNA fragments map near the gene of interest, as determined by *in situ* hybridization to polytene chromosomes. If such a DNA fragment is found, it can be used as the start point of a *chromosomal walk*. In the chromosomal walk, a *Drosophila* genomic library with the appropriate DNA fragment is screened in order to identify and isolate an overlapping set of DNA segments that contain that fragment.

The screening is then repeated, this time using the DNA fragments that lie farthest from the original starting point as hybridization probes. Thus, at each step is taken, it generates overlapping DNA segments that extend in both directions along the chromosome. The extent of the walk can be monitored by *in situ* hybridization, which reveals when the gene of interest is included in the cloned DNA fragments. Because each band plus the interband region in a polytene chromosome contains an average 30,000 bp, and because approximately 15,000 bp can be isolated in a single step, it not very difficult to cover large regions of the *Drosophila* genome. Furthermore, if no DNA fragments are found near the gene of interest, it is possible to dissect a single band from the polytene chromosomes using a scalpel and forceps.

The DNA segments isolated in this way can then be used as starting points for several chromosomal walks, which can be jointed

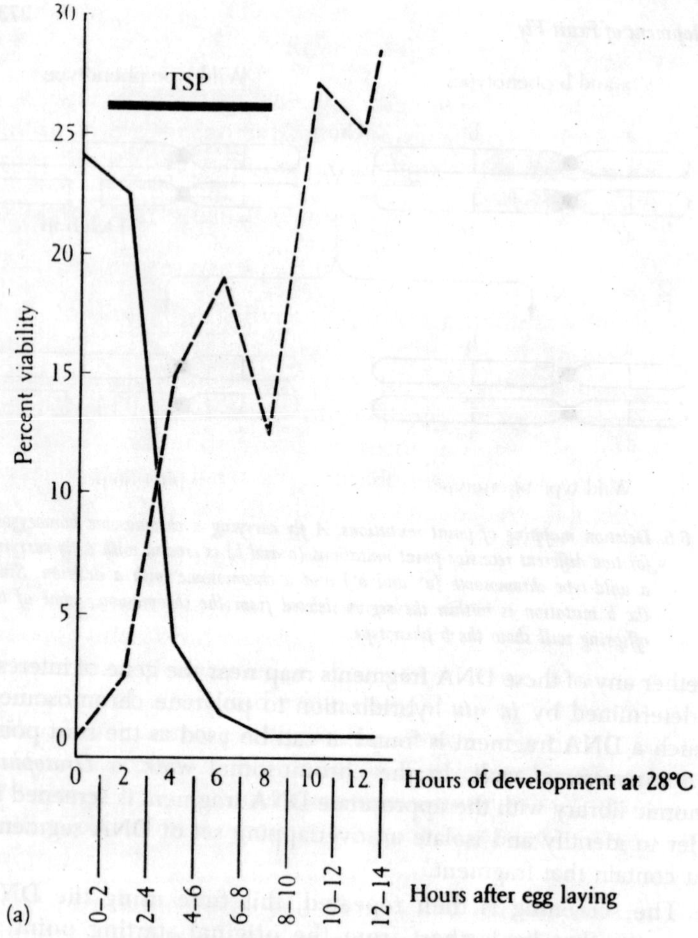

Fig. 8.7. Use of a temperature-sensitive lethal mutation to determine when a gene product is needed for viability. Embryos carrying a temperature-sensitive allele of ftz will live at 18°C. In this experiment, such mutant embryos were either shifted up in temperature from 18°C to 28°C (dashed line) or down in temperature from 28°C to 18°C (solid line) at different times after egg laying. They were then maintained at that temperature and the number of viable flies was determined. If the temperature is high early in development, all of the embryos die, but by six-ten hours after egg laying many embryos can survive at high temperature. The intersection of these curves defines the temperature-sensitive period (TSP) when the gene product is needed for survival.

to cover the region containing the gene. With these powerful techniques, in which the polytene chromosomes are always used as

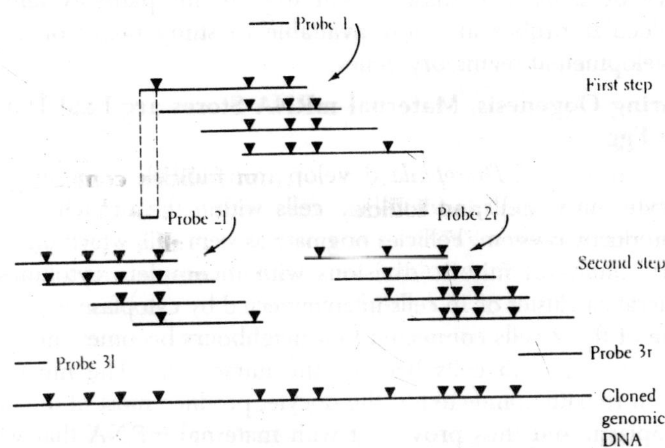

Fig. 8.8. Chromosome walking. A radioactive DNA probe is used to start the walk (probe 1). This is used to screen a bacteriophage lambda library of randomly cleaved Drosophila genomic DNA. Several overlapping phage inserts that can hybridize to the probe are isolated from the library and restriction mapped; this is the first step. The most distal fragments are then used for the next round of screening (probe 2l and 2r) to extend the walk another step. By piecing together the restriction map of all the overlapping phage inserts, one can reconstruct the restriction map of the Drosophila genomic DNA in that region. The lines represent Drosophila DNA segments, and the arrows represent restriction enzyme cleavage sites.

a reference point, any gene can be isolated that has been cytologically mapped, regardless of the function of that gene. This feat is not possible in any other higher organism.

Summary

The *Drosophila* life cycle follows the typical insect pattern: embryo, larva, pupa and adult. Larval moulting and metamorphosis are controlled by three hormones–PTTH, ecdysone, and juvenile hormone–via their effects on gene expression. The larvae have large polytene chromosomes in their salivary glands that are used to study gene location, function, and chromosomal aberrations. Many mutations are known to affect different stages of development in *Drosophila,* and these are classified by their phenotypic effects. Furthermore, virtually any gene in *Drosophila* can be molecular cloning.

Early Stages of Drosophila Development

In this section we shall see how maternal gene control the earliest stages of embryogenesis how that basic pattern of development is elaborated by the zygote genome to establish the body plan of the adult fly. In addition, we shall see how mutant have been used to dissect each step in this pathway and how molecular probes are now available to study many of the key developmental regulatory genes.

During Oogenesis, Maternal mRNA Stores are Laid Down in the Egg

The egg of *Drosophila* develop from follicles containing the oocyte, nurse cells, and follicle cells within the branched female reproductive system. Follicles originate as stem cell, which undergoes four rounds of mitotic divisions with incomplete cytokinesis to generate a cluster of 16 cells interconnected by cytoplasmic channels. One of the 2 cells connected to 4 neighbours becomes the oocyte, and the other 15 cells become the nurse cells. The nurse cells, which remain connected to the oocyte, produce most of the oocyte cytoplasm and thus provide it with maternal mRNA that will be translated during the early stages of embryogenesis. A single layer of follicle cells surround the oocyte and nurse cells. These follicle cells collect yolk proteins and other cytoplasmic elements from the hemolymph (the insect equivalent of blood) and pass them into the oocyte. They also produce the vitelline membrane surrounding the oocyte cytoplasm and the complex, multilayered chorion (eggshell).

In early Embryogenesis, all the Cleavage Nuclei Occupy a Single Cells

The initial 13 zygote nuclear division in *Drosophila* are very rapid (five to ten minute per division) and proceed without cytoplasmic division. The nuclei remain centrally located until the eighth nuclear division, when they begin to migrate in a stepwise manner toward the egg surface to form the *syncytial blastoderm*. Approximately 200 nuclei remain in the center to form the yolk, and 17 to 18 nuclei reach the posterior pole of the egg to form the pole cells, which develop into the germ cells. After the thirteenth mitosis, two hours after egg laying, cellular membranes begin to form around each nucleus, creating the *cellular blastoderm*. Gastrulation begins three hours after egg laying with the invagination of the mesoderm, which is followed by the invaginations of the anterior midgut and the posterior midgut. The pole cells invaginate at the posterior midgut at about five and a half hours. The region of mesodermal invagination, called the *germ band*, now begins to extend, or migrate, around the central yolk

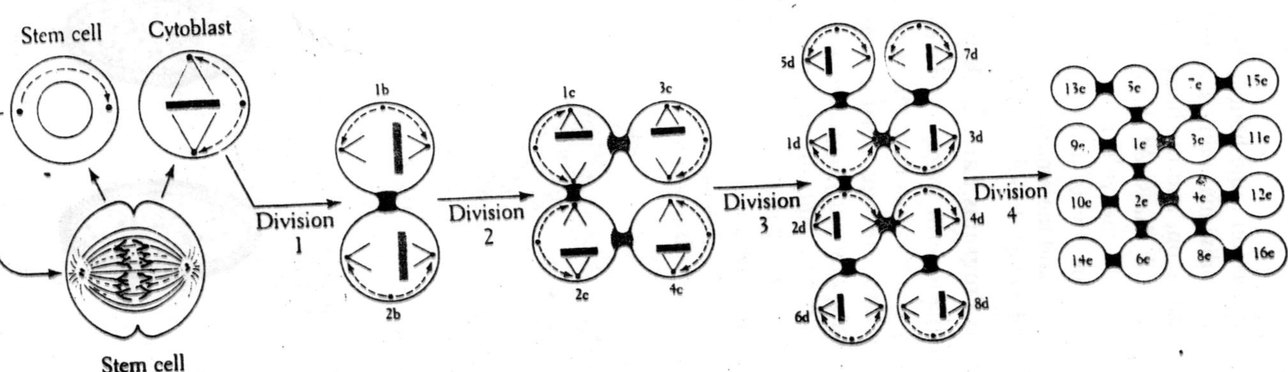

Fig. 8.9. Cell division leading to the formation of the egg chamber. The stem cell divides into a stem cell and a cytoblast. The cytoblast divides 4 times, with incomplete cytokinesis, and generates 14 cells connected by ring canals. Of these, either cell 1e or 2e becomes the oocyte; the others become nurse cells.

mass. At the maximum extension (nine hours), the germ band can be seen to divide into the individual body segments. Germ band extension now reverses direction, and the body segments begin to extend dorsally over the yolk mass. At this same time, the gut cells begin to migrate around the yolk mass internally, and the head assumes the larval shape. By 14 to 15 hours after egg laying these processes are complete and the embryo has assumed the larval form. The larva is fully developed and hatches at 24 hours.

The Maternal Genome Programs Early Embryogenesis

Before the cellular blastoderm stage, the development of the embryos is controlled by components from the maternal cytoplasm that are incorporated into the egg during oogenesis. These components can be identified by studying maternal effect mutations, where the phenotype of the mutant offspring depends not on its own genotype but rather on the genotype of the mother. One such cytoplasmic determinant that has a very early effect on development

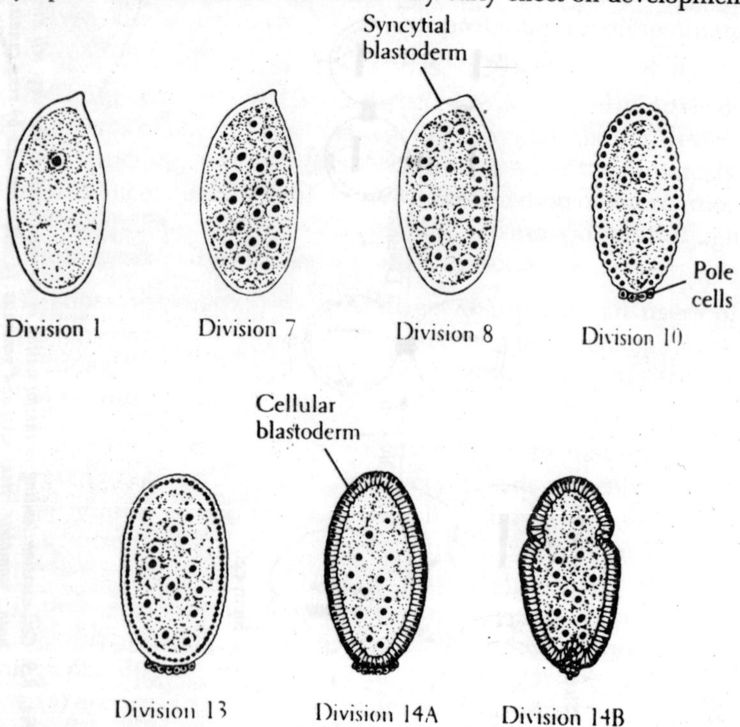

Fig. 8.10. Early embryogenesis.

is composed of *polar granules*, cytoplasmic particles that are found at the posterior end of the egg. These granules are incorporated into the posterior cells at the cellular blastoderm stage and direct those cells to become pole cells. The maternal effect mutation *grandchildless* perturbs this process. Mutant embryos have abnormal polar granules that do not migrate properly; thus the germ cells do not form and the resultant fly is sterile. No other region of the egg has visible cytoplasmic bodies with determinant properties. Although the polar granules appear to be composed of ribonucleoproteins, their exact biochemical composition is not yet known.

Some Maternal Effect Genes Dictate the Polarity of the Embryo

The best-studied maternal contribution to early embryogenesis is the establishment of polarity in the three axes of the embryo: anterior-posterior, dorsal-ventral, and left-right. Mutations such as *bicaudal* and *dicephalic* affect the anterior-posterior axis of the embryo. Some eggs laid by *bicaudal* females develop into embryos with two

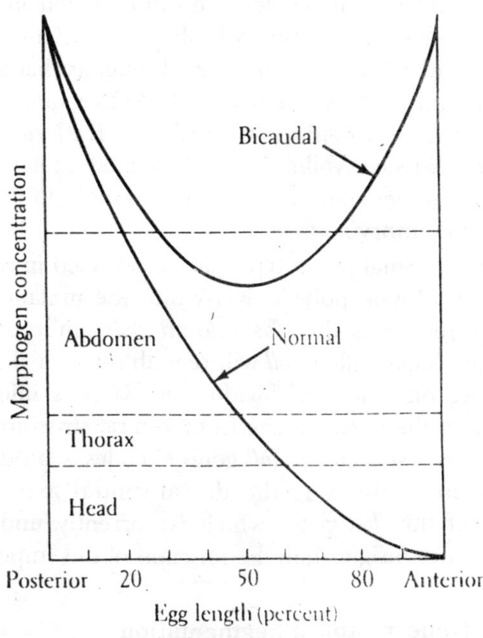

Fig. 8.11. A hypothetical morphogen gradient defining anterior-posterior structures in normal and bicaudal embryos. The horizontal scale show the anterior-posterior axis. The solid line represents a monotonic morphogen gradient in normal Drosophila embryos; higher concentrations of morphogen dictate more abdominal structures. The dashed line represents a U-shaped gradient in a bicaudal mutant embryo.

posterior end. Similarly, *dicephalic* female lay eggs that have anterior structures at both poles and no posterior ends. These unusual mutant phenotypes can be explained in terms of maternally derived *morphogen gradients* that run along the anterior-posterior axis. (A *morphogen* is a substance that dictates a particular developmental response–in this case, anterior or posterior structures.

A monotonic gradient of a hypothetical morphogen that specifies posterior structures could be present in wild-type eggs. *Bicaudal* eggs, on the other hand, could have a U-shaped gradient of this morphogen, thus generating two mirror image anterior ends. The same model could also explain the *dicephalic* phenotype. In this case, however, there is morphological evidence to support the gradient model. In wild-type female flies, the nurse cells are attached to the anterior end of the oocyte. In *dicephalic* females, however, nurse cells are evenly distributed on both sides of the oocyte. Thus, whereas a normal gradient is generated if the anterior-specific morphogen is fed only into the anterior of the oocyte, the nurse cells of *dicephalic* females feed the morphogen into both anterior and posterior ends, creating a U-shaped gradient and the resultant *dicephalic* embryo. Similar analysis of mutants that affect embryonic dorsal-ventral polarity has revealed the first biochemical evidence in favour of a morphogen gradient. Kathym Anderson and Christiane Nusslein-Volhard have identified a total of 11 genes that, when mutant, generate dorsal structures on both dorsal and ventral sides of the embryo.

The abnormal phenotype can be reversed in some mutants by injecting wild-type poly(A) RNA into the mutant embryos. Even more interesting is the observation that embryos mutant at the dorsalizing locus called *Toll* will reestablish their dorsal-ventral axis depending on where the wide-type RNA is injected. The cells proximal to the injection site from ventral structures. These results strongly suggest that the *Toll* gene encodes a product that plays a key role in establishing the dorsal-ventral axis. The molecular analysis of the *Toll* gene, which is currently under way, should provide more insights into the function of this important regulatory ene.

otic Gene Establish Segmentation

he zygotic genome is first expressed in *Drosophila* at the l blastoderm stage, at which time it elaborates the basis tablished by the maternal genome. To identify the zygotic genes that play a role in embryogenesis. Nusselin-Volhard and Eric Wieschaus first collected a large number of genes that cause embryo lethality. They have identified 147 loci that affect the larval cuticular pattern, which should represent more than 90 percent of the genes involved in this pathway. Of these, 24 loci affect the segmentation pattern of the embryo. Like all arthropods, *Drosophila* is segmented. The adult head consists of atleast three segments (called C1, C2 and C3, for short), the thorax consists of three segments (T1-T3), and the abdomen consists of at least eight segments (A1-A8).

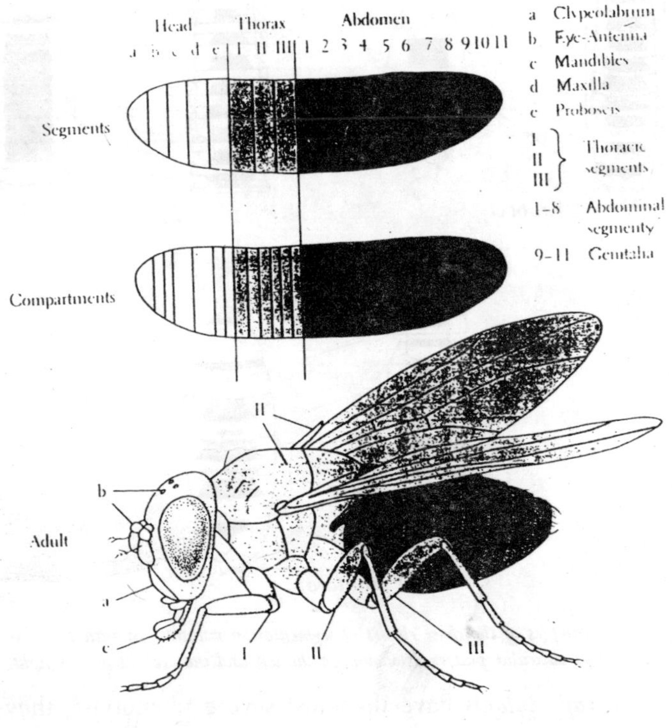

Fig. 8.12. The segmental organization of Drosophila.

The larval cuticular pattern shown in fig. 8.12 clearly depicts the ventral ***denticle belts*** that are characteristic of each segment. This patter is established early in embryogenesis and continues through the larval and adult stages. The 24 segmentation genes identified by Nusslein-Volhard and Wieschaus can be arranged into three classes according to there effect. Thus three different units of spatial organization appear to be involved in establishing this repeating motif.

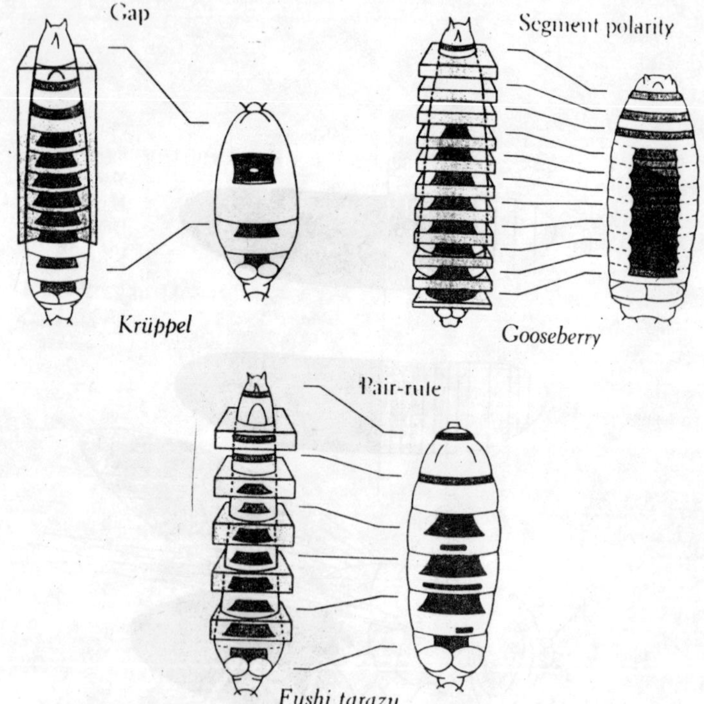

Fig. 8.13. Examples of the three classes of segmentation mutants. In each case, the wild-type cuticular pattern is shown on the left and the mutant on the right.

The *gap mutants* have the most severe phenotype; they are missing groups of adjacent segments in the thoracic and abdominal region. One group mutant, *Kruppel,* has only the A6, A7, and A8 abdominal segments, with a mirror duplication of the A6 segment. The gap genes are among the first of the zygotic genes to act and probably play a role in deciphering the anterior–posterior positional information present in the egg. Unlike the gap genes, the other two classes of segmentation mutants act uniformly along the length of the embryo, affecting the segmentation pattern with a periodicity of either one or two segments. Thus, ***segment polarity mutants*** are missing a portion of each segment and have the remaining region duplicated with mirror-image symmetry.

On the other hand, ***pair-rule mutants*** have a pattern deletion in alternating segments resulting in half the normal number of segments. The pair-rule gene that has been most extensive studied is *fushi tarazu* (*ftz*). Homozygous *ftz* mutant embryos are missing the posterior half of every other segment (starting with C2) and the anterior half of the adjacent. This gene has recently been cloned and characterized. In encodes a 1900 nucleotide mRNA that is first expressed at the syncytial blastoderm stage. The most striking feature of *ftz* transcription is that the RNA is expressed in rows of blastoderm nuclei that clearly reveal the segmented arrangement of the embryo, even prior to cellularization.

Furthermore, the regions of the embryo that express *ftz* RNA correspond to the region that are deleted in homozygous mutant embryos. These combined genetic and molecular data suggest that *ftz* plays a curcial role, along with other segmentation genes, in establishing the segmentation pattern of the embryo. To determine the spatial distribution of transcription for specific genes requires a powerful technique that has an important place in the study of *Drosophila* development. This technique utilized *in situ* hybridization, much as described earlier for polytene chromosomes, except that there radioactive–cloned DNA probes are used to determine where the complementary RNAs are located in thin (8-10 mm) histological sections of *Drosophila.* In this manner, the spatial distribution of transcripts of any cloned *Drosophila* gene can be determined during development.

Homeotic Genes Determine How Each Segment Develops

We have seen how the basic segmentation pattern of the fly is established during embryogenesis. Now the major question is, How do different cells within those segments become determined to form the elaborate structures found in the adult fly? A cell must fulfill two criteria to be determined. First, it must develop autonosmously. For example, if a cell determined to form a part of the leg is dissociated from tis neighbours and mixed with cells of a different type, it will still form leg structures. Second, a determined cell must transmit its determined state to its progeny. E. Hadron studied determination in the cells of imaginal discs by using the technique of *imaginal disc transplantation*. In this technique, portions of imaginal discs are removed from larva and cultured in the abdomen of female flies, where the discs continue to grow without differentiating to form adult structures.

The disc fragments can be cultured for months or years by transferring them to fresh adult hosts, or they can be tested for their ability to differentiate by introducing them into a larval host where they will undergo metamorphosis. Hardon observed that tissue derived from one disc was still capable of forming adult structures characteristic of that disc after months in culture, indicating that determination is very stable. After a long period of culture, however, a certain percentage of cells did change their determination and formed structures characteristic of different discs. These changes were termed *transdetermination*. Transdetermination involves several important features. First of all, it is not entirely random, because it occurs in specific directions at specific rates.

For example, genital cells never transdetermine directly into wing but do transdetermine into leg and then from leg to wing. Second, there is no difference in transdetermination between cells from a freshly isolated disc and cells that have transdetermined to that disc type. Thus genital cells that have transdetermined into a leg can further transdetermine in the same manner as normal leg disc cells. Last, only groups of cells transdetermine, not single cells. This observation, combined with the frequency of transdetermination, suggests that this phenomenon is not the result of spontaneous mutation. The experiments of Hadron and his coworkers raise intriguing questions: How are determination and transdetermination controlled? Is determination regulated by specific genes? If so, then mutations in these genes should lead to abnormal patterns of determination is imaginal discs. Such mutations, called *homeotic mutations,* have in fact been found and have been found and have been the subjects of intensive study.

Homeotic mutations produce a dramatic phenotype in which one portion of the larval or adult body is replaced by another. They do not affect the basic segmentation pattern of the fly but instead determine which structures develop within a particular segment. The similarities of homeotic mutant phenotypes to transdetermination phenotypes suggest that transdetermination may be caused by the occasional improper expression of homeotic genes.

Over 30 homeotic loci have now been identified. Most of these loci are clustered into two groups–the *antennapedia complex* and the *bithorax complex,* both on the right arm of the third chromosome. Mutations in the antennapedia complex affect the head and thoracic segments, whereas mutations in the bithorax complex affect the thoracic and abdominal segments. An example of a mutation in the antennapedia complex, called *Antennapedia* (*Antp*) transforms the antennae into legs.

Bithorax Genes Control Patterning in the Thorax and Abdomen

The bithorax complex has been extensively studied by E. Lewis at Cal. Tech for over 30 years. This complex is divided into three functional groups based on the complementation pattern of lethal bithorax mutants. The *Ultrabithorax* (*Ubx*) domain specifies the identity of the region encompassing the posterior half of the second thoracic segment (T2), the entire third thoracic segment (T3), and the anterior half of the first abdominal segment (A1). The other two functional groups, called *abdominal-A* (*abd-A*), and *abdominal-b* (*abd-B*), specify the identities of the abdominal segments. The *Ubx* domain is, tin turn, subdivided into four functional domains defined by the effects of four classes of recessive mutations: *abx, bx, pbx,* and *bxd.* For example, a fly homozygous for *bx* has the anterior portion of T3 transformed into the corresponding anterior portion of T2. A *pbx* fly has the posterior portion of T3 transformed into a posterior T2. The effect of these mutations are additive, so a *bx-pbx* double-mutant fly has two copies of the second thoracic segment and no third segment. Some of these mutants survive to adulthood to reveal the dramatic phenotype of a fly with four wings. Lewis has proposed a model to account for how each gene in the bithorax complex normally functions to maintain the state of determination in one region of the body.

According to this model, all cells in the second thoracic through the abdominal segments begin with the same original state of determination, which is to form T2. This ground state information is then suppressed in each segment posterior to T2 by the additive action of individual bithorax gene products. Thus, the *bx* and *pbx* functions suppress the T2 state in T3 to generate the appropriate T3 structures in the adult fly. Likewise, another set of bithorax genes that suppress the T3 state are activated in A1 and together with *bx* and *pbx,* dictate appropriate A1 development. This

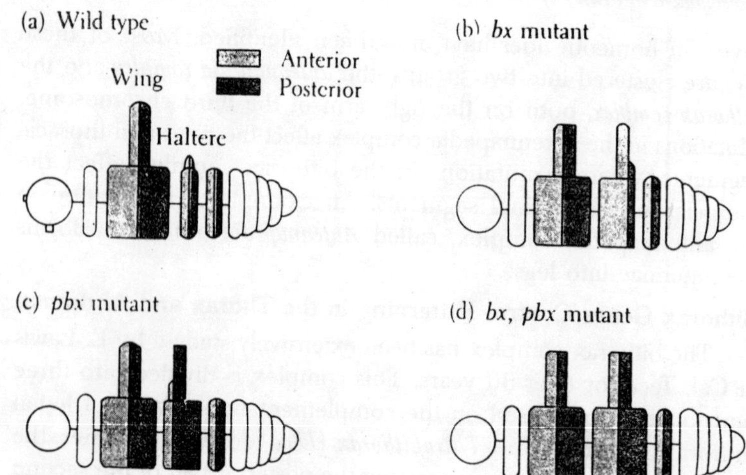

Fig. 8.14. Body segmentation pattern in (a) wild-type, (b) bx mutant, (c) pbx mutant, and (d) bx, pbx mutant flies.

combinatorial scheme proceeds to the last abdominal segment, where it is postulated that all functions of the bithorax complex are turned on. Molecular analysis of the bithorax region has revealed a complex pattern of transcription that mirrors the complexity of bithorax genetics. Much work remains to be done; even so, several significant features have emerged from our current knowledge of bithorax gene structure.

The *Ubx* domain can be divided into two unusually long transcription units, 75,000 and 25,000 bp in length, that correspond to the *Ubx* and *bxd* functions, respectively. In turn, each of the primary transcripts produced from these genes are differentially spliced to form several mRNAs of a more normal size. Interestingly, the cytoplasmic *bxd* RNAs do not contain any long, open reading frames and, therefore, may not be translated into proteins. More recently other homeotic genes have also been found to be unusually long.

In particular, the *Antp* gene spans a remarkable 100,000 bp. It is not yet known why these genes are so long, although this length necessarily imposes a minimal time in which the gene can be transcribed, because RNA polymerase moves at a constant rate of approximately 1000 bp per minute. This is one way in which the organism could control the timing of its gene expression. The distribution of *Ubx* transcripts has been determined by *in situ*

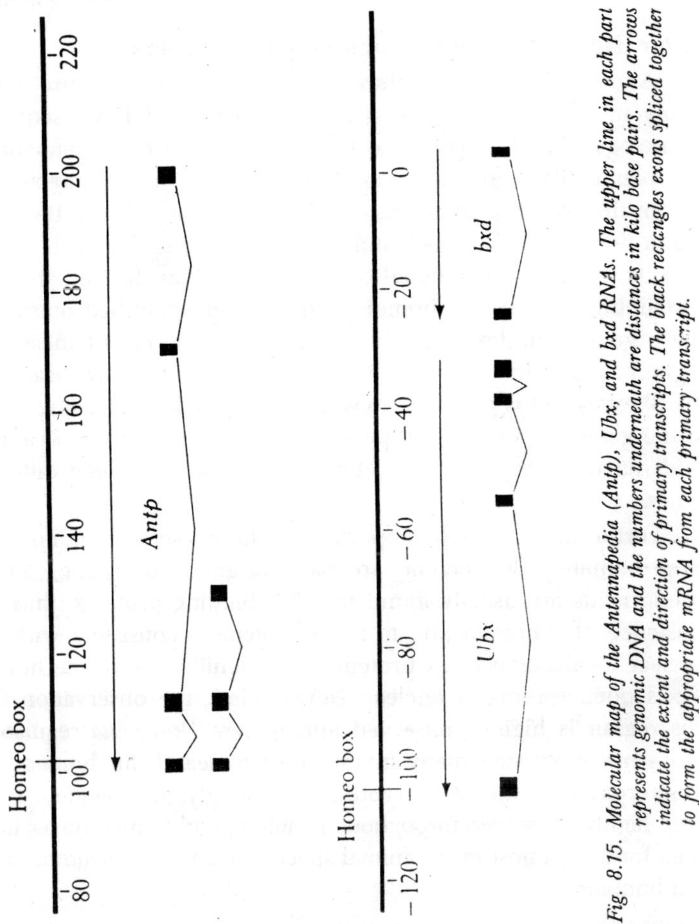

Fig. 8.15. Molecular map of the Antennapedia (Antp), Ubx, and bxd RNAs. The upper line in each part represents genomic DNA and the numbers underneath are distances in kilo base pairs. The arrows indicate the extent and direction of primary transcripts. The black rectangles exons spliced together to form the appropriate mRNA from each primary transcript.

hybridization of radioactive *Ubx* DNA probes to histological sections of embryos. As was the case when the *ftz* gene was studied by this technique, the region in which the *Ubx* gene is transcribed corresponds to the region affected by *Ubx* mutations. This method has recently been taken one step further by using fluorescently labelled antibodies to search out sites where the proteins encoded by *Ubx* are localized. As expected, this procedure revealed a distribution of *Ubx* proteins coincident with the regions in which the gene is transcribed. Significantly, the *Ubx* proteins are localized in the nucleus, suggesting that their regulatory properties may be mediated by their ability to bind specific DNA sequences.

Some Homeotic Genes Contain a Homeo Box

Another interesting feature of homeotic gene structure is that many of these genes share a 180-bp segment of DNA sequence homology, called a *homeo box.* Five genes within the antennapedia complex contain homeo boxes. There are only three homeo boxes within the bithorax complex, one located within each of the three major domains (*Ubx, abd-A,* and *abd-B*), suggesting that each homeo box may define a functional domain. More than five other genes containing homeo box homology have been identified outside of these gene complexes. Later we shall discuss one of these, the *engrailed* gene. In all cases studied, the homeo box is located in the 3' exon and is coextensive with the translational reading frame of the gene. As much as 80 percent to 90 percent of the 60 amino acids encoded by the homeo box are identical between different genes.

Furthermore, a high proportion of the conserved amino acids (approximately 30 percent) are basic (arginine or lysine). These amino acids are usually found in DNA-binding proteins; thus the existence of a homeo box in the *Ubx* gene is consistent with the nuclear localization of its proteins. The significance of the homeo box sequence remains unclear. Nevertheless, the observation that this region is highly conserved among key *Drosophila* regulatory genes has motivated many investigators to search for homeo box counterparts in other eukaryotes, interestingly, this sequence has been highly conserved throughout evolution, and homeo boxes have been fond in almost every animal species examined, including mice and humans.

Summary

The body plan of the adult fruit fly is established by the activities of three classes of genes: maternal effect genes, segmentation genes, and homeotic genes. The polarity of the egg is formed during oogenesis by maternally encoded gene products. This pattern is elaborated during embryogenesis, when the zygotic segmentation genes establish the basic metameric repeated pattern that is evident in the larval and adult fly. Finally, the identity of each segment is provided by the homeotic genes, which read the positional information along the anterior-posterior axis and accordingly dictate the formation of structures and tissues appropriate for that region. Thus, the rudiments of the adult fly are assembled in a stepwise manner during oogenesis and early embryonic development. The genes that comprise these three classes have been identified by mutational analysis and are now being characterized at the molecular leve. In general, the distribution of specific transcripts within the developing embryo correlates with the regions affected by mutation. Furthermore, some of the homeotic genes are unusually long, and some contain a 180 bp DNA sequence homology called a homeo box.

DETERMINATION IN DROSOPHILA

Fate Mapping with Genetic Mosaics

Important questions can be asked about the determination decisions made at the cellular blastoderm stage. For example, does determination occur in single cells, each of which then gives rise to a particular tissue, or does determination occur in groups of blastoderm cells? Also, how are these determined cells arranged in the blastoderm? Unfortunately, it is not easy to observe the cells that become determined to form the adult structures early in development. Instead, much important information has been obtained indirectly, from studies of male-female sex mosaics called *gynandromorphs*. Because sex determination in *Drosophila* depends on the number of X chromosomes in the cell, the loss of one of the X chromosomes from a female cell (XX) produces a male cell (XO) whose mitotic descendants will all be male cell. Individuals in which such a loss has occurred as part male and part female and are known as *gynandromorphs*.

By genetically marking the X chromosome so that its loss causes a change in, for example, cuticular pigmentation, the patches of male and female tissue can be distinguished later in development. These studies use a special X chromosome that is lost at a high frequency very early in development. As a result, this technique allows researchers to focus on developmental decisions made during early embryogenesis. For example, the observation that all of the structures derived from imaginal discs in gynandromorpha can be male-female mosaics reveals that imaginal discs originate form groups of cells and not individual cells. Furthermore, the smallest size of a mosaic patch observed in structures derived from an imaginal disc indicates the number of cells comprising that disc primordium. The greater the distance between two cells in the blastoderm, the more likely that the structures arising from those cells will be of different sexes in a gynandromorph. Thus, by studying the distribution of mosaic tissue in a large number of gynandromorphs, a fate map can be constructed of those tissues in the cellular blastoderm.

The fraction represented by the number of times two structures are of one sex relative to the number of times those structures are mosaics gives a numerical value that indicates the distance between their precursor cells in the blastoderm. The unit of these measurements is called a *sturt* in honour of the person who developed the technique. A, Sturtevant. A *blastoderm fate map* can be derived from these data by plotting the different distances; it is striking that the arrangement of precursor cells determined in this manner looks very similar to the arrangement of these cell based on their morphology.

Segments are Subdivided into Compartments

Because the X chromosome in gynandromorph studies tends to be lost very early in development, this technique is useful only for studying early developmental decisions. To study developmental decisions made at later times, investigators can induce recombination between chromosomes during mitosis by irrdiation at any time during development. By using chromosomes that carry a viable recessive mutation, such *mitotic recombination* creates a cell that is homozygous for the mutation and thus displays the appropriate phenotype.

Meanwhile, surrounding cells that have not undergone recombination are heterozygous for the mutation and do not display the phenotype. As discussed regarding fate maps of corn, all the cells in one patch are a clone, the descendants of a single cell. If a determined cell undergoes recombination, then its mitotic progeny will also be determined and thus be restricted to forming a particular portion of the adult structures. If, however, an undetermined cell undergoes recombination, its progeny may be differently determined. Hence, by irradiating at different times during development and then looking at the relationship of normal and recombinant clones in the adult flies, it is possible to determine when development commitments are made. Antonio Garcia-Bellido, working in Spain, examined the development of the imaginal wing-discs using induced mitotic recombination and discovered that the range of structures which a cell can form changes as development proceeds.

Recombination induced within 24 hours or egg laying generates clones that can form any structure in the anterior part of the wing

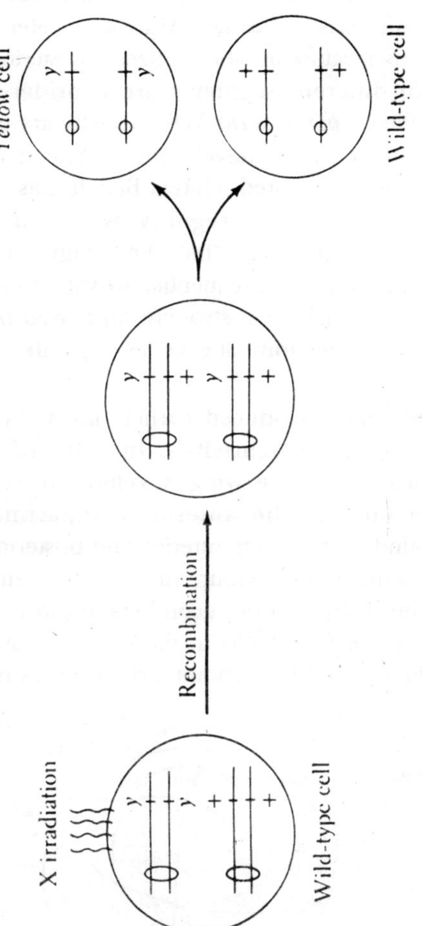

Fig. 8.16. X irradiation of a cell in mitosis causes recombination between homologous chromosomes. The recombined chromosomes segregate with mitosis to form a yellow mutant cell and a wild type cell. These cells then divide to form clones of yellow or normal tissue.

or in the posterior part, but never in both. Irradiation slightly later (24 to 48 hours) results in clones in the anterior and posterior areas that form either dorsal or ventral wing structures, but not both. These restrictions on position result from sharp boundaries that the clones can contract for hundreds of cells without crossing. Garcia-Bellido called such restrictions *compartment boundaries,* and the regions of the disc that they delimit, *compartments.* Compartment boundaries are most likely present within each segment of the fly, dividing in into anterior and posterior compartments. Thus, we can now see that it is in fact compartment boundaries that are recognized by homeotic mutations such as *Ubx* and not just some indiscriminate border within each segment.

The boundaries defined by mutations in the bithorax complex have motivated Alphonso Martinez-Arias and Peter Lawrence to propose a new nomenclature in which adjacent posterior and anterior compartments of different segments are considered a discrete functional unit called a *parasegment.* What exactly are compartments and what role do they play in development? The answers to these questions are still being debated. Garcia-Bellido has proposed that compartments represent the regions of activity of specific *developmental control genes,* and that the progressive nature of compartmentalization reflects a sequential activation of these control genes. This hypothesis has been strongly supported by the finding that the activity of some homeotic genes is limited to particular compartments.

The *engrailed* gene introduced earlier has a direct effect on compartment boundaries themselves. In *engrailed* mutants, the posterior compartment of the wing develops structures that are normally found only in the anterior compartment and the compartment boundary between anterior and posterior wing tissue is absent. This homeotic behaviour can also be seen in clones of *engrailed* cells induced by mitotic recombination in a normal wing. However, clones of *engrailed* cells in the anterior compartment of the wing obey the compartment boundary, whereas mutant clones

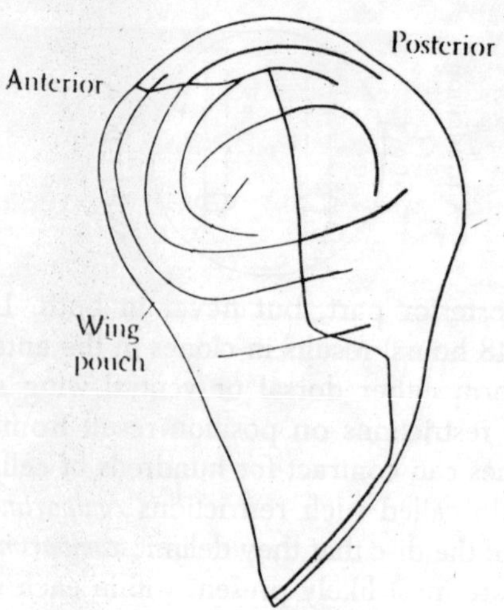

Fig. 8.17. Schematic figure showing the boundary of the anterior and posterior compartments in the wing disc.

in the posterior compartment do not recognize the boundary and cross into the anterior compartment. This suggests that *engrailed* function is essential for proper anterior–posterior compartmentalization and that the product of this gene must be present in posterior cells to distinguish them from anterior cells. In agreement with this prediction, *in situ* hybridization of *engrailed* DNA probes to RNA in intact imaginal discs clearly reveals that *engrailed* RNA is present only in the posterior compartment of those discs.

Summary

Commitments made early in the development of *Drosophila* can be studied by creating male-female mosaic flies called gynandromorphs. Analysis of gynadromorphs is used to determine approximately how many progenitor cells form a particular adult structure and where those cells form a particular adult structure and where those cells lie on the blastoderm fate map. Later developmental commitments can be studied by using X-ray-induced mitotic recombination. This technique has revealed compartment boundaries within each segment. The *engrailed* gene plays a key role in established anterior and posterior compartments.

Gene Regulation in Drosophila

In this section we shall review some newly developed techniques that greatly increase our ability to study gene regulation in *Drosophila*. In particular, we shall see how genes can be reintroduced into *Drosophila* in order to study their role in development.

Transposable Elements Allow Genes to be Cloned and Then Reintroduced into the Drosophila Genome

As we saw in maize, transposable elements provide a valuable tool for studying the genome. In *Drosophila* there are four classes of transposable elements. We shall discuss two of these here: *copia elements* and *P elements*. *Copia,* the first transposable element found in *Drosophila*, was identified because of its high reiteration in the genome. In spite of the low frequency of transposition of *copia*, its insertion has caused a variety of spontaneous mutations.

In such cases, a cloned *copia* DNA probe can be used to recover those mutant genes from a genomic DNA library. Thus, the presence of a *copia* element within any gene in *Drosophila* provides a molecular tag by which that gene can be isolated. However, this technique of *transposon tagging* is clearly limited by the low frequency of *copia*

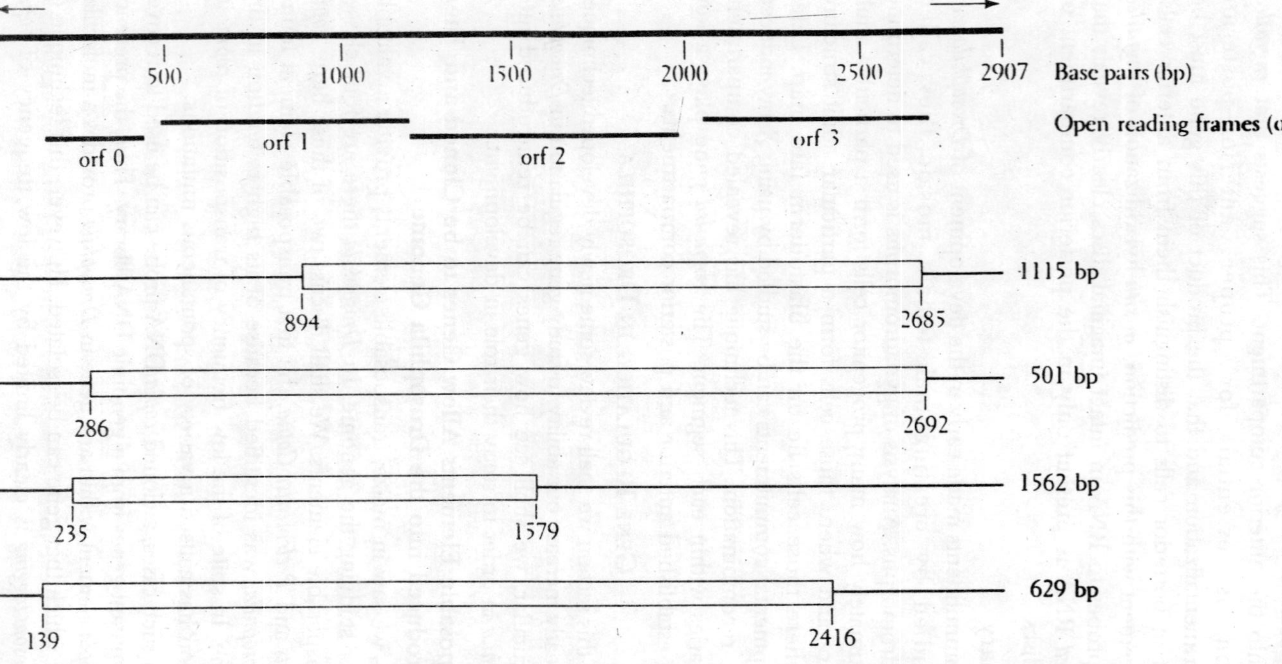

Fig. 8.18. The full-length 2907 base pair (bp) P element is shown on top, with the locations of the 31 bp inverted terminal repeats (arrows) and translational open reading frames (orf). Four defective P elements are shown below, with the deletions denoted by an open box. Note that these defective P elements have retained their terminal repeats and thus are capable of further transposition.

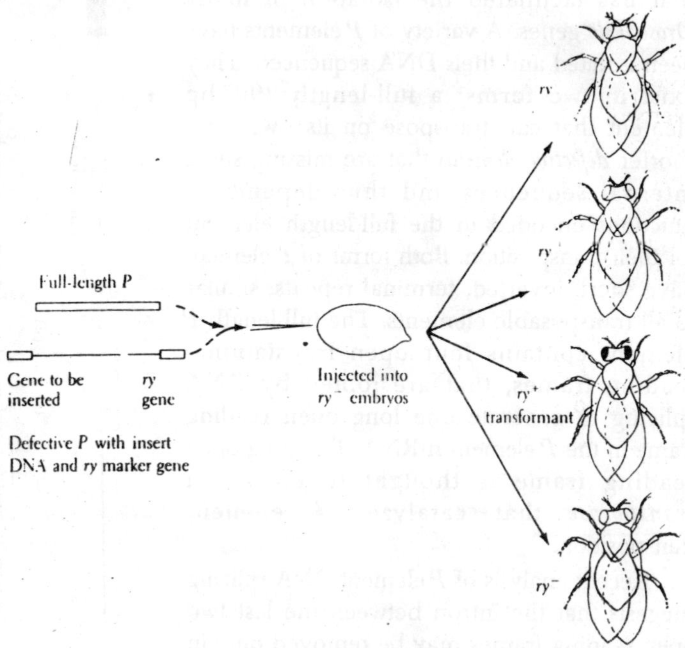

Fig. 8.19. P-element-mediated fly transformation.

transposition. For ***transposon tagging*** to be most useful, a transposable element that move with a high frequency under defined conditions would be ideal. *P* element provide such a system.

P transposable elements were discovered by G. Rubin and his coworkers during their study of the molecular basis of ***hybrid dysgenesis***. The phenomenon of hybrid dysgenesis occurs when a male fly carrying *P* transposable elements in his genome (P strain) mates with a female fly that has no such elements (M strain). Under these conditions, the *P* elements are destablized and transpose with a high frequency in the germ line of the zygote, generating a high rate of mutation. The mutant offspring are P strain in character, and the induced dysgenic mutations remain stable until the fly is crossed to an M strain female. Thus, by inducing hybrid dysgenesis and screening the offspring to identify individuals with the appropriate mutant phenotype, it is possible to determine when a *P* element has jumped into any gene of interest.

The gene can then be isolated as described earlier for mutations associated with *copia* insertion. This is a very powerful technique that has facilitated the isolation of many *Drosophila* genes. A variety of *P* elements have been isolated and their DNA sequenced. They exist in two forms: a full-length 2907-bp element that can transpose on its own, and shorter *defective elements* that are missing some internal sequences and thus depend on functions encoded in the full-length element for their transposition. Both forms of *P* element have short, inverted, terminal repeats, similar to all transposable elements. The full-length *P* element contains four open translational reading frames, that are joined by RNA splicing to generate one long open reading frame in the *P* element mRNA. This long open reading frame is thought to encode a *transposase* that catalyzes *P* element transposition.

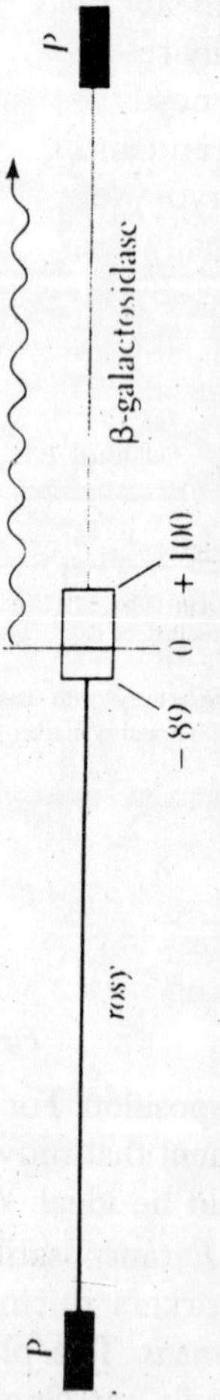

Recent analysis of *P* element RNA splicing suggests that the intron between the last two open reading frames may be removed only in the germ line. This elegant example of regulating gene expression in a tissue-specific manner would account for why hybrid dysgenesis occurs only in the germ line and never in somatic tissue. The observation that defective *P* elements can transpose in the presence of a full-length *P* element suggests to G. Rubin and A. Spradling that *P* elements provide a means of introducing DNA stably into the *Drosophila* germ line. This is done by inserting the DNA of interest into a defective *P* element while maintaining the inverted terminal repeats that are essential for its transposition. This DNA construct is then injected near the pole cells of early *Drosophila* embryos along with full-length *P* element DNA. The full-length element provides the transposase needed for the defective *P* element construct to transpose into the germ cell DNA. This technique is highly efficient: approximately 10 percent of the injected embryos carry the defective *P* element construct in their germ line and pass that new genetic information on to their offspring.

Fig. 8.20. P element carrying rosy gene (to identify transformants) and a segment of the heat-sock promoter fused to the E. coli β-galactosidase gene.

In addition to a high frequency of insertion, this technique usually results in the insertion of a single element into the genome. The newly integrated gene can then function normally, both in time and space, throughout development. In this case, the injected embryos were mutant at the *white* locus. This mutation causes the normally red eyes of the adult fly to loose all pigment and become white. By using *P* elements to insert a normal copy of the *white* gene into the *white* mutant genome, it is possible to rescue this mutation and generate flies with normal eye colour. In fact, such a rescue experiment is the best way to show that a cloned gene corresponds to a particular genetic locus.

Marker Genes Simplify the Study of Gene Expression

Transformation has also been used to study how gene expression is regulated in *Drosophila*. This puffing is known to be caused by a dramatic increase in the rate of transcription of the heat shock genes. *P*-element-mediated fly transformation has been sued to determine which regulatory DNA sequences flanking the promoter of the heat shock genes are necessary for controlling transcription. This experiment was facilitated by dissociating the heat shock promoter from its normal coding region and fusing it to another coding region whose protein product can be easily assayed.

9

DEVELOPMENTAL ASSEST IN INSECTS

In insects the developmental assest is called *diapause.* The diapause may be defined as a temporary interruption of development in a specific stage of ontogenesis which, in a particular species, would occur even if the insects were kept under optimum conditions. A span of opposing conditions of a definite character is often necessary before development can recommence. The opposite of diapause is quiesence, where inhibition of development is the immediate consequence of adverse conditions, at the end of which development is instantly renewed. It has been mentioned that the beginning of a 'true' diapause is independent of current environmental conditions; this does not mean that environmental conditions play no part at any point in the development of a given species in determining the occurrence of diapause in the most unfavourable conditions for the development of species, for example the low temperatures of winter or summer drought, and its adaptive value for the species distinctively demonstrate the role played by environment in the evolution of a given speices.

Apart from this common phylogenetic adaptation, is so-called 'facultative' diapause, the ambient conditions during earlier stages of life-history of the species concerned have been found to decide the occurance of the genetically conditioned 'facultative' diapause. More information on the principle of diapause led to the discovery of two types of deviations from the above 'classic' definition. The first includes cases in which a state corresponding to 'true' diapause may be induced within a long phase of ontogenesis, as in the IIIrd and VIIth larval instar of *Dendrolimus pini*, or during the larval development of *Lucilia*, when any opposing conditions such as unduly dry or moist food and wrong temperature induce arrested growth. This state is merged with and determined by various hibernating habits, as in *Eurygaster*, *Leptinotarsa* or Coccineilidae. In the second type, typical diapause can be disturb not only by adverse conditions but also artificially, e.g., by the action of long photoperiods, as in *Pyrrhocoris.* This can be easily confirmed by the recent discovery of the neurosecretory mechanism of diapause. As early as 1936, Wigglesworth showed that removal of the brain (by decapitation) at any point of postembryonic development induced in *Rhodnius prolixus* a state corresponding to diapause in every respect. In fact, all transitions between the two extremes, i.e. true diapause and quiescence, are found in various species, but those of the latter type are much more abundant, common, because of high selective valued.

Diapause Development

The span of adverse conditions (coldness) necessary for the recommencement of development varies with the species from two or three weeks to several (4 to 5) months. This period of arrested morphological development does not involve a complete rest in the physiological sense of the world; it is some what the time required for specific physiological changes to occur, which makes it possible for the insect organism to restart development. It is therefore called the period of diapause development (cf. Andrewartha, 1952). The diapause development of a particular species has a definite minimum length at certain temperature. In most species where this aspect has so far been studied, the maximum temperature for diapause development is around 5°C. With both decrease or increase of temperature there is a remarkable increase in the length of diapause development, ranging from several weeks to several years as the temperature approaches the 'living' optimum (20 to 25°C), while development practically stops at temperatures below 0°C.

The Stage at which Diapause Occurs

The occurance of diapause is generally at the most varied stages of ontogenetic development in different insects. For any one species, however, it is joined with a particular stage of development. There are many cases of embryonic diapause occuring at early embryonic stages (as in *Orgyia antiqua* or *Bombyx mori*), near the middle of embryonic development (as in *Locusta migratoria*, other Orthoptera and some of the Lepidoptera), or at the end of embryogenesis in a fully developed larva prior to hatching (as in *Lymantria dispar*). In the same manner, diapause can occur in any instar of larval development, for example the first instar in *Cacoecia fumiferana*, the second in *Euproctis chrysorrhoea*, the third in *Gastropacha quercifolia*, the penultimate in *Gryllus campestris* or the last larval instar in *Macrothylacia rubi*. Very often diapause occurs in the pupal instar, as in Satiurnia pyri, Hyphantria cunea, etc. Imaginal diapause mainly in-volves those parts of the body which continue to develop during adult life, primiarily the ovaries of female insects (e.g. *Anopheles maculipennis* or *Pyrrhocoris opterus*). There are rave cases when diapause occurs at two separate stages within one generation (e.g. in Denrolimus pini or Cephalcia abietis; see later).

Obligatory or Facultative Diapause

In few diapause manifests itself in each generations with out any interference of the environmental conditions the animals experience. In others, it occurs only in some generations, when specific environ-mental conditions prevail or have prevailed at an earlier stage. The former type mainly occurs in insects with a long period of post- embryonic development lasting one or more years. When it happens in each generation, the diapause is termed obligatory. The second type occurs in insects which have more than one generations a year, and appears at the appropriate stage only in those generations which bridge the period of unsuitable conditions, e.g. the overwintering generations. In such instance diapause is facultative and is dependent on the appropriate conditions in the period of its determination. From the genetic point of view, either type of development (with or without diapause) has the character of a hereditary modification, as known among the castes of the social insects, for example, or in many plants, e.g. the dandelion (*Taraxacum officinale*).

Special mechanisms are usually developed is some of the stages occuring before diapause when it is probably decided by specific conditions of, for example, photoperiod and temperature, Whereas obligatory diapause is mainly associated with univoltine species, facultative diapause occurs in polyvoltine ones, i.e. bivoltine, trivoltine, tetravoltin or multivoltine. Alterations between obligatory and facultative diapause are known also where development can continue without a break for many generations under suitable conditions, but may easily be interrupted when unsuitable conditions appear. When induced in such species, the delay in development has the character of a true diapause, but may easily be broken by specific environmental conditions, e.g. long photoperiod, and thus occupies an intermediate position between diapause and quiescence. This kind of intermediate position between diapause and quiescence. This kind of evolution is descrnible in lucilia erythrocephala and *Dendrolimus pini.*

The Physiology of Diapause

The significant feature of diapause is the inhibition of growth, is connected with many biochemical and physiological changes in the insect organism. These depict the adaptive nature of diapause, as they increases the ability of the insect to survive the adverse conditions (winter, drought). Some physiological characteristics are similar in all types of diapause:

Metabolism

Overall metabolism

The beginning of many types of diapause is correlated with a marked decrease in oxygen consumption. The respiration curve fails to a characteristic very low level - about one-tenth of the normal oxygen consumption of the particular stage - and this level is maintained practically throughout diapause. Diapause is generally associated with the bottom of the oxygen consumption curve for non-diapausing insects. Many entomologists have studied the respiratory metabolism during diapause and the influence of the metamorphosis hormones. The oxygen consumption in diapausing adult females of *Pyrrhocoris apterus* was found to be similar with that occuring in normal (non-diapausing) specimens after removal of both ca and cc. It also corresponded with that in normal untreated males, whereas it was definitely higher in the females which had only been allatectomized. Typical low diapause respiration with a low water content and an increased amount of fat was found by Tombes (1964) in the aestivating curculionid, *Hypera postica* (Gyll). DNA synthesis during metamorphosis and diapause was examined by the incorporation of tritiated thymidine in *cecropia* and *cynthia* silkworms.

In diapausing pupae, the incorporation of thymidine was limited to spermatogonia, haemocytes, and a few other cells. In the diapausing epidermal cells, DNA synthesis appeared only after a large epidermal injury was limited to a immediate vicinity of the wound. However, a generalized incorporation was observed after the termination of diapause at the beginning of adult development. This is assumed to be the only proof for differentiating between the increase in metabolism of normally developing specimens and those which have been affected by injury (Bowers and Williams, 1964).

The respiratory enzymes

Many researches have demonstrated a series of changes in the activity of the respiratory enzymes. From the observed increase in the resistance to HCN and CO as well as a resistance (seventy times higher) to organic poisons which block the cytochrome system (i.e. pilocarpin and diphtheria toxin), it was concluded that the cytochrome system is inactivated during diapause and replaced by a particular cytochrome, b_5. But, further experimentation has led to a different conclusion. Investigations on the resistance to cyanide and CO of the heart = beat of diapausing *cecropia pupae*, which continues for $5^1/_2$ hours in the complete absence of oxygen, showed a large surplus of cytochrome oxidase in comparison with cytochrome c, producing the observed greater resistance. Kurland and Schneiderman (1959) attempted to generalize from these results to other types of diapause, including embryonic diapause, on the basis of their studies on the respiration of diapausing pupae of several saturniid silkworms (*Cecropia, Cynthia, Promethea, Polyphemus*). They assumed that cytochrome c would be the factor constraining respiration during diapause, and that changes in respiratory metabolism would be the result of the lack of this factor as a direct or indirect consequence of MH.

The distribution of cytochrome oxidase and succinic oxydase in adults of the Colorado beetle (Leptinotarsa decemlineata) during the period of the diapause period was studied comprehensively detail by Ushatinskaya (1957b). The results of her work and those of other Soviet authors are included in her monograph on the cold-hardiness of insects. The physiology of diapause in sawflies was studied in detail by Siama (1959a-c) in a series of research papers. The physiology of imaginal diapause in two coccinellid species (*Semiadalia undecimnotata* Schneid., and Coccinella septempunctata L.) was studied by Hodek and Cerkasov (1958, 1959).

The utilizatiion and accumulation of reserves

A significant physiological feature of diapause is the reduced utilizatioin of reserve materials (protein and fats) correlated with the reduced metabolism. Prebble (1941) showed that, in the sawfly filpinia polytoma, no more that about 2 per cent of the reserve materials (dry matter) was consumed during the whole winter period. Similar results have been obtained with other species. This is probably one of the most important selective advantages of diapause. Where reduced metabolism occurs during suitable living conditions so that the diapausing insect is able to accept food, as in the so-called pre-diapause of the imaginal diapause, there is an accumulation of fats and other storing substances in the body of insect.

Water balance

An increase in the ratio of dry matter is another significant feature of diapausing insects making them more resistant to cold. Normal water balance is usually restored following the resum- ption of growth. The lack of required humidity may markedly prolong diapause in some cases. In summer diapause, as occurs in locust eggs, and less frequently in winter diapause, as in coding moth, diapause can be broken by immersing the insect in water. It is important to note that changes in the water balance as well as the afore-mentioned features of diapause seem to be the result of changes in the humoral balance of one particular endocrine factor, the AH, which controls, amongst other things, the water balance of the body.

Digestion during diapause

Many insects have unrestrained movement and are able to accept food during diapause. This is specially true of imaginal diapause, in so far as it occurs during suitable environmental conditions. It has been observed, however, that most insects incapable of utilizing the food, either for the development of the gonads or for restoring their fat-body reserves, so that they die of exhaustion within a few days. This can be understood as an AH-deficiency syndrome, now that its role in the activity of the digestive enzyme has been confirmed by Thomsen and Moller (1959) and by Strangways–Dixon (1960, 1961). Since the last edition of this book there have been many new discoveries on various aspects of the physiological changes connected with diapause (e.g. Maslennikova, 1972). Although it is not fearble discuss them here, it should be mentioned that they should be regarded as consequences of the same principle changes in the insects neurohormonal balance.

Determination and Photoperiodism

The determination of diapause

It has been affirmid that diapause coincides closely with the period of environmental conditions, adverse for a given species. In multivoltine species, a special method is required to ensure such synchronization. Usually the insect has a hereditary structure which will respond with diapause to particular environmental conditions which normally precede diapause, such as a certain day-length (photoperiodism), or a change in temperature. Since diapause itself is closely associated to a particular stage of development, so is its period of determination. It is therefore heriditary. As indicated by Novak (1958b), the phylogenic origin of this genetic mechanism can be explained as the result of selection for a tendency which perhaps occurs in all insects to a very small extent at certain stages of development. This tendency to respond to various inclement conditions, by a temporary break in development at some subsequent stage, may originally have no specific direction (e.g. for a long or short photoperiod) and only its value for the insect's survival under the particular environmental conditions may finally induce strengthening of another kind of response. Therefore, it is proved in bivoltine races of *Bombyx mori* that optimal conditions during the embryonic period and during the first part of larval development (Ist and II instar) tend to produce egg diapause in the next generation.

The same conditions during the IIIrd instar producse no effect on diapause. While, in the second half of larval development (IVth and Vth instar) these same conditions (long photoperiod and higher temperature) tend to induce development without diapause and the converse also applies (cf. Emme, 1953). By specialization through selection for the early instar effect short-day photoperiodism results, while specialization on the latter effects result in long-day photoperiodism. The role of selection in the evolution of this type of dependence seems obvious from the markedly adaptive character of all the mechanisms. The analysis of photoperiodism in connection to diapause has induced much interest since its importance was first stressed by Ernst Scharrer (1952, 1959) that it can now be regarded as a social branch of research on diapause. Basic work in the field has been carried out by Danilevski's team since 1945. Even here, the neuroendocrine basis of all these relationships is in undoubtedly clear.

Different in degree and yet similar to photoperiodism is the action of temperature during the determining period of facultative diapause. It is interesting that the influence of temperature in a particular species works in the same direction as photoperiodism: in short day species a low temperature has a positive effect in same way as a short photoperiod while higher temperatures induce diapause; in long-day species the temperature and photoperiodism act in the opposite response has a clear adaptive value, e.g. a low temperature together with a short photoperiod during embryonic development causes the deposition of non-diapausing eggs in the bivoltine race of *Bombyx mori*. It is widely accepted that the first spring generation of silkworms hatches in the early spring (March or April in China) when low temperatures and relatively short days are normal conditions. The females of the second generation thus deposit non-diapausing eggs of the second generation are which hatch in about June, when the days are long and the average temperature is high. In this way the females of the second generation are accustomed to lay without diapause and higher humidity winter period, whereas a hatch in autumn would naturally caused death through cold and lack of food.

The interaction of the effects of photoperiod and temperature when the conditions are varied experimentally is extremely complese (1949, Danilevske.) Along with the effects of photoperiod and temperature, an influence of humidity on diapause has also been discovered in many species e.g. *Bombyx mori*. Again the effect of humidity is aligned with the influences of the other two factors; with short-day species, conditions of lower humidity induces development without diapause and higher humidity induces diapause. The effect of humidity in the summer type of diapause in insects from tropical and subtropical countries appears to be rather more involved, though and yet fully understood. The influence of food and its composition on the determination of diapause is also doubtful. Many results in the literature have proved the described cases to be unreliable when repeated with a controlled photoperiod. In other cases it is still unclear whether observed effect is specific or merely very generally affects development, e.g. suboptimal nutrition might make the insect more susceptible to factors inducing diapause.

There is, however, a definite case of an influence of want of food on diapause determination in blood-sucking insects. Here a special mechanism has developed which synchronizes development with periods of food abundance and induces a state corresponding to a diapause during periods of starvation. Shortage of food also seems to be a stimulus inducing diapause in some plant-feeding insects, e.g. in *Diatraea lineolata*, a subtropical maizeborer.

Termination of Diapause

As earlier mentioned, the degree of temperature during diapause development may produce a profound effect on the length of diapause. Never-theless, no change of temperature within ecological limits is able to break diapause at any given moment and re-induce development. But there are a number of other factors, including experimental treatments, which may be able to induce normal development in animals already determined for diapause. Perhaps the best known are the various treatments which suppress diapause in diapausing Bombyx mori eggs. It is important to note that most of these factors are only likely to break diapause during a definite, relatively short sensitive period; this is usually just before or at the beginning of diapause. For instance a short immersion of silkworm eggs examined for diapause in concentrated HCI can stop the process of diapause completely if performed during the first 48 hours after oviposition, i.e. before the eggs have their normal grey-blue colour. After this moment the acid treatment is completely ineffective.

An effect on diapause similar to the HCI treatment can be obtained by increasing the temperature to about 46°C for to min in the same sensitive period. This procedure is such a potent stimulus that it can induce development, and hence artificial parthenogenesis, in unfertilized Bombyx eggs, the state which can thus be compared with true embryonic diapause. The process of breaking diapause by increased photoperiod was studied by Williams and Addission (1964) in *Antherea pemyi*. They realized that long-day conditions (15 to 18 hours light promoted temination of the diapause whereas short-day conditions) (4 to 12 hours) prevented it. The experiments showed that the brain itself is the site of the action of light, most probably by nervous control of AH production. In this connection, the transparent window in the pupal integument immediately above the brain seems to be of primary importance, as shown by Shakbazov (1961). A photoperiodic acceleration of the termination of diapause was also found by McLeod and Beck (1963) in *Ostrinia* (=*Pyrausta*) *nubilalis.* A similar acceleration was observed following administartion of large doses of ammonium acetate or other ammonium compounds. Syzergism in accelerating diapause development was found between the action of NH_4 ions and a long photoperiod.

It has been shown that the effect of a long photoperiod in preventing diapause can be replaced by an additional short photophase (5 to 30 min) applied daily at about the middle of the dark period, even when combined with a very short photoperiod (e.g 10 hours) and ecology of photoperiodism in insects was summarized. Diapause in larvae of the fly *Lipara lucens* can be disturbed by a pin prick or by contact with growing pin. Similar stimuli was also readily break the relatiavely feeble diapause in flies of genus Lucilia. Diapause in eggs of the grasshopper *Melanopus* can be broken at any time in its duration by immersing the eggs in xylo or other wax solvents. It has been demonstrated that effect of xylo is to remove the waterproof wax layer (which covers the hydrophyle during diapause, so preventing water penetration) which is fundamental cause of diapause. Removing both cuticular envelops (the yellow and white cuticle) and immersing the eggs in insect-Ringer has the effect similar to the same experiment done by xylo. It may be stated that the effect of strong mineral acids on breaking diapause, e.g. in silkworm eggs, lies in producing a particular temperature on the surface of the eggs, whilch is hte main cause of the breaking of diapause. The major cause for the diapause break in various examples of post-embryonic diapause by stimulation of the nervous system probably lies in the renewed secretory activity of the cerebral neurosecretory cells. The only treatment which is able to break postembroynic diapause at any stage of its development to induce experimentally an active concentration of AH (by implantation of an active brain) or MH (by implantation of prothoracic glands or by the injection of ecdysone).

The mechanism of the effects of photoperiodism and other external factors

There is some proof concerning the way in which external stimuli are transmitted to the neurosecretory activity. One of the earliest research papers relevant to this problem was written by Dixon (1949). He suggested, on the basis of his investigations on diapause in the oriental fruit-moth *Laspeyresia molesta,* that a series of hypothetical two-phase chemical reactions formed the basis of the response to photoperiod. One of these phases requires light, whereas the other needs darkness. Ushatinskaya (1961c) stated that in this context cytochrome oxidase might play a significant role its inhibition by CO is prevented by light, this being a photosensitive reaction of a type which would be the basis of such a mechanism. The first hint on this question appears to come from a paper by van der Kloot (1955). He studied the sequence of changes in the electrical activity of the brain, i.e., the titre of cholinesterase (ChE), electrical activity (EA) and the amount of cholinergic substances in connection with the observed AH production in H. Cecropia. He discovered that all these changes are closely connected with diapause. Thus electrical activity drops together with AH and cholin- esterase activity, remaining practically at zero throughout diapause with a steep rise afterwards. The cholinergic substances, however, follow an almost exactly opposite pattern. The amount steadily increases during diapause development and reaction its maximum at the point when diapause ends. About this time, electrical activity reappears and rapidly increases together with the titre of the cholinesterase.

It seems that the observed accumulation of cholinergic stubstances forming a substrate for cholinesterase is the direct internal cause of the effects of chilling, because it is optimal at

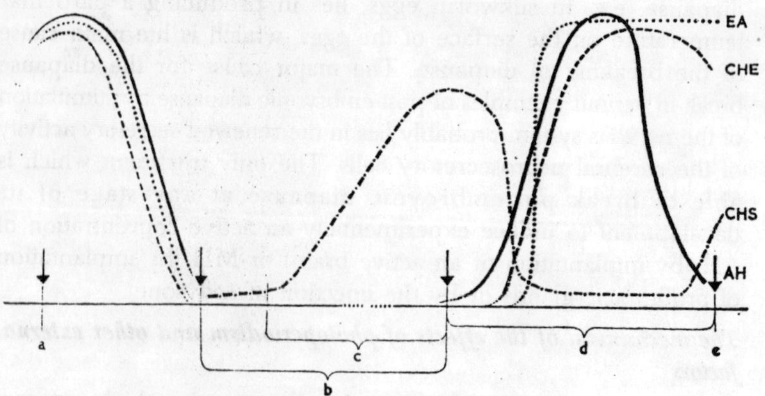

Fig. 9.1. The interrelationship in activation hormone production (AH), electrical activity of brain (EA), choline esterase content (CHE), cholinergic substances (CHS) in Hyalophora cecropia during the pupal instar (a) beginning of cocoon spinning; (b) period of diapause; (c) period of low electrical activity; (d) period of development; (e) imaginal moulting (ecdysis).

low temperatures. It is confirmed that such a surplus of cholinergic substances as a substrate for the renewal of cholinesterase activity is necessary for production of AH and the renewal of electrical activity. In the brain of the non-diapausing *Galleria mellonella*, the cholinesterase content and the amount of cholinergic substances is stable during metamorphosis. In the medial neurosecretory cells of *Bupalus piniarius*, neurosecretory material seems to be produced and correspondingly the spontaneous electrical activity of the brain persists during pupal diapause (Schoonhoven, 1962). The existence of photoperiod by *Leptinotarsa decemlineata* has been investigated in a series of papers by de Wilde et al. (1959).

The Colorado beetle is a typical long-day species, diapausing in a photoperiod of less than 14 hours of light. The threshold intensity of white light is less than 0.1 lux. Experiments aimed at localizing the site of perception have not as yet yielded clear-cut results. Removal of the eyes or interference with their function does not affect the photoperiodic reactions. The eyes are perhaps not the only sites of perception. Unlike Dixon and Ushatinskaya, de Wilde indicated a possible effect of feedin, perhaps connected with a rhythm of feeding activity. A review on the problem of photoperiodism in insects has been produced by Lees (1959). A monograph by Danilevski (1961) on the photoperiodism of insects is also significant.

The Hormonal Principle of Diapause

A succession of many other theories has been formulated to explain the mechanism and the origin of diapause. The important ones can be divided into three groups:

1. Diapause is hormonally conditioned, either as a result of the absence of metamorphosis hormones indispensable for development, or, in the case of embryonic diapause, the result of the action of a special humoral factor. A suggestion that such is one role of the brain hormone can be found in Kopec's (1917, 1922) original papers based on his experiments with *Lymantria dispar*. The first clear formulation of this concept was given by Wiggleworth (1936a) in his study of the hormonal system of the blood-sucking bug *Rhodnius prolixus*.
2. Diapause is the result of a decrease in the water content of the tissues caused among other things by the accumulation of reserve materials, primarily fats. This leads to the reduced metabolism and to the other features of diapause. Diapause is,

however, much more probably the cause rather than the effect of all these changes, as Andrewartha (1960) has stated, or perhaps the result of one and the same cause as these changes.

3. Diapause is caused by the accumulation of a specific inhibitory substance `acting through the inhibition of the life processes' which is slowly decomposed during diapause development. This category also includes the concept of the general diapause hormone proposed by linton (1953). However, no clear proof of a substance with such characteristics has so far been acquired leaving the findings on the humoral conditioning of embryonic diapause in Bombyx.

Every experimental evidence which has been collected unambiguously supports the hormonal theory of diapause. The most significant work. But the alterations in the hormonal system, should not be understood as a primary cause autonomously determining the onset of diapause in any individual. It should instead be considered the mechanism which realizes the effect of the environment on the principal functions of the insect organism. The environmental conditions affect these functions either directly in the same generation by means of the nervous system and neurosecretory cells, or indirectly through previous phylogenetic stages, via the genetic mechanism under the perpetuae pressure of natural selection. Three major types of diapause can be distinguished on the basis of their hormonal mechanisms: (1) Diapause caused by a lack of AH and (therefore) of MH. To this category belong all cases of larval and pupal diapause known so far and, doubtless, also late embryonic diapause. (ii) Diapause caused by lack of AH and JH; this mechanism is concerned in imaginal diapause. (iii) Diapause caused by the action of the neurosecretory factor produced by the suboesophageal ganglia of the female, and affecting the development of eggs. This category comprises early embryonic diapauses.

The diapause effected by a deficiency of the activation and moulting hormones is the most common type of diapause, and the type so far most thoroughly investigates. It consists of an arrest of growth and development due primarily to the absence of MH, itself a consequence of an inhibition of AH production. It was originally supposed by most authors that the lack of MH was the only internal cause of this type of diapause. Recent findings on the role of AH in digestion and water balance, however, make it very probable that, besides the indirect effect of AH in activating the prothoracic glands, there also persists a direct effect of the hormone, the lack of which causes an important part of the diapause syndrome.

Pupal Diapause

The initial and comprehensive research on of pupal diapause was carried out by William and his school, using the giant silkworm *Hyalophora (= platysamia) cecropia*. The main facts about this type of diapause may be briefly summarized as follows.

In this particular species, diapause begins at the same time as the ecdysis to the pupa. The remarkablely low metabolism of diapause begins somewhat later, however, at the trough of the U-shaped curve. Diapause metabolism, unlike that of developing insects, is highly resistant to the metabolic poisons HCN and CO and to toxins of bacterial origin such as diptheria toxin, and differs from normal metabolism in a variety of other feathers which disappear when development is renewed. Pupal diapause is often on ligatory as in *H. cecropia* and the Middle-Eastern saturniids (Saturnia piri, S. Spini Eudia pavonia), perhaps even be facultative as in the bivoltine *Antherea pernyi.* Experiments with *extirpation*, transplantation, sectioning pupae, injection of extracts, etc. have clarified that the primary internal cause for the inhibition of development in pupal diapause lies in an interruption of the secretory activity of the neurosecretory cells of the pairs intercerbralis, and possibly includes entails a blocking of the passage of neurosecretory granules into the haemolymph. By implanting an active brain into the diapausing insect it is possible to break diapause at any stage of diapause development. The effect of chilling on the termination of diapause is especially concerned with the brain. This is demonstrated the fact that the implantation of the brain alone from a chilled specimen is able to break diapause in another unchilled pupa, whereas the chilling of a decerebrated pupa is ineffectual on diapause. Adding the AH source is, however, only effective when the host specimen retains its own prothoracic glands or when they are implanted at the same time.

The implantation of one or more active brains into an isolated abdomen has no effect, whereas the implantation of active prothoracic glands breaks diapause and development continues. A similar impact is produced by injecting MH in the form of ecdysone prepared by the method used by Butenandt and Karlson.

Little is known about which of the other known characteristics of diapause are the direct effects of the lack of MH, which are caused by the lack of AH, and which are perhaps an indirect consequence of growth inhibition. The results from the work on H. ceropia were similar to that obtained by other entomologists using diapausing pupae of other insects. Thus in Hyphantria cunea, Anterea pernyi and Eudia pavonia the implantation of either an active brain or prothoracic glands results in breaking diapause.

A exhaustive histological research on diapausing pupae of the moth Mimaltiliae was carried out by Highnam (1958a, b). He demonstrated that the neurosecretory cells of the brian are inactive in the diapausing pupa at room temperature, but that they produce intracellular material during the first 3 weeks of chilling (at 3-C). The material (granules) passes to the corpora cardiaca and towards the end of diapause development the neurosecretory cells again become inactive. A similar period of activity was observed in the corpora allata. The removal from specimens of both corpora cardiaca and corpora allata, however, had no effect on the duration of diapause. Such specimens had a notably thicker epidermis and their fat bodies contained a much smaller number of inclusions than the controls. The author supposes that the corpora allata play some role in maintaining diapause, perhaps by controlling the metabolism of the fat body in some way.

No signs of secretory activity were observed while the subsequent development at 25°C was taking place. A number of research work has been published on the effects of AH and MH deficiency on metabolic enzyme during pupal diapause. Apart from there work mentioned above, Shapirio and Williams (1957a, b) studied the cytochrome system in various tissues of late larvae and pupae of *H. cecropia* using low temperature spectroscopy and a DU spectrophotometer. They report the presence in last instar larvae of a complete cytochrome system with moderate to high concentrations of cytochromes a, a_3, b, c, and b_5, whereas in diapausing pupae no b and c were detectable by the methods used. When development continued after the termination of diapause, the activity of the original complete system reappeared. The explanation of the observations given by the researchers must, however, be compared with the later findings of Williams and Harvey (1958) and of Kurland and Schneiderman (1958), which have been earlier stated.

It has been stressed that a specific inhibitory factor, inactivated by cold, participated in the induction of a diapause state in diapausedetermined *Pieris brassicae* pupae. It was shown, that such insects can recover from the chilling, and their pupae continue to develop even if the obtain is extirpated at the outset of diapause. Reactivation in the absence of brain AH might, however, be attributable to the amount of AH produced by the ventral nerve cord ganglia, but gernerally this does not acquire an active concentration. Pupal diapause can take place at any point during the pupal instar. Such process in a given species is bound to the same phase of the pupal period and in most species it is genetically fixed. An example of a very late pupal diapause is the pronymph of the sawfly *Cephalcia abietis.*

Larval Diapause

We do not know much about the various example of larval diapause. Based on what is known it is apparently clear that larval diapause does not differ fundamentally from pupal diapause. It has been observed that in diapausing, penultimate instar larvae of *Gryllus campestris* the histology of the prothoracic glands shows reduced secretory activity and the brain appears to be responsible for these changes. The brain and prothoracic glands have also been found to control diapause in the larvae of *Sialis lutaria.* Larval diapause in ants of the genus Myrmica and its relation to caste polymorphism has been examined comprehensith by Brian (1959, 1961, cf. 1960). Much of the research work dealing with diapause in larvae have been concerned with the diapause which occurs in sawflies.

However, this is not a true larval diapause, but rather the diapause of an instar in metamorphosis occurring between the true larva and pupa. The existence of neurosecretory cells in the brain of the sawfly *Diprion pini* has been demonstrated by L'Helias (1953a-c). Church (1953, 1955) has shown that diapause in *Cephus-cinctus* is caused by a temporary interruption of the necurosecretory activity of the larval brain, and that the renewal of this activity requires a certain period of chilling. The same author later showed that certain prothoracic glands described by him are necessary to terminate diapause. Two periods of diapause, separated by a period of limited interdiapause development, have been found in the pre-pupal stages, the eonymph and pronymph of another sawfly, *Cephalcia abietis* (Pamphilidae). Normally, this insect's span of development is two years. The first diapause period, which occurs at the eonymph stage, needs a considerably longer period of chilling (not less than four months) than the second period. This corresponds to the pre-pupal stage at an advanced stage of the moulting process.

It generally occurs in the second winter, and is shorter under the same conditions as the first (eonymph) diapause, the preceding winter. The interdiapause development normally occurs in the summer months between the first as second hibernations. The pronymphal diapause can be broken by the implantation of an active brain. Ligaturing experiments have demonstrated a clear reliance on the brain and prothoracic centers in this sawfly, just as in other species. Like pupal diapause, larval diapause can occur at any point during larval development. The IInd larval instar of *Euproctis chrysorrhoea*, which hibernates in specific common nests with over 200 caterpillars, can best stated as an example of a very early larval diapause.

Late Embryonic Diapause

There is no clear indication concerning the diapause which occurs in more or less completely formed larvae before hatch-ing, as is found in *Lymantria despair*. The fact that active neurosecretory cells and prothoracic glands can be demonstrated in embryos makes it clear that the hormonal mechanism at work at this stage is the same as that in larval diapause, though this is not the case is younger embryos.

Adult Diapause

Imaginal diapause is caused by the clack of activation and juvenile hormones. A quite similar neurosecretory mechanism operates as in the aforementioned cases of true diapause, but with the difference that it is a lack of JH, not of MH, which is the main factor in adult insects. Moreover, the neurosecretory mechanism of JH is not fundamentally different from that of MH. The hormonal mechanism of imaginal diapause has been most thoroughly studied in Coleoptera, for example in Dytisus marginalis and Lepitnotarsa decemlineata and in Diptera, for example Calliphra erythrocephala and in the mosquitoes *Anopheles macullpennis* Culex pipiens, C. molestus and Aedes aegypii Similar methods functions in the imaginal diapause of Heteroptera, as has been shown by Johnson (1958) with *Oncopeltus fasciatus* and by Slama (1962) with *Pyrrhocoris apurus*.

Little is known about imaginal diapause in Lepidoptera, such as that which occurs in many common butterflies of moderate geographic zone like Nymphalidae or Pieridae which overwinter as adults; but a similar mechanism may well be accepted. Morphologically, imaginal diapause is only distinct in the interruption of growth of the ovaries and in the suppression of the functions of the accessory glands of both males and females. As has been shown by the experiments of E. Thomsen (1952) with Calliphora erythrocephala, this seems to involve a complex syndrome of both AH and JH deficiency. The implantation of corpora allata can cause the ripening of some of the eggs, but a quantitatively complete termination of diapause is only possible by the simultaneous action of both JH and AH. Besides this, the presence of AH seems indispensable for the corpora allata to function normally. The most important physiological effect of the lack of AH in diapausing adults is that on proteinase activity in the gut, which is probably also the cause, either directly or indirectly, of a variety of other physiological changes in species with imaginal diapause. The connection of the corpora allata to imaginal diapause is clear from the necessity, first shown by Wigglesworth (1935, 1936) of JH for the ovarian follicle cells. Experimental evidence of the dependence of imaginal diapause on a temporary break in ca activity as suggested by Detinova (1945) for Anopheles maculipennis, was provided by Joly (1945) in his detailed study of Dytiscus marginalis and has since been confirmed by a number of other authors in various groups of insects which undergo imaginal diapause, e.g. by de Wilde et al. (1954-1961) in a series of papers on Leptinotarsa decemlineata; Hodek and Novak (1961) in Eurygaster, Larsen and Bodenstein (1959) in *Aedes aegypti* and other mosquito species, etc.

Here, it is enough to say that the primary internal cause of imaginal diapause lies in the neurosecretory system of the pars intercerebralis, which appears to be part of the mechanism of action of external stimuli. The absence of AH, of course, produces a series of effects in the body, in addition to the inhibition of the ca, as shown by the papers of E. Thomsen and I. Moller (1959), Strangways-Dixon (1960) and others. This is because the state after allatectomy is not identical with the natural imaginal diapause and why diapause also occurs in males, the gonads of which are not dependent on the presence of JH. This has been clearly demonstrated by Slama (1963) in his experiments on the effects of allatectomy, cardicectomy and of the natural diapause on the oxygen consumption in *Pyrrhocoris apterus*. A stage, often stated by entomologist as `predia-pause' has been observed before hibernation in most examples of imaginal diapause in insects.

It is accepted to be as a period of preparation for hibernation, during which various biochemical changes occur in the body to make the insect more resistant to cold and other adverse conditions, e.g. an increase of reserve materials (fat), lack of water content in the tissues, thus increasing their resistance to freezing, etc. Individuals of diapausing generations grow less quickly than those with continuous development, and related to this is the fact that they usually reach a greater weight and size during their periods- of development. Many authors have attempted to explain these changes as caused by an internal factor (conditioned by environmental conditions) which induces diapause. This supposition was made, e.g. in the so-called 'food mobilization' theory. There is, however, weighty evidence against this supposition. In the majority of insects undergoing imaginal diapause, diapauses starts without such earlier changes in the organism, and appears to be fundamentally identical with `prediapause'. Also AH or JH can terminate diapause following a preparative prediapause. Easier definition of this stage is possible: there is no difference in physiological conditioning between `prediapause' and the subsequent diapause; both being due to a deficiency of AH and JH. When this deficiency occurs within normal environmental conditions, feeding and the other functions of the body continue, but the ovaries and accessory glands are inactive. Food utilization is there- fore limited to energy and accessory glands are inactive. Food utilization is there-fore limited to energy and basal metabolism, with the result that surplus food accumulates in the form of reserve materials.

Other changes are probably caused by the lack of AH activity on the gut proteinases. All the continuing processes interrupted during prediapause are arrested by the adverse conditions of winter. Prediapause is therefore undoubtedly true diapause, where what is regarded as proper diapause of the ovary and accessory gland tissues combined with quiescence of all the other parts of the body, particularly the subsequent quiescence of the whole organism. The latter changes from normal quiescence, however, in that no hormones exist in the body at the time. Inhibition of the neurosecretory system is nevertheless over in that period. The results obtained by Hodek (1962) suggest, however, that in 'prediapause' the syn-dromes of AH and JH deficiency seem to be combined with inherited instincts for hibernating- such as the instinct for flight to hibernation sites in Coccinellidae, or in Eurygaster intergriceps. This result is confirmed by recent findings on the direct effects of the metamorphosis hormones on imaginal instincts, as shown by Zdarek (1970) and others. More experimental evidence will be required, however, before the relationship between the release of these instincts and the onset of diapause can be explained.

It may consist in a parallel response to the same stimulus (e.g. a change of photoperiod, or one may excite the other (diapause the instincts, or vice versa, e.g. by keeping the insects in the dark, without food). These conclusions on the nature of 'pre-diapause' corroborate with the fact that the occurrence of a prediapause is almost entirely limited to imaginal diapause and only affects a small part of the body while other tissues continue their functions. However, it should also be noted that in many insects diapause ends before the start of winter, the cold days and especially the cold nights of autumn providing insufficient chilling. By contrast, diapause may be terminated at the beginning winter simply by placing the insects in optimal conditions for further development. Inspite the state of such insects in winter corresponds in most respects to true diapause, it differs by corresponding with quiescence in that it may be broken by transferring the insects to laboratory conditions.

As regards the other biochemical characteristics of imaginal diapause, apart from the direct effect of the lack of AH on proteinase activity, only the diapause of the ovarian tissues corresponds to a diapause caused by AH-MH deficiency but their state influences the tissues that develop the body. Apart from water diapause, a summer dormancy in *Leptinotarsa decemlineata* has been described by Ushatinskaya (1961) as occurring in a proportion of the beetles surviving from the previous year. The beetles creep into the soil in the hot summer months (June, July) and remain there for about one month, after which they continue to feed. Possibly summer diapause has the same hormonal mechanism as the winter one. In the autumn some of the `old' beetles of the previous year enter another 'winter diapause' which differs to some extent from that of the first year.

Early Embryonic Diapause

Early embryonic diapause is connected with the diapause hormone (DH), a neuro-hormone of the suboesophageal ganglion. The major distinctive feature between embryonic diapause (except late embryonic diapause mentioned earlier) and the two earlier categories is that in the present type determination usually takes place in the females of the previous generation, so that it must be assumed that it is transmitted via the eggs. The other important feature of this type of diapause is that it occurs at a stage of ontogeny before any of the metamorphosis hormones become active. The relatively small number of species in which the hormonal mechanism of embryonic diapause has been adequately studied precludes any broad generalization from what has been observed. However, the biochemical features of embryonic diapause are quite similar to that of diapause in later stages of ontogeny. For instance, diapausing embryos of the grasshopper *Melanopus differentials* show not only a U-shaped respiration curve which is quite similar to that in the pupa, but also a reduced susceptibility to HCN and CO which again disappears with the onset of development.

Similar characters have been found by Wolsky (1941) in diapausing eggs of Bombyx mori. Diapause in the eggs of migratory locusts, as of *Locustana pardalina*, is of a markedly different character, and it probably resembles more to the type associated with AH-MH deficiency. We are gratified by the work of Fukuda (1952a-c, 1953a-c), and Hasegawa (1964) and their colleagues for the first trustworthy information on the hormonal mechanism of embryonic diapause. These papers are, however, not concerned with diapause itself, but with the way it is determined in the mother organism. Since long time it has been accepted that the occurrence of diapause in the bivoltine races of the silkworm is determined by the environmental conditions under which the egg of the parent generation developed. But, the environmental conditions do not, as might perhaps be conjectured affect the rudiments of the gonads directly, but indirectly through the secretory functions of the nervous system which is yet undeveloped when determination occurs.

The correct stimulus to the ovaries occurs not till the subsequent pupal period, as Umeya (1936) demonstrated for the first time in his ovary transplantation experiments. The complex mechanism whereby this transmission occurs was first discovered, however, by the systematic research of the two aforementioned authors. The following paragraph states the summary of their findings. It can be demonstrated that the incidence of diapause in eggs is conditioned by the endocrine activity of special neurosecretory cells in the suboesophageal ganglion of the mother specimen. The extirpation of this ganglion at the beginning of the pupal instar causes eggs, which were determined for diapause through their mother, to develop without diapause. Similarly univoltine females can be caused to lay non-diapausing eggs, a thing which they never do under normal conditions. Conversely, if, up to a critical point in time, a female pupa which has been conditioned to lay non-diapausing eggs receives a suboesophageal ganglion from a female which would have laid diapausing eggs, it also will lay diapausing eggs. The same effect can be produced by the implantation of suboesophageal ganglia of either females or males of a univoltine race. That the relevant DH is nonspecific for species, genus or family was shown by implantations of the suboesophageal ganglia from other Lepidoptera.

The capacity to induce diapause was found not only in the ganglia of other species with an embryonic diapause such as *Antherea yamamayi*, *Dictyoploca japonica* and *Lymantria dispar*, but also in the ganglia of Antherea pernyi, a species with a pupal diapause. Ganglia from pupae of *Philosamia cynthia* and *Dendrolimus pini* did not have the similar. Yet more complex is the mechanism of transmission, complicated. A female conditioned to lay non-diapausing eggs can be made to deposit diapausing eggs by decerebration or imply by cutting the circumoesophageal connectives before a certain time. From these kind experiments it was affirmed that the endocrine activity of the suboesophageal ganglion, which conditions diapause in the eggs, is constricted by a nervous stimulus from the brain transmitted through the circumoesophageal connectives. Hasegawa (1957) stated that he has succeeded in isolating an effective DH extract from Bombyx mori pupae which had been conditioned to lay diapausing eggs (univoltine race).

The suboesophageal ganglia together with the brains of about 15-000 pupae were dissected and homogenized in 80 per cent methanol. After contrifugation, the supernatant fluid was evaporated under vacuum, diluted with water and washed with ether. The aqueous layer was then extracted with chloroform. The evaporated chloroform extract yielded a fluid which showed a high diapause inducing activity when it was injected into female pupae conditioned to lay non-diapausing eggs. An extract was obtained by the same method, from the heads of adult females. Isobe et al. (1973) recently succeeded in isolating two DH components from the heads of two million mated adult Bombyx males. They were separated from a complex lipid fraction by gel permeation chromatography, using Sephadex LH-20 and organic eluants. Both the components, A and B, seem to have a relatively low molecular weight (between 2000 and 4000) and they both seem to be complex lipids with a number of peptide bonds among several types of amino acids and various other components.

Table 9.1. Isolation of the diapause hormone from Bombyx mori.

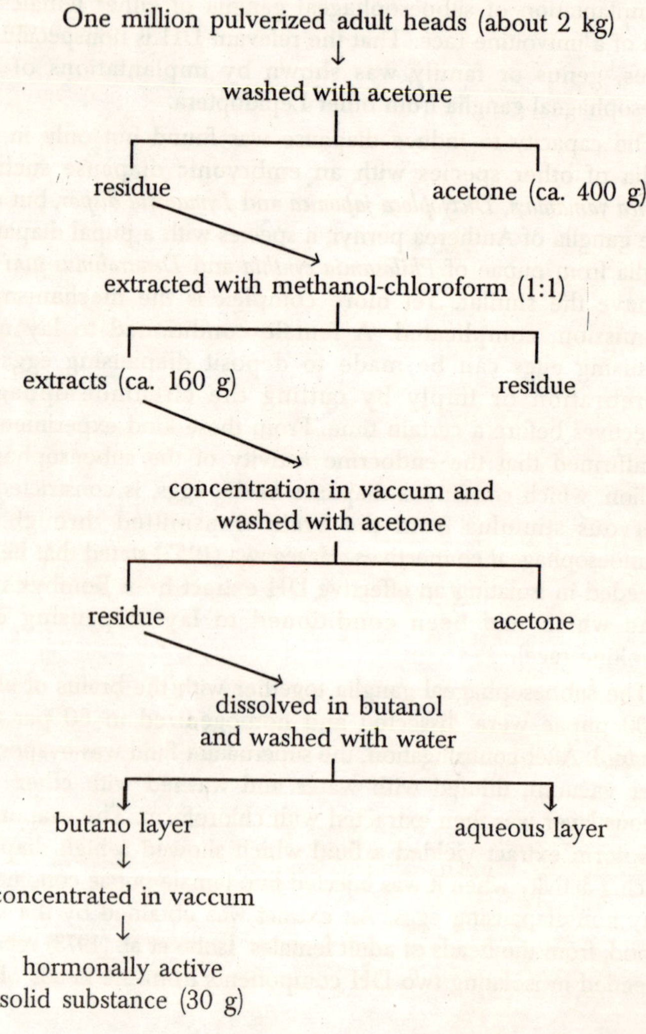

The process of extracting the hormone was described (the active principle was found in the aqueous fraction after ether extraction and was then removed from this with chloroform). Its physicochemical and chromatographic properties and its acid, base, heat and light stability were determined. Most research was done by Hasegawa (1963) and Yamashita and Hasegawa (1964a, b) who further examined the mode of action of the diapause hormone in the silkworm (Bombyx mori). The fate of the hormone after injection into a female pupa and its effect on the penetration of 3-hydroxy- kynurenine into the developing egg are discussed. The latter substance in the precursor of the serosa pigment (ommochrome) determining the colouration of the diapausing eggs; it is much less abundant in non-diapausing eggs. More experiments where done to know about, the effect of the diapause hormone on the glycogen content of the ovaries and the blood sugar level of the female pupae.

It has been confirmed that one of the actions of the hormone was to stimulate the penetration of blood sugars into the pupal ovaries and their deposition therein the form of glycogen, thus reducing their content in the haemolymph. Further break through in the investigation of the mechanism of embryonic diapause development in silkworms was made by Chino (1961). It was shown that almost all the glycogen present in the eggs of Bombyx mori before the onset of diapause was rapidly decomposed by a complicated enzymatic reaction, the end products of which were the two polyhydric alcohols sorbitol and glycerol. A resynthesis of glycogen started at the moment when diapause ended. It was further shown that sorbitol is formed by dehydrogenases catalysing the following two reactions:

$$\text{glucose TPNH} \rightarrow \text{sorbitol} + \text{TPN} \qquad (5)$$

$$\text{glucose (or fructose)-6-phosphate} + \text{TPNH} \rightarrow \text{sorbitol-6-phosphate} + \text{TPN}^{+} \qquad (6)$$

Glycerol is formed by enzymes catalysing the following reactions:

$$\text{dihydroxyacetone phosphate DPNH} \rightarrow \text{a-glycerol phosphate} + \text{DPN}^{+} \qquad (5')$$

$$\text{glyceraldehyde (or dihydroxyacetone)} + \text{TPNH-glycerol} + \text{TPN}^{+} \qquad (6')$$

But these polyhydric alcohol dehydrogenases were not, the factors which induced diapause, for they were also present in developing eggs in which an increased amount of polyhydric alcohols never occur. The adaptive character of the accumulation of such substances which increase the resistance of diapausing eggs against freezing is clear.

It was discovered that if the crude diapause hormone extract was injected into female *Bombyx mori* pupae at the appropriate time, it not only induced diapause, but also assisted in the progress of glycogen accumulation in the eggs and distinctly stimulated trehalose activity in the haemo-lymph, resulting in a high glycogen level in the ovaries and its reduction in the fat body (Hasegawa and Yamashita, 1970). In another study, Kai and Hasegawa (1973) found that esterase A in Bombyx mori eggs was concerned with the diapause termination mechanism. When esterase A, separated by electrophoresis from non-diapause eggs or from eggs in which diapause had terminated, was added to an *in vitro* culture of *yolk* from diapausing eggs, the *yolk* cells were extremely affected (their cell membranes were dissolved, releasing the contents into the medium).

The Synchronization of Diapause in Parasites with that of their Hosts

Most of the researchers have confirmed that the endoparasitic larvae of entomophagous insects have a life-cycle closely synchronized with that of the host. With small entomophagous parasites such as some Braconids and Tachinids, if the host larva is to survive and develop to an adult, such synchronization appears to be necessary. A special internal mechanism must be supposed in cases where both parasite and host undergo diapause. In this connection several authors have suggested an intervention by some of the metamorphosis hormones. This possibility has been studied experimentally by Schoonhoven (1962) in *Bupalus piniarius* L. (Geometridae) and its parasite Eucarcelia rutilla Vill. (Tachinidae).

A series of experiments involving chilling, decerebration and parabiosis of diapausing host puape has provided evidence that one of the two metamorphosis hormones associated with pupal diapause (AH and MH) is responsible for activating the parasite when the imaginal development of the host starts. Schoonhoven explained why this hormone must be MH. Apparently, the parasite is more easily affected by the hormone than the host. Maslennikova (1961) obtained similar conslusions while working on the development of the Chalcid, *Pteromalus puparum*, and its host *Pieris brassicae* and *Pieris rapae.* Maslennikova concluded that the physiological influence of the host on the parasite larva is exerted via the metamorphosis hormones, possibly MH.

Artificial Diapause

The most remarkable research in this field was done by Wigglesworth (1936), using Rhodnius prolixus, who was the first to show that the artificial elimination of the source of AH results in a state which in many respects is similar to the true natural diapause. For instance decapitated R. prolixus nymphs, though unable to move or accept food, may live very much longer than normal nymphs. The same is true for decerebrated *Hyalophora cecropia* pupae and isolated pupal abdomens, as has been shown by Williams (1947, 1952). The physiology of such an experimentally produced diapause in Bombyx mori pupae is identical with natural diapause in many respects, although no natural pupal diapause occurs in Bombyx. There is a significant decrease in oxygen consumption and an increase in resistance of HCH although to a lesser degree than in the natural diapause of the saturniid *Eudia pavonia* with which it was compared (Novak, 1959, 1961). It may be concluded that the enzyme system working in natural diapause is also operating in the experimentally induced case, even though it may be much less developed. This supports the theory that the origin of diapause was the result of a gradual evolution, under the action of natural selection and concludes in an adaptation to adverse condition of environment.